D1433886

Profiles of Drug Substances, Excipients, and Related Methodology

EDITORIAL BOARD

Profiles of Drug Substances, Excipients, and Related Methodology

Volume 31

edited by

Harry G. Brittain

Center for Pharmaceutical Physics
10 Charles Road
Milford, New Jersey 08848, USA

Founding Editor

Klaus Florey

2004

Amsterdam • Boston • Heidelberg • London • New York • Oxford
Paris • San Diego • San Francisco • Singapore • Sydney • Tokyo

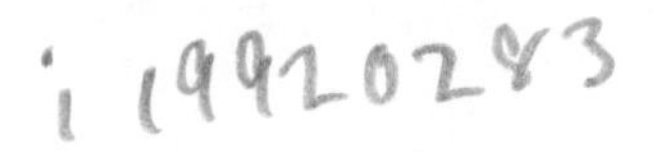

ELSEVIER B.V.
Sara Burgerhartstraat 25
P.O. Box 211, 1000 AE
Amsterdam, The Netherlands

ELSEVIER Inc.
525 B Street, Suite 1900
San Diego, CA 92101-4495
USA

ELSEVIER Ltd
The Boulevard, Langford Lane
Kidlington, Oxford OX5 1GB
UK

ELSEVIER Ltd
84 Theobalds Road
London WC1X 8RR
UK

1st edition 2004

Library of Congress Cataloging in Publication Data
A catalog record is available from the Library of Congress.

British Library Cataloguing in Publication Data
A catalogue record is available from the British Library.

ISBN: 0-12-260831-3

The paper used in this publication meets the requirements of ANSI/NISO Z39.48-1992 (Permanence of Paper).

Transferred to digital printing 2005

CONTENTS

Preface vii
Affiliations of Editor and Contributors ix

Profiles of Drug Substances

1. **Acetylcholine Chloride:**
   - **1.1 Physical Profile** 3
   - **1.2 Analytical Profile** 21

   Abdullah A. Al-Badr and Humeida A. El-Obeid

2. **Benazepril Hydrochloride: Comprehensive Profile** 117
   F. Belal, H.H. Abdine and A.A. Al-Badr

3. **Ciprofloxacin:**
   - **3.1 Physical Profile** 163
   - **3.2 Analytical Profile** 179
   - **3.3 Drug Metabolism and Pharmacokinetic Profile** 209

   Mohammed A. Al-Omar

4. **Dipyridamole: Comprehensive Profile** 215
   A. Khalil, F. Belal and Abdullah A. Al-Badr

5. **Mefenamic Acid: Analytical Profile** 281
   Hadi Poerwono, Retno Widyowati, Hajime Kubo, Kimio Higashiyama and Gunawan Indrayanto

6. **Nimodipine**
   - **6.1 Physical Profile** 337
   - **6.2 Analytical Profile** 355
   - **6.3 Drug Metabolism and Pharmacokinetic Profile** 371

   Mohammed A. Al-Omar

Related Methodology Review Articles

7. **Evaluation of the Particle Size Distribution of Pharmaceutical Solids** 379
   Harry G. Brittain

Cumulative Index 421

PREFACE

The comprehensive profiling of drug substances and pharmaceutical excipients as to their physical and analytical characteristics remains at the core of pharmaceutical development. As a result, the compilation and publication of comprehensive summaries of physical and chemical data, analytical methods, routes of compound preparation, degration pathways, uses and applications, etc., has always been a vital function to both academia and industry. Historically, the ***Profiles*** series has always provided this type of information, and has striven to publish the highest quality available.

As the science of pharmaceutics continues to grow and mature, the need for information similarly expands along many new fronts and the appropriate vehicles where investigators may find the information they need must develop accordingly. As a result, the content of the ***Profiles*** series has expanded so that chapters will therefore fall into one or more of the following main categories:

1. Physical characterization of a drug substance or excipient
2. Analytical methods for a drug substance or excipient
3. Detailed discussions of the clinical uses, pharmacology, pharmacokinetics, safety, or toxicity of a drug substance or excipient
4. Reviews of methodology useful for the characterization of drug substances or excipients

The expansion in scope of the ***Profiles*** series first appeared in Volume 30, and continues in the present volume. Several of the chapters in the current volume are comprehensive in nature, but others are more specialized. The volume also contains a methodology review article on particle size distribution. It is anticipated that future volumes of the ***Profiles*** series will contain similar methodology reviews, as well as other types of review articles that summarize the current state in a particular field of pharmaceutics. As always, I welcome communications from anyone in the pharmaceutical community who might want to provide an opinion or a contribution.

Harry G. Brittain
Editor, Profiles of Drug Substances,
Excipients, and Related Methodology
hbrittain@earthlink.net

AFFILIATIONS OF EDITOR AND CONTRIBUTORS

H.H. Abdine: Department of Pharmaceutical Chemistry, College of Pharmacy, King Saud University, P.O. Box 2457, Riyadh-11451, Kingdom of Saudi Arabia

Abdullah A. Al-Badr: Department of Pharmaceutical Chemistry, College of Pharmacy, King Saud University, P.O. Box 2457, Riyadh-11451, Kingdom of Saudi Arabia

Mohammed A. Al-Omar: Department of Pharmaceutical Chemistry, College of Pharmacy, King Saud University, P.O. Box 2457, Riyadh-11451, Kingdom of Saudi Arabia

F. Belal: Department of Pharmaceutical Chemistry, College of Pharmacy, King Saud University, P.O. Box 2457, Riyadh-11451, Kingdom of Saudi Arabia

Harry G. Brittain: Center for Pharmaceutical Physics, 10 Charles Road, Milford, NJ 08848, USA

Humeida A. El-Obeid: Department of Pharmaceutical Chemistry, College of Pharmacy, King Saud University, P.O. Box 2457, Riyadh-11451, Kingdom of Saudi Arabia

Kimio Higashiyama: Institute of Medicinal Chemistry Hoshi University, 4-41, Ebara 2-chome, Shinagawa-ku, Tokyo 142-8501, Japan

Gunawan Indrayanto: Faculty of Pharmacy, Airlangga, University Jl. Dharmawangsa Dalam, Surabaya 60286, Indonesia

A. Khalil: Department of Pharmaceutical Chemistry, College of Pharmacy, King Saud University, P.O. Box 2457, Riyadh-11451, Kingdom of Saudi Arabia

Hajime Kubo: Institute of Medicinal Chemistry Hoshi University, 4-41, Ebara 2-chome, Shinagawa-ku, Tokyo 142-8501, Japan

Hadi Poerwono: Faculty of Pharmacy, Airlangga University, Jl. Dharmawangsa Dalam, Surabaya 60286, Indonesia

Retno Widyowati: Faculty of Pharmacy, Airlangga University, Jl. Dharmawangsa Dalam, Surabaya 60286, Indonesia

Profiles of Drug Substances

Acetylcholine Chloride: Physical Profile

Abdullah A. Al-Badr and Humeida A. El-Obeid

Department of Pharmaceutical Chemistry
College of Pharmacy, King Saud University
P.O. Box 2457, Riyadh-11451
Kingdom of Saudi Arabia

PROFILES OF DRUG SUBSTANCES,
EXCIPIENTS, AND RELATED
METHODOLOGY – VOLUME 31
DOI: 10.1016/S0000-0000(00)00000-0

CONTENTS

1. **Description** 4
 1.1 Nomenclature 4
  1.1.1 Systematic chemical names 4
  1.1.2 Nonproprietary names 5
  1.1.3 Propietary names 5
 1.2 Formulae 5
  1.2.1 Empirical formula, molecular weight and CAS numbers 5
  1.2.2 Structural formula 6
 1.3 Elemental analysis 6
  1.3.1 Acetylcholine 6
  1.3.2 Acetylcholine hydroxide 6
  1.3.3 Acetylcholine chloride 6
  1.3.4 Acetylcholine bromide 6
 1.4 Appearance and color 6
2. **Physical Characteristics** 6
 2.1 pH 6
 2.2 Solubility characteristics 6
 2.3 Crystallographic properties 7
  2.3.1 Crystal data 7
  2.3.2 X-Ray powder diffraction 8
 2.4 Thermal methods of analysis 9
  2.4.1 Melting behavior 9
  2.4.2 Differential scanning calorimetry 10
 2.5 Spectroscopy 10
  2.5.1 UV/VIS spectroscopy 10
  2.5.2 Vibrational spectroscopy 10
  2.5.3 Nuclear magnetic resonance spectroscopy 11
 2.6 Mass spectrometry 14
3. **Stability** 14
4. **Acknowledgments** 17
5. **References** 17

1. DESCRIPTION

1.1 Nomenclature

1.1.1 Systematic chemical names [1, 2]

Ammonium, 2-(acetyloxy)-N,N,N-trimethylethyl, chloride.

Ethanaminium, 2-(acetyloxy)-N,N,N-trimethyl, chloride.

(2-Acetyloxyethyl)-trimethylammonium chloride.

(2-Acetoxyethyl)-trimethylammonium chloride.

O-Acetyl-choline chloride.

2-(Acetyloxy)-N,N,N-trimethylethanaminium chloride.

(2-Acetoxyethyl)-trimethylammonium hydroxide.

1.1.2 Nonproprietary names
Acetylcholine, Acetylcholine chloride, Chlorune D, Acetylcholinium chloratum, Acetylcholin, Acetylcholini chloridum [1–3].

1.1.3 Proprietary names
Miochol, Alcon, Acecoline, Arterocholine, Ovisot, Covochol, CooperVision, Coopervision, Acetylcholin ophthalmologic, Pragmaline, Tonocholin B, Miovision, Covochol [1–3].

1.2 Formulae

1.2.1 Empirical formula, molecular weight and CAS numbers [1]

1.2.1.1 Acetylcholine
$C_7H_{16}NO_2^+$ 146.21 [51–84–3]

1.2.1.2 Acetylcholine hydroxide
$C_7H_{17}NO_3$ 163.2 —

1.2.1.3 Acetylcholine chloride
$C_7H_{16}ClNO_2$ 181.68 [60–31–1]

1.2.1.4 Acetylcholine bromide
$C_7H_{16}BrNO_2$ 226.14 [66–23–9]

1.2.2 Structural formula

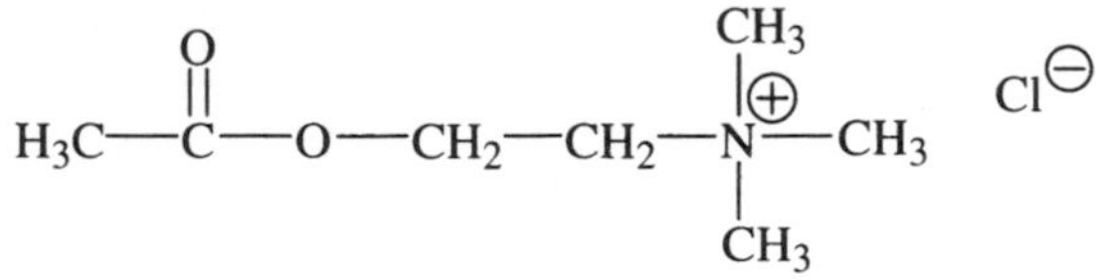

$$H_3C-C(=O)-O-CH_2-CH_2-N^{\oplus}(CH_3)_3 \quad Cl^{\ominus}$$

1.3 Elemental analysis

1.3.1 Acetylcholine

C 57.51%, H 11.03%, N 9.58% O 21.88%.

1.3.2 Acetylcholine hydroxide

C 51.51%, H 10.50%, N 8.58%, O 29.41%.

1.3.3 Acetylcholine chloride

C 46.28%, H 8.88%, Cl 19.52%, N 7.71%, O 17.61% (1).

1.3.4 Acetylcholine bromide

C 37.18%, H 7.13%, Br 35.34%, N 6.20%, O 14.15%.

1.4 Appearance and color

Acetylcholine chloride is obtained as white or off-white hygroscopic crystals, or as a crystalline powder. The salt is odorless, or nearly odorless, and is a very deliquescent powder. Acetylcholine bromide is obtained as deliquescent crystals, or as a white crystalline powder. The substance is hydrolyzed by hot water and alkali [1–5].

2. PHYSICAL CHARACTERISTICS

2.1 pH

Following reconstitution, commercially available injectable formulations are characterized by a pH range of 5–7.8 [4].

2.2 Solubility characteristics

Acetylcholine chloride is freely soluble in water, alcohol, propylene glycol, and chloroform, and is practically insoluble in ether. The solubility of 500 mg is complete in 5 mL alcohol, resulting in a colorless solution. Solutions of acetylcholine chloride in water are unstable. Likewise, acetylcholine bromide is very soluble in cold water; decomposed by hot water or alkalis, soluble in alcohol, and practically insoluble in ether [3–5].

2.3 Crystallographic properties

2.3.1 Crystal data

The crystal structures of acetylcholine bromide and acetylcholine iodide have both been investigated by Sorum [6, 7]. The structure was solved in the $P2_1$ space group, with two crystallographically different molecules in the asymmetric unit. The crystal structure of acetylcholine chloride was reported by Allen [8].

Acetylcholine chloride is very hygroscopic and extremely soluble, giving a solution of high viscosity. For this reason, it is difficult to prepare crystals appropriate for X-ray analysis. At 25°C, a saturated solution contains approximately 0.962 g/mL of solute, and has a density of 1.19 g/mL. When a solution, nearly saturated at room temperature, is cooled to approximately −10°C, the anhydrous salt crystallizes slowly as clusters of needles. There was no evidence of hydrate formation. Single crystals, mounted and sealed in glass capillaries were examined about the needle axis [*a*] by oscillation and Weissenberg methods using filtered copper radiation, and by the precession method using molybdenum radiation. Provided that the sealing was satisfactory these mounted crystals remained unchanged for a period of several weeks.

The crystal data obtained were as follows:

Acetylcholine chloride	$(CH_3)_3\overset{\oplus}{N}CH_2CH_2OCOCH_3 . Cl^{\ominus}$
Orthorhombic	Formula Weight: 181.5
$a = 6.28$ Å	Unit cell volume: 952 Å^3
$b = 9.93$ Å	
$c = 15.26$ Å	

Space group: P $2_12_12_1$.

Systematic extinctions observed:

$h00$ for $h = 2n$ $0k0$ for $k = 2n$ $00l$ for $l = 2n$

Density: Observed 1.202 g/mL; Calculated 1.27 g/mL

Molecules per unit cell: 4

The density of the crystal was estimated by flotation in benzene/chloroform mixtures. It is recognized that the value is not very accurate and is almost certainly lower than the true value because of the difficulty of freezing the crystal completely from the viscous mother liquor. However, the method is sufficient to confirm the number of molecules per unit cell [8].

2.3.2 X-Ray powder diffraction

The X-ray powder diffraction pattern of acetylcholine chloride was obtained using a Philips PW-1710 diffractometer system with a single crystal monochromator and using copper K_α radiation. The powder pattern is shown in Figure 1. The scattering angles, *d*-spacings, and relative intensities were automatically obtained for a pure sample of acetylcholine chloride, and are shown in Table 1.

Canepa [9] suggested that one of the two molecular forms of acetylcholine bromide reported by Sorum [6, 7] was incorrect. Dunitz [10], on the other hand, observed some unusual systematic absences and suggested that the crystal used was a twinned $P2_1/c$ rather than a single $P2_1$ specimen. In response to these observations, Svinning and Sorum [11] refined the structure of acetylcholine bromide based on previous photographic intensities corrected for twinning by Canepa, Pauling and

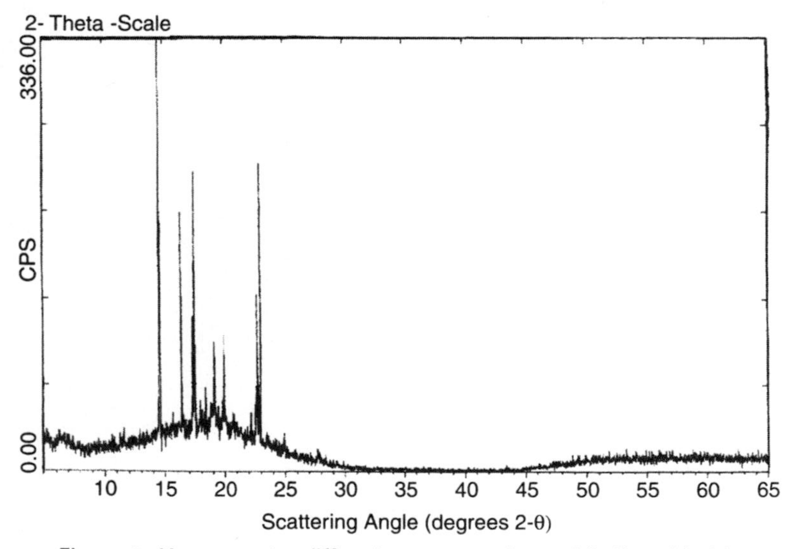

Figure 1. X-ray powder diffraction pattern of acetylcholine chloride.

Table 1

Crystallographic Data from the X-ray Powder Diffraction Pattern of Acetylcholine Chloride

Scattering Angle degrees -2θ	*d*-Spacing (Å)	Relative intensity (%)	Scattering Angle degrees -2θ	*d*-Spacing (Å)	Relative intensity (%)
7.054	12.5219	6.67	11.606	7.6184	6.69
14.592	6.0653	100.00	14.881	5.9484	12.13
16.435	5.3890	37.06	17.497	5.0643	68.38
17.967	4.9329	20.01	18.428	4.8104	20.93
19.149	4.6310	38.98	19.955	4.4459	35.48
22.734	3.9083	37.49	23.011	3.8618	34.16
24.906	3.5721	8.25	27.755	3.2115	5.05

Sorum [12] to a final R of 0.10. Thus, the reinvestigation of the structure of acetylcholine bromide, $[C_7H_{16}O_2N]^+Br^-$, with X-ray diffraction intensities collected from two untwinned crystals showed that the crystals are monoclinic, and are characterized by a space group of $P2_1/n$, with $a = 10.966$ (4), $b = 13.729$ (7), $c = 7.159$ (4) Å, $\beta = 108.18$ (7)°, and $Z = 4$. The structure was refined by full-matrix least squares calculations using 1730 observed reflections, and anisotropic temperature factors for all non-hydrogen atoms. The final R was found to be 0.041. Atomic coordinates, thermal parameters, bond lengths, and angles were compared with those from a previous work on acetylcholine derivatives.

2.4 Thermal methods of analysis

2.4.1 Melting behavior

Acetylcholine chloride, when dried earlier at 110°C in a capillary tube for one hour, melts between 149 and 152°C [5]. Acetylcholine bromide melts at 143°C [3].

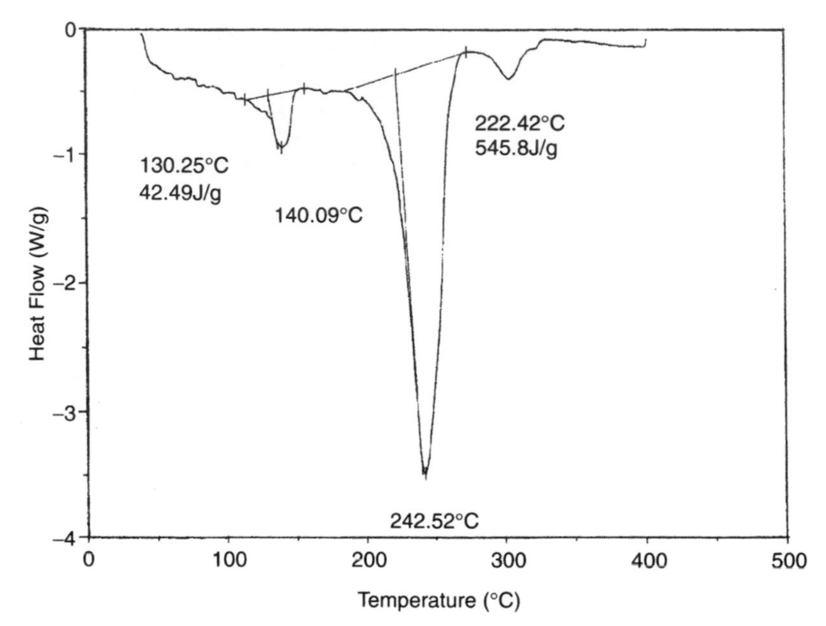

Figure 2. Differential scanning calorimetry thermogram of acetylcholine chloride.

2.4.2 Differential scanning calorimetry

Differential scanning calorimetry of acetylcholine chloride was performed using a Dupont DSC Model TA 9900 thermal analyzer system, interfaced to a Dupont data collection system. Figure 2 shows the DSC thermogram, which was recorded from 50 to 350°C.

2.5 Spectroscopy

2.5.1 UV/VIS spectroscopy

Owing to its lack of chromophores, acetylcholine chloride does not exhibit any absorption bands in the ultraviolet or visible region.

2.5.2 Vibrational spectroscopy

The infrared spectrum of the substance that is shown in Figure 3 was recorded on a Pye Unicam SP infrared spectrophotometer (KBr pellet method). The major observed bands have been correlated with the functional groups, and these are 2970 and 2920 cm^{-1} (CH_3 and CH_2 stretching), 1755 cm^{-1} (C=O stretching) and 1465 cm^{-1} (CH deformation). Clarke reported the principal peaks for acetylcholine bromide at wavenumbers of 949, 1089, 869, 1231, 1000, 1740 cm^{-1} (KBr pellet) [3].

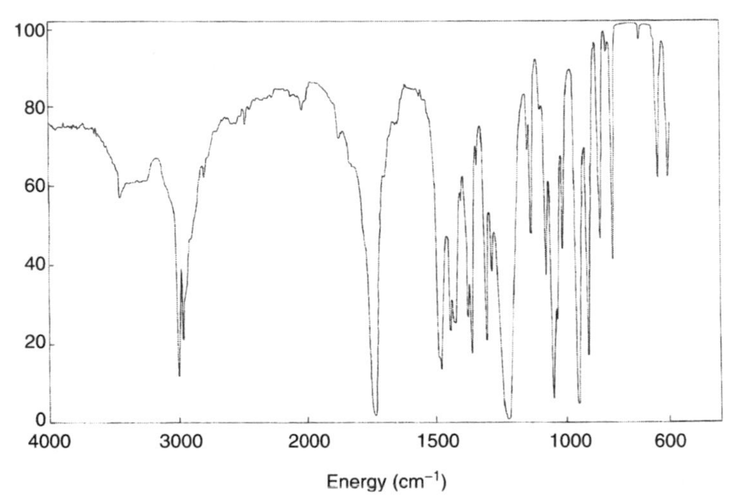

Figure 3. Infrared absorption spectrum of acetylcholine chloride.

Srivastava *et al.* carried out a complete study of the vibrational spectrum of acetylglycine, and evaluated its relation to the spectrum of acetylcholine [13]. A normal coordinate analysis of acetylglycine was carried out using the Wilson's GF matrix method. Vibrational frequencies were assigned, and the infrared spectra of acetylglycine and acetylcholine compared. Conformation-sensitive modes of acetylcholine were identified, and a transferable Urey-Bradley force field was also obtained.

2.5.3 Nuclear magnetic resonance spectroscopy

2.5.3.1 ^{1}H-NMR spectra

The ^{1}H-NMR spectrum of acetylcholine chloride in $CDCl_3$ was recorded on a Varian XL200 NMR spectrometer using TMS as the internal reference. The simple proton (Figure 4) and HOMCOR (Figure 8) spectra were used to determine the exact chemical shifts and proton coupling. The HETCOR spectrum (Figure 9) was used to assign the protons to their respective carbons in the molecule. The resulting proton chemical shifts are assigned as listed in Table 2. ^{1}H-NMR studies [14] on acetylcholine were used in determining its conformation.

2.5.3.2 ^{13}C-NMR spectra

The ^{13}C-NMR spectrum of acetylcholine chloride was recorded in $CDCl_3$, using TMS as the internal reference, on a Varian XL 200 NMR

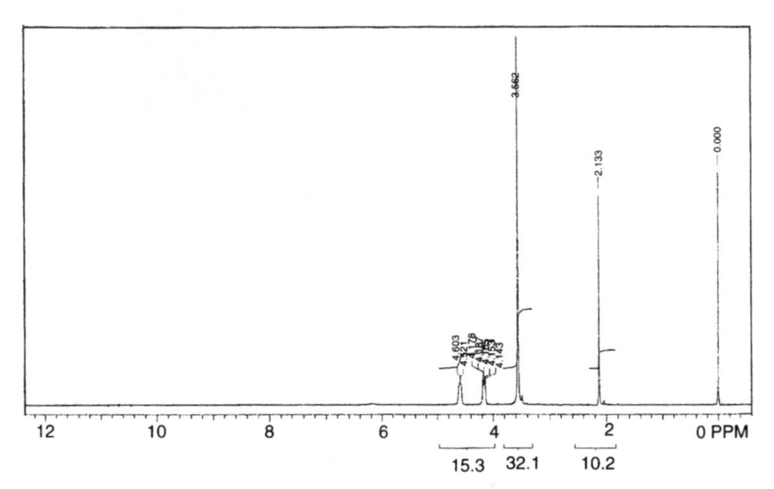

Figure 4. ^{1}H-NMR spectrum of acetylcholine chloride in $CDCl_3$.

Table 2

^{1}H-NMR Chemical Shift Assignments for Acetylcholine Chloride

Chemical shifts (ppm, relative to TMS)	Number of protons	Multiplicity	Assignment
2.13	3	singlet	$-\underset{\underset{O}{\|}}{C}-\underline{C}H_3$
4.14	2	multiplet	$\equiv\underset{+}{N}-\underline{C}H_2-$
4.56	2	multiplet	$-O-\underline{C}H_2-$
3.55	9	singlet	$\equiv\underset{+}{N}-\underline{C}H_3$

spectrophotometer. Figures 5, 6, and 7 show the S2PUL pulse sequence, and the APT and DEPT spectra, respectively. The assigned chemical shifts are listed in Table 3.

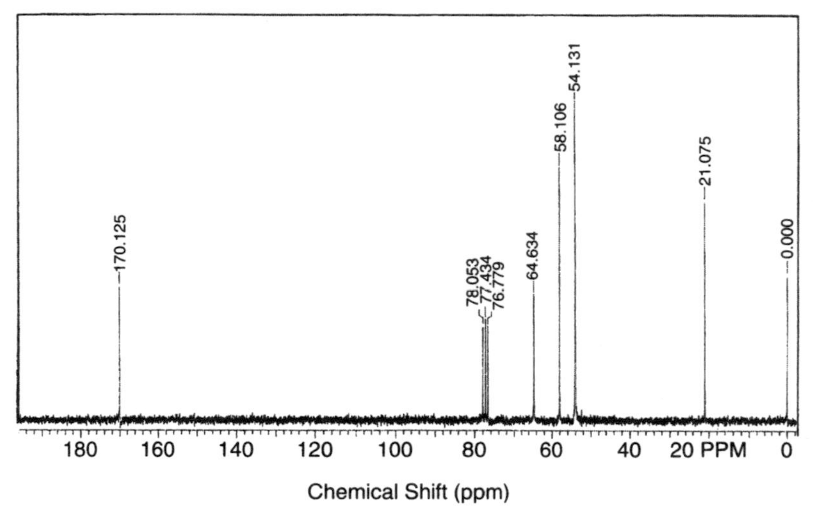

Figure 5. ^{13}C-NMR spectrum of acetylcholine chloride in $CDCl_3$.

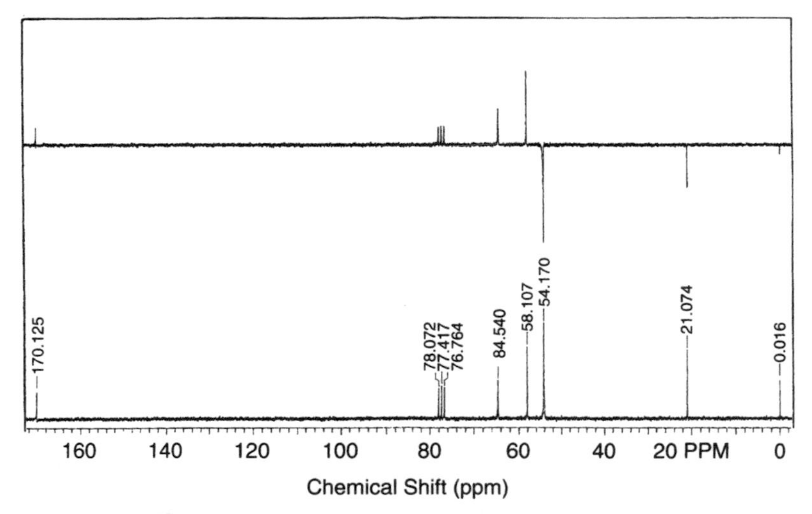

Figure 6. ^{13}C-NMR spectrum (APT) of acetylcholine chloride in $CDCl_3$.

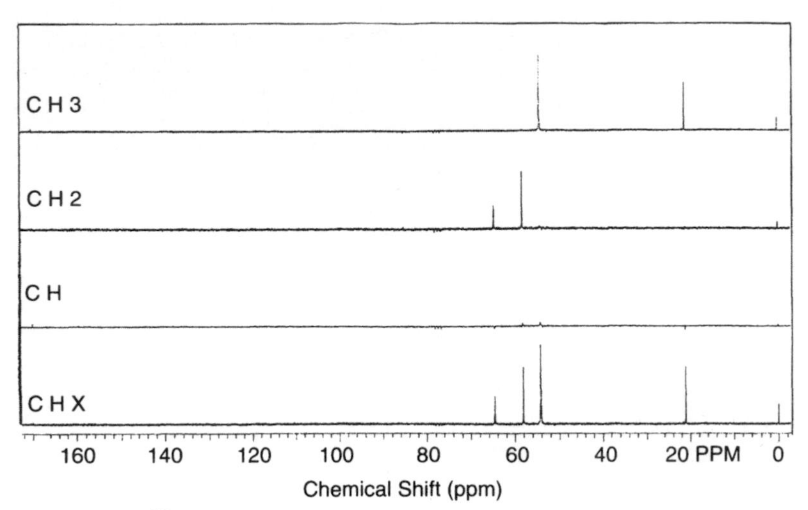

Figure 7. ^{13}C-NMR spectrum (DEPT) of acetylcholine chloride in $CDCl_3$.

The ^{13}C chemical shifts of acetylcholine chloride were assigned by Murari and Baumann [15]. The results of their ^{13}C-NMR studies [16] on acetylcholine chloride were used for analyzing the conformational isomers of the molecule.

2.6 Mass Spectrometry

The mass spectrum of acetylcholine chloride was obtained utilizing a Shimadzu PQ-5000 mass spectrometer, with the parent ion being collided with helium carrier gas. The mass spectrum is shown in Figure 10, and Table 4 shows the mass fragmentation pattern. Clarke lists the principal mass spectral peaks at m/z 58, 43, 57, 149, 71, 42, 41, 55 [3].

Mass spectrometry was used in combination with chromatographic methods for the analysis of acetylcholine in biological systems, with the mass spectral peak at m/z 58 being most frequently used for detection. Quantitation of acetylcholine and its related compounds was successfully performed using GC/MS [17–26] and HPLC/MS [27–29].

3. STABILITY

Aqueous solutions of acetylcholine chloride are found to be unstable. Such solutions are decomposed by heat, and are incompatible with alkalis

Table 3

^{13}C-NMR Chemical Shift Assignments for Acetylcholine Chloride

$$H_3\underset{1}{C}-\underset{2}{\overset{\overset{O}{\|}}{C}}-O-\underset{3}{C}H_2-\underset{4}{C}H_2-\overset{\overset{CH_3}{|}}{\underset{\underset{\underset{5}{C}H_3}{|}}{N^{\oplus}}}-CH_3 \quad Cl^{\ominus}$$

Chemical shift (ppm, relative to TMS)		Carbon Assignments
Found	**Reported [15]**	
21.07	21.3	1
54.17	54.8	5
58.11	59.2	3
64.64	65.2	4
170.12	173.8	2

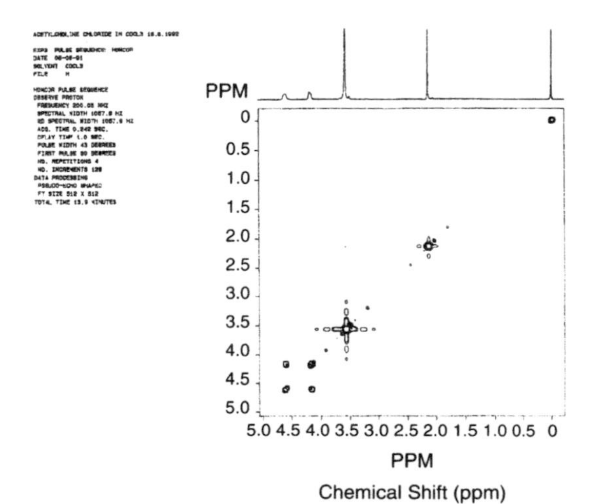

Figure 8. HOMCOR-NMR spectrum of acetylcholine chloride in $CDCl_3$.

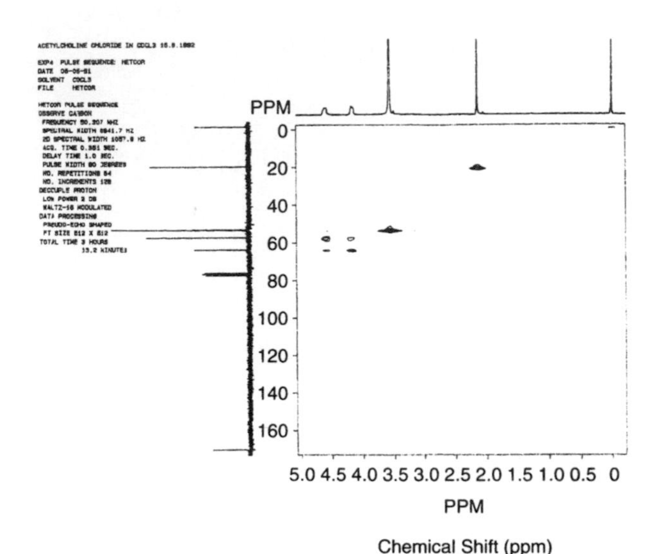

Figure 9. HETCOR-NMR spectrum of acetylcholine chloride in $CDCl_3$.

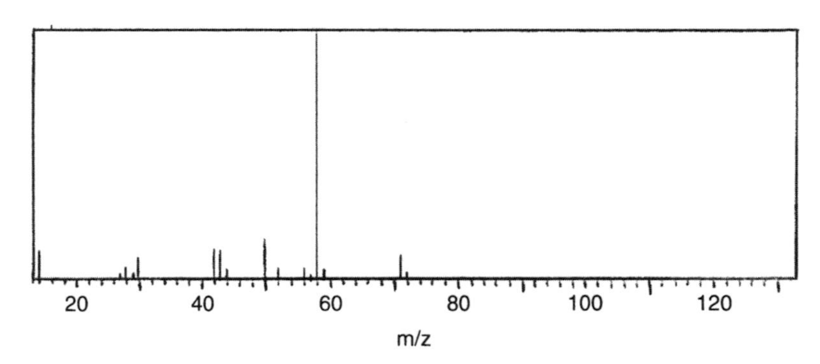

Figure 10. Mass spectrum of acetylcholine chloride.

and acids. Some reports revealed that the solution is stable for 2–3 h following reconstitution, after which breakdown products are observed. The reconstituted solution should be used immediately, and any unused portion discarded. The unconstituted two-chambered vials of acetylcholine chloride for injection should be stored in air-tight containers at 15–30°C, and protected from freezing [1, 4, 5, 30].

Table 4

Assignments for the Fragmentation Pattern Observed in the Mass Spectrum of Acetycholine Chloride

m/z	Relative intensity (%)	Fragment
71	8	$[H_3C{-}N(CH_3){-}CH{=}CH_2]^{+\bullet} \rightleftharpoons H_3C{-}\overset{+}{N}(CH_3){=}CH{-}\dot{C}H_2$
58	100	$H_3C{-}\overset{+}{N}(CH_3){=}CH_2$
43	10	$H_3C{-}C{\equiv}\overset{+}{O}$
42	10	$[H_2C{=}C{=}O]^{+\bullet}$

4. ACKNOWLEDGMENTS

The authors wish to thank Mr. Tanvir A. Butt, Department of Pharmaceutical Chemistry, College of Pharmacy, King Saud University, Riyadh, Saudi Arabia, for typing this profile.

5. REFERENCES

1. ***The Merck Index***, 11th edn., Budavari, S., ed., Merck and Co., Inc., Rahway, N.J., p. 14 (1989).
2. ***Index Noninium: International Drug Directory***, compiled by the Scientific Documentation Center of the Swiss Pharmaceutical Society, S. Budavari, ed., Zurich, p. 8 (1987).

3. ***Clarke's Isolation and Identification of Drugs in Pharmaceuticals Fluids and Post-Mortem Materials***, 2nd edn., A.C. Moffat, ed., The Pharmaceutical Press, London, p. 317 (1986).
4. ***AHFS Drug Information***, G. K. McEvoy, ed., The American Society of Health-System Pharmacists, Maryland, USA, p. 1924 (1995).
5. ***The United States Pharmacopoeia (USP XX)***, United State Pharmacopoeial Convention, Inc., Rockville, MD, 20852, p. 1045 (1980).
6. H. Sorum, *Acta. Chem. Scand.*, ***10***, 1669 (1956).
7. H. Sorum, *Acta. Chem. Scand.*, ***13***, 345 (1959).
8. K.W. Allen, *Acta. Cryst.*, ***15***, 1052 (1962).
9. F.G. Canepa, *Nature, Lond.*, ***201***, 184 (1964).
10. J.D. Dunitz, *Acta Chem. Scand.*, ***17***, 1471 (1963).
11. T. Svinning and H. Sorum, *Acta Cryst.*, ***B31***, 1581 (1975).
12. F.G. Canepa, P. Pauling, and H. Sorum, *Nature, Lond.*, ***210***, 907 (1966).
13. R.B. Srivastava, S.C. Agarwala, N.P. Roberts, and V.D. Gupta, *Indian J. Biochem. Biophys.*, ***12***, 238 (1975).
14. C.C.J. Culvenor and N.S. Ham, *Chem. Commun.*, 537 (1966).
15. R. Murari and W.J. Baumann, *J. Am. Chem. Soc.*, ***103***, 1238 (1981).
16. L. Cassidei and O. Sciacovelli, *J. Am. Chem. Soc.*, ***103***, 933 (1981).
17. O. Niwa, T. Horiuchi, M. Morita, T. Huang, and P.T. Kissinger, *Anal. Chim. Acta*, ***318***, 167 (1996).
18. R.L. Polak and P.C. Molenaar, *J. Neurochem.*, ***23***, 1295 (1974).
19. R.L. Polak and P.C. Molenaar, *J. Neurochem.*, ***32***, 407 (1979).
20. F. Mikes, *Chem. Rundsch*, ***33***, 1 (1980).
21. A.K. Singh and L.R. Drewes, *J. Chromatogr., Biomed. Appl.*, ***40***, 170 (1985).
22. D.J. Liberato, A.L. Yergey, and S.T. Weintraub, *Biomed. Environ. Mass Spectrom*, ***13***, 171 (1986).
23. M. Watanabe, A. Kimura, K. Akasaka, and S. Hayashi, *Biochem. Med. Metab. Biol*, ***36***, 355 (1986).
24. E.A. Pomfret, K.A. DaCosta, L.L. Schurman, and S.H. Zeisel, *Anal. Biochem.*, ***180***, 85 (1989).
25. A.V. Terry Jun, L.A. Silks, R.B. Dunlap, J.D. Odom, and J.W. Kosh, *J. Chromatogr*, ***585***, 101 (1991).
26. T.A. Patterson and J.W. Kosh, *Biol. Mass Spectrom*, ***21***, 299 (1992).

27. X.Y. Zhu, P.S.H. Wong, M. Gregor, G.F. Gitzen, L.A. Coury, and P.T. Kissinger, *Rapid Commun. Mass Spectrom.*, ***14***, 1695 (2000).
28. H. Ishimaru, Y. Ikarashi, and Y. Maruyama, *Biol. Mass. Spectrom.*, ***22***, 681 (1993).
29. L.D. Acevedo, Y.D. Xu, X. Zhang, R.J. Pearce, and A. Yergey, *J. Mass. Spectrom.*, ***31***, 1399 (1996).
30. ***Product Information: Miochol***®, Acetylcholine, Cooper, Vision Pharmaceuticals, San German, Puerto Rico (1995).

Acetylcholine Chloride: Analytical Profile

Abdullah A. Al-Badr and Humeida A. El-Obeid

Department of Pharmaceutical Chemistry
College of Pharmacy, King Saud University
P.O. Box 2457, Riyadh-11451
Kingdom of Saudi Arabia

PROFILES OF DRUG SUBSTANCES,
EXCIPIENTS, AND RELATED
METHODOLOGY – VOLUME 31
DOI: 10.1016/S0000-0000(00)00000-0

CONTENTS

1. **General Methods of Analysis** 23
1.1 Identification 23
1.2 Reviews on methods of analysis 24
1.3 Assay methods 25
1.3.1 Method 1: Percentage of acetyl 25
1.3.2 Method 2: Percentage of chlorine 25
2. **Titrimetric Methods of Analysis** 25
2.1 Acidimetric methods 25
2.2 Potentiometric methods 26
3. **Electrochemical Methods of Analysis** 26
3.1 Conductimetric methods 26
3.2 Polarographic methods 27
3.3 Voltammetric methods 27
3.4 Coulometric flow titration methods 28
3.5 Flow injection method 66
3.6 Amperometric methods 66
4. **Spectrophotometric Methods of Analysis** 67
4.1 Infrared spectrophotometric method 67
4.2 Photometric methods 67
4.3 Mass spectrometric method 68
4.4 Colorimetric methods 68
4.5 Fluorimetric methods 70
4.6 Chemiluminescence methods 72
5. **Biological Assay Methods of Analysis** 74
6. **Chromatographic Methods of Analysis** 75
6.1 Thin layer chromatography (TLC) 75
6.2 Gas chromatography (GC) 75
6.3 High performance liquid chromatography (HPLC) 79
6.4 Liquid chromatography (LC) 80
6.5 Cation-exchange liquid chromatography 97
6.6 High performance liquid chromatography–mass spectrometry (HPLC/MS) 97
6.7 Gas chromatography–mass spectrometry (GC/MS) 98
6.8 Capillary electrophoresis 101
6.9 Radioassay methods 102
7. **Other Methods of Analysis** 105
7.1 Ion-selective method 105

7.2 Microwave methods 105
7.3 Microdialysis methods. 105
8. **Acknowledgment**. 105
9. **References** 106

1. GENERAL METHODS OF ANALYSIS

1.1 Identification

Vizi *et al.* described a simple and sensitive method for the identification and assay of acetylcholine [1]. A modification of the acetylcholine assay technique for guinea pig ileum has been combined with either minivolume gel filtration or high pressure liquid chromatography separation of the samples. In addition, labeled acetylcholine (^{14}C-ACh) was eluted with unlabeled acetylcholine, or with samples expected to contain acetylcholine, and their elution profiles were compared by bioassay combined with radioassay of eluted fractions. When the elution of both biological activity and label occurred in the same fraction, it was concluded that the substance assayed on guinea pig ileum was acetylcholine. This method was found to be sensitive to 0.5 ng (1.8 pmol) of acetylcholine, and its reproducibility was within 5%.

Wiegrebe and Vibig reported a method for the identification of choline and choline salts [2]. A critique of the methods of identification of acetylcholine chloride and choline chloride according to the Swiss Pharmacopoeia VI and of choline chloride according to the German Pharmacopoeia 8 was given and alternative procedures suggested. As alternatives, the pyrolytic degradation of acetylcholine to acetaldehyde, which can be identified by the Simon-Awe method, and of choline to formaldehyde followed by a reaction with chromotropic acid to give a deep violet color, are suggested.

Mynka *et al.* used infrared spectrophotometry for the identification and determination of acetylcholine [3]. Beer's law was obeyed from 1 to 10 μg/mL of acetylcholine, and the coefficient of variation was 2.7%.

Zhu *et al.* used an *in vivo* microdialysis and a reversed phase ion-pair liquid chromatography-tandem mass spectrometry method for the determination and identification of acetylcholine in rat brain [4]. A microdialysis probe was surgically implanted in the striatum of a rat, and this was perfused with Ringer's solution at a flow rate of 2 μL/min.

The resulting microdialysate samples were mixed (9:1) with acetyl-β-methylcholine (internal standard, 1 ng/μL), and a 950 μL portion was analyzed on a Discovery C_{18} column (50 cm × 2.1 mm i.d.). The method used 20 mM ammonium acetate/20 mM heptafluorobutyric acid in methanol/water (1:9, pH 3.2) as the mobile phase at 0.4 mL/min, and the analyte and internal standard were detected by tandem ion-trap mass spectrometry. The calibration graph was linear up to 500 fg/μL of acetylcholine, and the on-column detection limit was 200 fg acetylcholine.

A sensitive and specific bioautographic method for the identification of choline and its derivatives, was developed by Lewin and Marcus, utilizing a *Neurospora crassa* mutant [5]. As little as 0.03 γ of choline chloride is detectable. The applicability of the method for detection of choline derivatives in lipid hydrolyzate is discussed. A method for determining the rate of diffusion of compounds from paper chromatograms into bioautograph agar is described.

Unger and Ryan used a secondary-ion mass spectrometry method for the identification of choline and its esters [6]. Secondary-ion mass spectrometry provides the mass spectra of choline and its esters, from which molecular weights and structures may be readily obtained. Structure-spectral correlations were developed using analogies from other forms of mass spectrometry. Spectra can be recorded on metal or paper substrates, and individual components can be identified even in complex samples.

1.2 Reviews on methods of analysis

The recent developments in the determination of acetylcholine and choline, the advantages and the disadvantages of a variety of analytical methods used in the analysis, were reviewed and discussed by Maruyama *et al.* [7]. Hanin published an overview for the methods used for the analysis and measurement of acetylcholine [8]. Shimada *et al.* presented a review, including the applications of some reactors in the analysis of choline and acetylcholine. The immobilized enzyme reactors were used for detection systems in high performance liquid chromatography [9].

Tsai reviewed the separation methods used in the determination of choline and acetylcholine [10]. This review surveyed the array of analytical techniques that have been adopted for the measurement of acetylcholine or its main precursor/metabolite (choline), ranging from simple (bioassay, radio enzymatic assay, gas chromatography–flame ionization detection, gas chromatography–mass spectrometry, high performance liquid chromatography–electrochemical detection, high performance liquid chromatography–mass spectrometry) to the sophisticated (combination of microimmobilized enzymatic reactor, microbore high performance liquid chromatography, and modified electrode technology for the detection of ultralow levels), with particular emphasis on the state of the art techniques.

1.3 Assay methods

United State Pharmacopeia XX describes the following assay methods for acetylcholine as a reagent [11]:

1.3.1 Method 1: Percentage of acetyl

Weigh accurately about 400 mg of analyte, previously dried at 105°C for 3 h, and dissolve in 15 mL of water in a glass-stoppered conical flask. Add 40 mL of 0.1 N-sodium hydroxide VS, and heat on a steam bath for 30 min. Insert the stopper, allow to cool, add phenolphthalein TS, and titrate the excess alkali with 0.1 N sulfuric acid VS. Determine the exact normality of the 0.1 N sodium hydroxide by titrating an aliquot of 40.0 mL after it has been treated in the same manner as in the test. Each milliliter of 0.1 N sodium hydroxide is equivalent to 4.30 mg of CH_3CO. Between 23.2 and 24% of CH_3CO is found.

1.3.2 Method 2: Percentage of chlorine

Weigh accurately about 400 mg of acetylcholine chloride, previously dried at 105°C for 3 h, and dissolve in 50 mL of water in a glass stoppered 125 mL flask. Add with agitation 30.0 mL of 0.1 N silver nitrate VS, then add 5 mL of nitric acid and 5 mL of nitrobenzene and shake well. Thereafter, add 2 mL of ferric ammonium sulfate TS, and titrate the excess silver nitrate with 0.1 ammonium thiocyanate VS. Each milliliter of 0.1 N silver nitrate is equivalent to 3.545 mg of Cl. Between 19.3 and 19.8% of Cl is found.

2. TITRIMETRIC METHODS OF ANALYSIS

2.1 Acidimetric methods

Szakacs-Pinter and Perl-Molnar described a new and simple procedure for the determination of choline and its derivative. The method is based on the acidimetric determination of the volatile base produced by the $K_2S_2O_8$ oxidation of choline in an alkaline medium [12]. The procedure was carried out by distilling the trimethylamine formed continuously during the reaction. Optimum conditions for the determination are: 0.5 M alkali, 15 min distillation time, and a 10 fold excess $K_2S_2O_8$. Choline or acetylcholine could be determined in 0.005–0.1 N solution of pharmaceutical preparations and animal feed.

2.2 Potentiometric methods

Kakutani *et al.* described an ion-transfer voltammetry and potentiometry method for the determination of acetylcholine with the interface between polymer–nitrobenzene gel and water [13]. The PVC-nitrobenzene gel electrode was prepared as described by Osakai *et al.* [14]. The transfer of acetylcholine ions across the interface between the gel electrode and water was studied by cyclic voltammetry, potential-step chronoamperometry, and potentiometry. The interface between the two immiscible electrolyte solutions acted as the ion-selective electrode surface for the determination of acetylcholine ions.

Baum reported a potentiometric determination of acetylcholine activity using an organic-cation-selective electrode [15]. The performance of a liquid membrane electrode selective for acetylcholine (Corning No. 476.200) was investigated. Measurements of the potential difference at various concentration of acetylcholine were made against a calomel reference-electrode.

Nakazawa and Tanaka described a potentiometric titration of salts of organic bases in non-aqueous solvents with the addition of bismuth nitrate [16]. The base halide or hydrohalide (0.7 mequiv.) is dissolved in 40 mL of anhydrous acetic acid, and then 40 mL of 1,4-dioxane and 2.5 mL of 5% $Bi(NO_3)_3$ solution in acetic acid are added. The $Bi(NO_3)_3$ prevents interference by the halide. The solution is titrated to a potentiometric endpoint with 0.1 M $HClO_4$ (a blank titration is also carried out). Results of purity assays of acetylcholine chloride and other compounds were tabulated, and it was found that the coefficient of variation was 0.18%.

3. ELECTROCHEMICAL METHODS OF ANALYSIS

3.1 Conductimetric methods

An immobilized enzyme flow injection conductimetric system was used by Godinho *et al.* for the determination of acetylcholine [17]. Aliquots of acetylcholine standard solutions were injected into a carrier stream of 0.1 M phosphate buffer of pH 7.4 at 1.19 mL/min, and passed through a polyethylene tube (3.5 cm × 3 mm i.d.) containing acetylcholineesterase immobilized on glass beads. The solution was merged with a stream of 2 M H_2SO_4 at 1.19 mL/min, and passed through a PTFE membrane diffusion cell. The acetic acid that diffused through the membrane was carried by a stream of water to a conductimetric cell. The calibration graph was linear from 10 μM to 1 mM acetylcholine, and the method was applied to the analysis of acetylcholine in a pharmaceutical formulation.

3.2 Polarographic methods

A quantitative method was reported by Maslova for the determination of acetylcholine in biological tissues by polarographic analysis utilizing a rotating platinum electrode [18]. The principle of the method is based upon a polarographic analysis of the iron ions which remain after the formation of a specific Fe–acetylhydroxamic acid complex. Using this method, it was shown that 1 mL of peripheral blood of a healthy adult man contained 6.6 ng of acetylcholine.

3.3 Voltammetric methods

The ion transfer voltammetry and potentiometry of acetylcholine with the interface between polymer-nitrobenzene gel and water was reported by Kakutani *et al.* [13]. The polyvinyl chloride-nitrobenzene gel electrode was prepared by Osakai *et al.* [14], and the transfer of acetylcholine ions across the interface between the gel electrode and water was studied by cyclic voltammetry, potential step chronoamperometry, and potentiometry. The interface between the two immiscible electrolyte solutions acted as the ion selective electrode surface for the analysis of acetylcholine ions.

Matsue *et al.* [19] recommended an electrochemical determination method using Nafion coated electrodes for the determination of electroinactive medicines. Acetylcholine was determined voltammetrically with the use of a Nafion 117-coated vitreous carbon working electrode and a saturated calomel electrode. The method is based on competition between (ferrocenylmethyl) trimethylammonium ion and the drug cations for entry into the Nafion layer. The presence of acetylcholine caused a decrease in the peak current for (ferrocenylmethyl) trimethylammonium ion, from which the acetylcholine concentration is determined. The method was described to have permitted the determination of 1 μM–1 mM of acetylcholine.

Osakai *et al.* used a microcomputer-controlled system for the ion transfer voltammetry procedure [20]. The system used is based on a NEC PC-9801 microcomputer, which was designed by using a polarizable oil–water interface as an ion-selective electrode surface. The system was applied to the determination of acetylcholine ion by cyclic, differential pulse, and normal pulse voltammetry at the PVC–nitrobenzene gel electrode. The amperometric measurement was carried out with voltage pulses of short durations and constant amplitude.

Hale *et al.* reported the use of an enzyme-modified carbon paste for the determination of acetylcholine [21]. The sensor was constructed from a carbon paste electrode containing acetylcholineesterase and choline oxidase, and the electron transfer mediator tetrathiafulvalene. The electrode was used for the cyclic voltammetric determination of acetylcholine in 0.1 M phosphate buffer at +200 mV versus saturated calomel electrode. Tetrathiafulvalene efficiently re-oxidized the reduced flavin adenine dinucleotide centers of choline oxidase. The calibration graph was linear up to 400 μM acetylcholine, and the detection limit was 0.5 μM.

Kawagoe *et al.* described an enzyme-modified organic conducting salt microelectrode, comprised of a carbon fiber electrode (20 μM), having a recessed tip into which crystals of the conducting salt tetrathiafulvalene-tetracyanoquinodimethane were galvanostatically deposited, followed by glucose oxidase cross-linked with glutaraldehyde for the detection of glucose [22]. For the detection of acetylcholine, acetylcholineesterase and choline oxidase were co-immobilized in the recess instead of glucose oxidase. The micrographic and cyclic voltammetric characterization of the recessed electrode was described. The acetylcholine microelectrode responded to 5 μM or 800 μM acetylcholine within four seconds (Table 1).

By using the electrolysis at the interface between two immiscible electrolyte solutions, Marecek and Samec determined acetylcholine by a differential pulse stripping voltammetric method [23].

3.4 Coulometric flow titration methods

Gyurcsanyi *et al.* studied the determination of acetylcholine after enzymatic hydrolysis by triangle-programmed coulometric flow titration [24]. A sample containing acetylcholine was pumped through an enzyme reactor containing acetylcholineesterase immobilized on controlled-pore glass. The resulting acetic acid was titrated against KOH with continuous

Table 1

Biosensors for the Detection and Analysis of Acetylcholine

Biological system	Biosensor design	Results and remarks	Reference
Acetylcholineesterase and choline oxidase	Carbon paste electrode containing AChE and ChO and the electron transfer mediator tetrathiafulvalene. The electrode was applied to the cyclic voltammetric determination of ACh in 0.1 M phosphate buffer at + 200 mV versus SCE.	The calibration graph was linear up to 400 μM ACh and the detection limit was 0.5 μM. The tetrathiafulvalene-based sensor operated efficiently at low applied potential (0 to − 100 mV versus SCE).	[21]
Acetylcholineesterase and choline oxidase	Carbon-fiber electrode having a recessed tip into which crystals of the conducting salts tetrathiafulvalene–tetracyanoquinodimethane were galvanostatically deposited followed by AChE and ChO which were co-immobilized in the recess.	No interference from uric acid. The electrode responded to 5 μM or 800 μM ACh within 4 s. Kinetic studies of the biosensor were reported.	[22]

(continued)

Table 1 (continued)

Biological system	Biosensor design	Results and remarks	Reference
Acetylcholineesterase (AChE)	Glass microelectrode with AChE immobilized on active surface.	Nernstian response. Detection limit is pH-dependent. At pH 8.5 detection limit was 10 µM. In cerebrospinal fluids it was 0.1 mM.	[63]
Choline oxidase and acetylcholineesterase	Enzymes immobilized on a nylon net attached to H_2O_2-selective amperometric sensor. ChO is used for choline and AChE and ChO for acetylcholine.	Rectilinear response in the range of 1–10 µM. Response time 1–2 min. Interferences occur from ascorbic acid, primary amines, and most seriously from betaine aldehyde.	[64]
Nicotinic receptor from *Torpedo californica*	Receptor was fixed into a cross-linked poly(vinylbutyral) membrane covering the gate of an ISFET which was mounted in a sample cell with a reference ISFET and Ag–AgCl reference electrode.	Sensitive to ACh, H^+, Na^+ and K^+. Steady state reached after 2 min. Initial rate of change of the voltage was a rectilinear function of log ACh concentration was a from 0.1 to 10 µM. When receptor was immobilized in the membrane by using lecithin an amplified response was obtained which was due to Na^+ flux across receptor channel.	[65]

Acetylcholine receptor from electric organ of *Torpedo* sp.	Receptor protein noncovalently bound on the surface of a planar interdigitated capacitative sensor.	Response was concentration dependent and specific for ACh and inhibited by (+)-tubocurarine, amantidine and α-neurotoxin.	[66]
Acetylcholineesterase	*Anal Abst.*, 1988, 50, 6J174 Sensor previously described.	Determination of ACh and Ch in rats brain extracts. Rectiliner calibration graphs were obtained for 12–190 μM acetylcholine. Response times, 2 min. Recoveries were 93–105% of ACh. Used for determination of ACh in ophthalmic preparations.	[67]
Acetylcholineesterase and choline oxidase	Sensor was constructed by attaching 4 mm^2 of film, mounted between Nephrophan membrane (25 μm) to a platinum electrode using a manual polarograph or a flow through system.	Response (amount of H_2O_2 produced) was rectilinear for 1–660 μM ACh. Electrode was stable for 20 days or 700 determinations.	[68]
Acetylcholineesterase and choline oxidase	Enzymes were entrapped in a photo-cross-linkable polymer. Detection of H_2O_2 liberated from the enzyme reaction.	The range of application was 20–750 μM for ACh. Response time was 2–4 min. The immobilized enzyme membranes were very stable, and stored in a dry state.	[69]

(*continued*)

Table 1 (continued)

Biological system	Biosensor design	Results and remarks	Reference
Acetylcholineesterase	pH sensitive ISFET (based on a silicon-SiO_2-Ta_2O_5) by coating its tantalum pentoxide layer with a gelatin membrane containing chloineesterase immobilized with use of glutaraldehyde. The sensor was operated versus silver–silver chloride in saturated KCl.	Suitable for determining 0.1–10 mM ACh. A response of 61 mV per decade of the substrate concentration at pH 7.5 was obtained.	[70]
Acetylcholineesterase and choline oxidase	Enzymes were co-immobilized on chemically preactivated immundyne polyamide membrane (thickness 120 μm, size cut-off 3 μm). Applying 20 μL of 1% ChO solution in 0.1 M-phosphate buffer (pH 6). Assay was based on electrochemical detection of the generated H_2O_2.	The response time was 2 min. Detection limit was 50 nM and the response was rectilinear up to 20 μM.	[71]

Acetylcholineesterase	Enzyme was covalently immobilized on a bovine serum albumin-modified H^+-selective coated wire electrode. The sensor was used in 5 mM phosphate buffer pH 7 at 30°C.	The response time was 3–10 min for 0.1–10 mM ACh. pH change of 6–8 had little effect. Coefficient of variation was 5.7 and 5.8%.	[72]
Acetylcholineesterase	A nereistoxin sensor is prepared from a combination of AChE, ChO and an O electrode.	Good linearity was observed. Used for determination of ACh and phosphate.	[73]
Acetylcholineesterase and choline oxidase	Carbon fiber electrode (7 μm) dipped into poly(vinyl alcohol)-quaternized stilbazole-ChO-AChE solution, drying in air and exposing to fluorescent room light for 1 h.	Calibration graph was rectilinear from 0.1 to 1 mM ACh. Detection limit was 0.05 mM, of ACh and detection time is 5 s.	[74]
Acetylcholineesterase and choline oxidase	Electrode was developed by co-immobilization of AChE with ChO–H_2O_2. Disposable sensors constructed by co-immobilization in a gelatin membrane on Pt electrode or by immobilizing AChE in polyurethane on a thick-film metallized Pt electrode.	A kinetically controlled bioenzyme sensor was also used at a low activity of ChE for determining inhibitors.	[75]

(continued)

Table 1 (continued)

Biological system	Biosensor design	Results and remarks	Reference
Acetylcholineesterase and choline oxidase	Prepared by mounting a carbon fiber (200 μm diam) in a glass capillary with silver paste and epoxy resin, electrochemically pretreating the electrode from 0 to 1.2 V for 15 min, and dipping the electrode in 11% PVA-Styryl pyridinium solution containing AChE and ChO.	The calibration graph was rectilinear from 0.2 to 1 mM of ACh. The response time was 0.8 min.	[76]
Acetylcholineesterase and choline oxidase	Enzyme membrane in H_2O was treated with 11% solution of PVA-SbQ (polyvinyl alcohol) with styryl pyridinium groups. Mixture was spread on a cellulose nitrate membrane and air dried. The membrane was exposed to UV radiation for 3 h and stored at 4°C. The enzyme membrane was fixed with a Pt electrode. Sample was dissolved in phosphate buffer and measured.	The best results were obtained at pH 8 and at 30°. The calibration graph was rectilinear for 5 mM ACh. The storage stability of the dry membrane was excellent.	[77]

Acetylcholineesterase and choline oxidase	Immobilization of horseradish peroxidase in the redox polymer; poly (4-vinylpyridine-chlorobis-(2,2′-bipyridyl) osmium cross-linked by means of polyoxyethylene 400 diglycidyl ether on polished vitreous carbon electrodes.	Response time of 30 s and maintained its sensitivity for 24 h at low substrate concentration. Responses were rectilinear up to 10 mM H_2O_2 for 100 μM choline with detection limit of 10 nM and 1 μM.	[78]
Acetylcholineesterase and choline oxidase	Carbon-fiber micro-electrode for determination of ACh and choline, mounted in a glass capillary tube. Enzyme was immobilized on the surface of the carbon fiber with albumin. Electrode was dip-coated with Nafion. Evaluation of selectivity and dynamic range, at a fixed potential of 1.2 V versus Ag–AgCl.	Calibration graph for ACh and Ch was rectilinear between 0.1 and 3 mM. Interference from ascorbic acid was not observed in the range 0.1 to 0.3 mM.	[79]
Enzymes	Enzymes immobilized in a polymer solution. Acetylcholine and choline are determined at the μM level with a precision of 3.5–7%.	Detection limits were 0.1 and 1 μM of Ch and ACh, respectively. Calibration graphs were rectilinear for 0.1–100 and 1–100 μM of choline and acetylcholine. Storage stability of the sensing element is satisfactory. Response time was 3–10 min.	[80]

(continued)

Table 1 (continued)

Biological system	Biosensor design	Results and remarks	Reference
Acetylcholineesterase and choline oxidase	Coating a polished vitreous-carbon electrode with 5 μL of a 0.5% solution of choline oxidase which was dried by heating. Residue was coated with 5 μL of a 1:10 mixture of Nafion and water. The dried electrode (ChO activity 0.382 iu/cm^2) was used at 0.25 versus SCE for the amperometric determination of choline in 0.1 M phosphate buffer of pH 6.5 containing 1 mM $K_3Fe(CN)_6$ as redox mediator. ACh sensor, 0.5 μL of a 0.1% solution of AChE to choline oxidase coating and heat drying.	Calibration graphs were rectilinear for 1–13 mM-choline (sensitivity 48 nA/mM), 0.04–0.06 mM-arsenocholine (0.44 μA/mM) and up to 5 mM acetylcholine. Detection limits were 50 μM, 10 μM and 43 μM, respectively. There was no interference from anionic and easily oxidized serum constituents.	[81]

Acetylcholineesterase and choline oxidase	Carbon fiber was mounted in a glass capillary with Ag paste and epoxy resin, washed. Electrochemical pretreatment at 1.2 V dipped into an aqueous solution of ChO and BSA containing polyvinyl alcohol/styryl pyridinium polymer, exposed to fluorescent light (300–460 nm), dip-coated in 5% Nafion solution. Measurements for AChE were made at 30° and 1.2 V in 0.1 M-2-(N-morpholino) ethanesulfonic acid buffer of pH 7.1 containing ACh, chloride.	Calibration graph was linear from 0.05 to 3 mM of acetylcholine chloride. A steady-state response was reached in 2 min. Calibration graph for purified AChE was linear up to 50 iu/mL. Results agreed with the spectrophotometric method. Sensor was stable for 1.5 months on storage at 5°C.	[82]
Acetylcholineesterase and choline oxidase	Co-immobilized enzyme (AChE and ChO) on 7 μm diameter carbon-fiber electrode entrapped with polyvinyl alcohol quaternized stilbazole. Sensor was used as an amperometric detector for ACh.	Linear response in the range 0.1–1 mM of ACh. Response time was 5 s.	[83]
Acetylcholineesterase and choline oxidase	Immobilized by cross linking with glutaraldehyde vapor on the exposed end (diameter 0.2 mm) of a silica-sleeved Pt electrode.	Detection limit for ACh and Ch were 20 and 10 pmol, respectively. Sensor was used successfully in the FIA and for detection in HPLC.	[84]

(continued)

Table 1 (continued)

Biological system	Biosensor design	Results and remarks	Reference
Acetylcholineesterase, urease, glucose oxidase and butyryl chloinesterase.	Immobilized enzyme on to the sensor chip by corsslinking with glutaraldehyde and BSA. Conductivity changes, produced by the enzyme-catalyzed hydrolysis of ACh were measured for the analysis.	Detection limits for ACh was 0.07 mM with corresponding sensitivity of 5.6 ± 0.2 μS/mM. The device could be also used for apparent Michaelis constant determination.	[85]
Acetylcholineesterase	Miniaturized multichannel transductor with planar Au electrode which was first covered with a choline-selective liquid membrane made from 66% PVC-polyvinyl acetate (PVA), 33% 2-nitrophenyl octyl ether plasticizer and 1% ion-pair choline phosphotungstate. A second layer of 2% AChE in the PVA-polyethylene dispersion was spread on the top. The electrode was used as working electrode versus Ag/AgCl for potentiometric measurement of Ch and ACh in 0.1 M Tris buffer at 7.4.	Optimum pH range for the sensor was 7–9. The calibration graph was linear from 0.02–10 mm ACh and detection limit was 5 μM. Response time was 3–5 min. Sensor was suitable for determination of ACh in biological fluids.	[86]

Acetylcholineesterase and choline oxidase	Co-immobilizing AChE and ChO on to a Pt disc microelectrode (200 μM diameter) with glutaraldehyde vapor. Sensor was used in flow-injection and LC system at a potential of 0.6 V versus Ag/AgCl/KCl (saturated) with mobile phase containing 0.1 M-phosphate buffer (pH 8) (for FIA) and 0.05 M phosphate buffer at pH 7.5 containing 0.03 mM SDS and 3 mM tetramethyl ammonium chloride (for LC). The LC separations were carried out on an ODS-5 column (25 cm × 0.5 mm i.d.).	The microsensor exhibited a linear response for acetylcholine and choline for 0.05–103 pmol.	[87]
Acetylcholineesterase and choline oxidase	A Cross-linkable polymer, poly (vinyl pyridine) derivatized at the N atoms with a combination of iron-linking and redox functionalities was used to immobilize the enzymes and to shuttle electrons. Enzymes were deposited with the polymer and deposited onto C electrode. For peroxide selectivity over ascorbate is achieved by incorporation of Nafion.	The microsensors if they can be successfully used *in vivo* will provide valuable information for brain diseases (Parkinsion's and Alzheimer's).	[88]

(*continued*)

Table 1 (continued)

Biological system	Biosensor design	Results and remarks	Reference
Acetylcholineesterase and choline oxidase	A rotating graphite-disc electrode was polished, defatted, cleaned and oxidized by immersion in 10% HNO_3/2.5%, $K_2Cr_2O_7$ at 2.2 V versus SCE for 10 s. AChE was covalently immobilized on to the electrode using a standard method. Measurements were made in 0.1 M-universal buffer of pH 7 at 0.8 V versus SCE.	Calibration graphs were linear from 0.6–10 μM substrate. RSD were 5% (n = 10). The detection limit was 10 μM acetylthiocholine, electrode response time was 15 s.	[89]

Acetylcholineesterase	A 350 μM diameter coherent imaging fiber coated on the distal surface with a planar layer of analyte-sensitive polymer that was thin enough not to affect the fiber's imaging capabilities. It was applied to a pH sensor array and an ACh biosensor array (each contain 6000 optical sensor). Fibers were coated with an immobilized layer of poly (hydroxyethylmethacrylate)-N-flurosceinylacrylamide and AChE-fluorescein isothiocyanate isomer poly (acryloamide-co-N-acryl oxysuccinimide), respectively.	The response time of the pH sensor was 2 s for a 0.5 unit increase in pH. The biosensor had a detection limit of 35 μM ACh and a linear response in the range 0.1 mM.	[90]
Acetylcholineesterase and choline oxidase	300 μL 0.1 M phosphate buffer (pH 6.5) containing 16 mg BSA and 1 mg each of ChO and AChE were mixed with 30 μL 25% glutaraldehyde diluted 10 fold with phosphate buffer. The solution was used to coat the surface of a Pt electrode. This enzyme electrode was used for the amperometric measurement of ACh and Ch.	Calibration graphs were linear upto 0.09 and 0.08 mM Ch and ACh, respectively. Detection limits were 0.1 μM of both Ch and ACh. Response time was 1 s for both Ch and ACh. The use of the sensor as detector for HPLC analysis for both Ch and ACh was demonstrated.	[91]

(continued)

Table 1 (continued)

Biological system	Biosensor design	Results and remarks	Reference
Acetylcholineesterase and choline oxidase	Au foil was treated with cystamine to produce a base layer of aminothiolate units, was derivatized by reaction of the amino group and disodium-4,4′-diisothiocyanato-trans-stilbene-2,2′-disulfonate. Enzymes were immobilized at the isothiocyanate group via thiourea link. The bifunctional sensor for ACh was prepared by stepwise immobilization of four layers of the enzyme ChO and three layers of AChE. Choline generated was detected amperometircally with the use of 2,6-dichlorophenolindophenol as a mediator in solution.	Electrical communication between the enzyme and the electrode is achieved either by the use of ferrocenecarboxylic acid as mediator in the assay buffer or by immobilization of [(ferrocenyl methyl)amino] hexanoic acid on the enzyme layer.	[92]

Nicotinic acetylcholine receptor	Receptor, isolated from *electrophorus electricus*, was incorporated into a polymeric film formed *in situ* on an electronic transductor containing two terminal 1.5 cm × 1.5 cm interdigitated gold electrode.	The response of the biosensor reached equilibrium within 5 s for ACh and remained stable up to 20 min. Detection limits ~ 25 ng in a 50 μL sample (*i.e.*, 0.5 μg/mL).	[93]
Acetylcholineesterase and choline oxidase	Coated wire enzyme electrode was prepared by coating a Ag wire with a homogenized solution of CH_2Cl_2 containing 0.28 gm of polyvinyl bytyral, 0.15 gm of di-n-amylphthalate, 10 mg of tridodecylamine and 2 mg of sodium tetraphenylborate.	Three different methods for immobilizing the enzyme on the coated electrode are described with use of AChE. The model electrode was used to determine ACh from 0.1 to 10 mM. Response time was 3–6 min.	[94]
Acetylcholineesterase	Enzyme membrane electrode was prepared galvanostatically on a stationary Pt and vitreous C disc electrode at a current density of 0.5 mA/cm^2 for 45 s in a solution containing 0.4 mL of enzyme solution, 0.02 of 25% glutaraldehyde, 0.02 mL ethanol and 0.02 mL of HCl.	Electrode response was linear for substrate over the range 0.1–1 mM. The RSD of the AChE electrode in the determination of 0.1, 0.5 and 1.0 mM ACh were 5.3, 5.0, and 4.8%, respectively.	[95]

(*continued*)

Table 1 (continued)

Biological system	Biosensor design	Results and remarks	Reference
Acetylcholineesterase and choline oxidase	Pt microelectrode, precoated with cellulose acetate, dipped in a buffered solution (pH 6.89) of choline oxidase or AChE containing glutaldehyde as cross linking agent.	The calibration graph was linear from 0.5 μM to 100 μM. Response time was 15–20 s. The optimum pH was 9–10.5. Electrode was stable over 41 days.	[96]
Acetylcholineesterase and choline oxidase	Enzyme immobilized over tetrathiafulvalene tetracyanoquinodimethane crystals packed into a cavity at the tip of a carbon-fiber electrode. The immobilization matrix consisted of dialdehyde starch/glutaraldehyde, and the sensor was covered with an outer Nafion membrane. The amperometric performance of the sensor was studied with the use of FIA system.	An applied potential of +100 mV versus SCE (Pt-wire auxiliary electrode) and a carrier flow rate of 1 mL/min. The Ch and ACh biosensors exhibited linear response upto 100 μM and 50 μM, respectively. Response times were 8.2 s.	[97]

Acetylcholineesterase	Screen-printed electrodes were prepared by disposition of a Ag conductive pad, a C pad and an insulating layer on a polyester film substrate. A graphite-based ink mixed with 0.5% Ru on activated C was used to print the working surface and the electrodes were heated at 110°C for 10 min.	Responses to H_2O_2 at +700 mV versus SCE were linear for 5–100 μM and the RSD was 6.7%. Acetylcholine determination was performed by adding AChE to the electrolyte. Best results were achieved using borate buffer of pH 9 containing 0.1 M KCl.	[98]
Acetylcholineesterase	Packing a thin layer of a glass bead mixture comprising AChE covalently immobilized on preactivated isothiocyanate glass and thymol blue indicator bound on aminopropyl glass on the tip of a bifurcated optical fiber sensor head. The sensor was incorporated into a flow cell and used to detect carbamate and organophosphate pesticides.	The calibration graph was linear for 2.5–25 mM acetylcholine and RDS were <2%; the response was reversible.	[99]

(*continued*)

Table 1 (continued)

Biological system	Biosensor design	Results and remarks	Reference
Acetylcholineesterase and choline oxidase	A dual syringe pump, one containing choline solution in 50 mM phosphate buffer of pH 8 in the presence of 0.5 mM EDTA. The other containing the sample solution consisting of a mixture of choline and acetylcholine in the same buffer solution. The two solutions were pumped via a microdialysis unit into an enzymatic reactor containing three separate compartments filled with ChO, catalase and AChE/ChO mixture, respectively.	The H_2O_2 generated in the first compartment by the enzymatic reaction of choline was consumed in the second compartment. The unreacted acetylcholine present in the sample reacted in the third compartment and the generated H_2O_2 was detected by an osmium-gel-HRP modified vitreous C electrode set at 0 V (versus Ag/AgCl). The calibration graph was linear up to 1 μM acetylcholine.	[100]

Acetylcholineesterase and choline oxidase	A glassy C electrode surface was modified with osmium poly (vinylpyridine) redox polymer containing horseradish peroxidase (Os-gel-HRP) and then coated with a co-immobilized layer of AChE and ChO. A 22 μL pre-reactor, in which ChO and catalase were immobilized on beads in series, was used to remove choline. The variation in extracellular concentration of ACh released from rat hippocampal tissue culture by electrical stimulation was observed continuously with the online biosensor combined with a micro-capillary sampling probe. Measurement of ACh and Ch was carried out by using a split disc C film dual electrode.	The detection limit was 4.8 nM, comparable to that obtained by LC with electrochemical detection combined with an enzyme reactor. In the online measurement of acetylcholine standard solutions, the sensitivity of the biosensor was 43.7 nA/μM at a flow rate of 16 μL/min.	[101]

(*continued*)

Table 1 (continued)

Biological system	Biosensor design	Results and remarks	Reference
Acetylcholineesterase	A 3 mm diameter, Pt disc electrode, either polished or black, was immersed for 20 min in a solution of avidin (100 μg/mL). After washing with PBS solution of pH 7.4, the electrode was immersed for 20 min in a biotin-labeled ChO solution (100 μg/mL). After treatment 10 times, the surface of the sensor was further modified with ChE in the same manner.	The amperometric response of the sensor was measured with a three-electrode cell at 0.6 V versus Ag/AgCl in 0.1 M phosphate buffer of pH 6.8. The response was linear from 1 μM to 1 mm of acetylcholine.	[102]

Acetylcholineesterase	A stock solution of 0.52 mg/mL of the pesticide trichlorophen in 10 mM phosphate buffer of pH 7.5 was diluted with buffer to various concentrations. The obtained solutions were then analyzed using an ACh biosensor based on the inhibition effect of trichlorophen on the function AChE which promotes the hydrolysis of the natural neurotransmitter, acetylcholine. The sensor was fabricated by immobilizing AChE onto the surface of an antimony disc electrode, which was then used in conjunction with a double junction Ag/AgCl (0.1 M-KCl) reference electrode with a 0.1 M lithium acetate salt bridge.	A linear calibration plot was obtained in the range of 10 ppb–10 ppm with a lowest detection limit of 0.1 ppb.	[103]

(*continued*)

Table 1 (continued)

Biological system	Biosensor design	Results and remarks	Reference
Acetylcholineesterase and choline oxidase	The detector consisted of two Pt electrodes (6 mm × 3 mm) sandwiched between Perspex sheets and separated by a 1 mm thick sheet of silicone rubber, and the carrier stream (0.5 mL/min) was 0.1 M phosphate buffer (pH 8.2). AChE and ChO were immobilized by glutraldehyde cross-linking to controlled-pore glass and packed into columns (3 cm × 2.5 mm) that were operated at 25°C.	Rectilinear calibration graphs for 10–100 μM choline and acetylcholine were obtained.	[104]
Nicotinic acetylcholine receptor	The biosensors were constructed with poly (vinylbutyral) membranes incorporating nicotinic acetylcholine receptor for the determination of ACh.	The detection range was 0.1–10 μM acetylcholine.	[105]

Acetylcholineesterase	Biosensors were fabricated from filter-supported solventless bilayer lipid membrane (BLM) and used for the analysis of the substrates of hydrolytic enzymes in a flow-through system. The codeposition of lipid (dipalmitoyl-phosphatidic acid) and protein solutions to form a BLM on a microporous glass fiber or polycarbonate ultra-filtration membrane disc was described. Enzyme was immobilized on the membrane by incorporation of protein solution into the lipid matrix at the air-electrolyte interface before BLM formation.	BLM containing AChE was used for the analysis of acetylcholine following injection of the substrate into a 0.1 M KCl/10 mM HEPES carrier electrolyte. The enzymatic reaction at the membrane surface caused changes in the electrostatic fields and phase structure of the BLM, resulting in ion current transients; the magnitude of these were linearly related to the substrate concentration down to the μM level. The RSD were 5%.	[106]

(*continued*)

Table 1 (continued)

Biological system	Biosensor design	Results and remarks	Reference
Acetylcholineesterase	The aqueous fermentation solution containing cholineesterase (90–100 mg/mL) was added to nitro-cellulose in butyl acetate (30 mg/mL). From 1 to 5 μL of the mixture was spotted onto a small area of FET and dried in the air. The FET was treated with 5% of glutaraldehyde for 5–10 min, washed with 10 mM borate buffer (pH 9.18) for 15 min, dried and stored at 4°C.	Acetylcholine iodide was studied in 15 mM NaCl and 13 mM borate buffer of pH 9–10. Detection limit for acetylcholine was 40 μM. Sensitivity losses was $\leq$ 40% over 20 days.	[107]

Acetylcholineesterase and choline oxidase	The ISE biosensors were fabricated from membranes containing 1.2% of 2,6-didodecyl β-cyclodextrin/65.6% O-nitrophenyloctyl ether/0.4% sodium tetrabis-3,5-kis-(trifluoromethyl)phenyl borate (TKB) and 32.8% PVC or polyurethane. The biosenser was prepared by coating a screen-printed C working electrode with 1,1′-bis-(methoxymethyl) ferrocene followed by application of a horseradish peroxidase/AChE/ChO solution.	A thin-film composed of 40% plyurethane/59.5% bis(1-butylpentyl)-adipate/4.8% 2,3,6-triethyl-β-cyclodextrin/0.6% TKB was cast over the enzyme layer. Acetylcholine was determined at 0.18 V versus Ag/AgCl. Detection limit was at the sub-pM level. Interference with other electroactive compounds was minimal.	[108]
Acetylcholineesterase and choline oxidase	A glassy carbon electrode (GCE) modified by electrodepositing sub μm Pt-black particles on the surface. ACh and Ch. micro biosensor arrays were fabricated based on immobilization of AChE–ChO or ChO by cross linking with gentaraldehyde on Pt-black GCE.	Significant enhancement in the performance of these biosensors was achieved. The chronoamperometric response of 1 μ biosensor array was linear from 29 to 1200 μM with detection limit of 8.7 μM acetylcholine.	[109]

(continued)

Table 1 (continued)

Biological system	Biosensor design	Results and remarks	Reference
Acetylcholineesterase	An acrylohydrazide-methacrylohydrazide pre-polymer was prepared and an aqueous 30% solution was added to a solution of AChE in phosphate buffer containing gelatin. The enzyme-copolymer solution was applied to the tip of a glass electrode, dried and cross-linked. Other sensors based on pH-sensitive Pd–PdO and Ir–IrO_2 electrodes were also prepared with the enzyme and copolymer.	Potential measurements were made with a Radiometer PHM 65 pH meter and recorder with use of HEPES buffer containing $MgCl_2$, NaCl, and gelatin as sample buffer and acetylcholine as substrate. The detection limits for nicotine and the organophosphorus compound were given.	[110]
Biological tissues	The electrode was constructed from a slice of pig muscle (0.9–1 mm thick), stuck over the flat end of a glass pH electrode and covered with a nylon net fixed with a rubber ring.	The calibration graph of potential versus concentration was rectilinear from 20 to 600 µg/mL of acetylcholine in 1 mM. KCl at 30°C. There was no interference from a wide range of biological compounds. Recovery was 98.2% and the coefficient of variation was less than 8% ($n = 8$).	[111]

Nicotinic receptor ligands	A silicon-based light addressable potentiometric sensor was applied to detect biotinylated α-bungarotoxin bound to carboxy fluorescein-nicotinic acetylcholine receptor conjugate. This conjugate (1–20 ng) was incubated for 1 min with 20 nM-biotinylated α-bungarotoxin in assay buffer solution containing 150 mM NaCl, 0.1% bovine serum albumin and 0.05% Triton X-100 in 10 mM Na_2H-NaH_2PO_4 (pH 7).	The sensor could be applied to determine agonists and antagonists. Allows the detection of nicotinic receptor ligands with a light addressable potentiometric sensor. After the addition of streptavidin solution, the mixture was filtered through the biotinylated nitrocellulose capture membrane.	[112]

(*continued*)

Table 1 (continued)

Biological system	Biosensor design	Results and remarks	Reference
Acetylcholineesterase	Bilayer lipid membranes were prepared by adding a solution of egg phosphatidylcholine and dipalmitoyl phosphatidic acid dropwise into the surface of aqueous 0.1 M KCl/10 mM HEPES, near the Saran Wrap partition of a two compartment plexiglass cell. A portion of AChE solution in 10 mM Tris hydrochloride buffer solution of pH 7.4 was applied. The electrolyte level was momentarily dropped below the orifice and raised to form a membrane. The membranes were used as transducers for the reaction of AChE with ACh. An external voltage (25 mV) was applied across the membrane between two Ag/AgCl reference electrodes.	Enzymatically generated hydronium ion causes transient current due to alteration of the electrostatic field by the ionization of dipalmitoyl phosphatidic acid. The response delay time was directly related to the substrate concentration where acetylcholine can be determined from 1 μM upto mM level.	[113]

Acetylcholineesterase	Immobilization of AChE on the surface of a glass electrode is achieved by coating the electrode with a AChE-bovine serum albumin mixture and gelatin, and successive treatment with a solution of glutaraldehyde and a mixture of glycine and lysine. The electrode was applied in the determination of ACh in phosphate buffer pH 8 at 25°C.	The acetylcholine concentration was found from a calibration graph which was rectilinear from 0.1 to 100 mM. The detection limit was 0.01 mM and the response time was 1 to 3 min. The electrode is intended for the application to biological materials.	[114]
Acetylcholineesterase and choline oxidase	Sensors were developed consisting of a Clarke oxygen electrode modified by superposition (over the polypropylene membrane of the electrode) of a dialysis membrane on which AChE and ChO were immobilized. Both formats are based on the conversion of the substrate to choline which is oxidized in the presence of choline oxidase causing a reduction in pO_2 which is detected by the electrode.	Free choline is measured with a similar electrode with only choline oxidase immobilized on the membrane. Calibration graphs were rectilinear upto 90 μM choline and upto 120 μM acetylcholine. Results were presented for the analysis of tissue extracts and pharmaceutical formulations.	[115]

(*continued*)

Table 1 (continued)

Biological system	Biosensor design	Results and remarks	Reference
Acetylcholineesterase and choline oxidase	Amniotic fluid was mixed at 27°C with phosphate buffer (pH 8) and the current was measured with an electrode consisting of an immobilized choline oxidase membrane at 650 mV versus Ag/AgCl. After reaching the steady state, acetylcholine or butyrylcholine was added and the current was measured.	Rapid and sensitive discriminating determination of AChE activity in amniotic fluid with choline sensor. The calibration graphs were rectilinear and the coefficient of variation was 2.5%.	[116]

Acetylcholineesterase and choline oxidase	Microdialysates were obtained from the brain of rats by fitting a cannula directly into the striatum and perfusing with Ringer's solution (2 μL/min) for 5 h. Fractions were automatically collected and analyzed by liquid chromatography. The eluate was passed through a 10 μM F 8903 enzyme reactor column containing AChE and ChO and the H_2O_2 produced from ACh by the enzymatic reactions was detected at vitreous C electrode operated at + 100 mV versus Ag/AgCl.	Detects the basal acetylcholine in rat brain microdialyzate. The detection limit obtained by this method was 10 fmol for acetylcholine. The average basal striatal concentration of acetylcholine was 31 fmol/5 μL dialysate.	[117]
Acetylcholineesterase	Acetylcholineesterase was immobilized on acetyl-cellulose activated by $NaIO_4$ to give an enzyme membrane which was bound to a pH sensing membrane containing 4,4′-bis-(N,N-didecylaminomethyl) azobenzene of an original electrode.	Average recovery was 99.5% with RSD of 2.18%. The static and dynamic linear response ranges were from 240 to 1300 and 300–1700 μg/mL acetylcholine, respectively. Optimum measurement medium was 1 mM KCl.	[118]

(*continued*)

Table 1 (continued)

Biological system	Biosensor design	Results and remarks	Reference
Acetylcholineesterase and choline oxidase	The sample was diluted with water and the resulting solution was pumped into a flow-through cell containing a rotating (1064 rpm) Teflon disc with AChE and ChO co-immobilized on the top surface of the disc. Measurements were made by stopping the flow for 60 s. The H_2O_2 produced from the enzymatic reaction was detected amperometrically at a Pt ring working electrode located above the rotating Teflon disc and held at +0.6 V versus Ag/AgCl.	Enzymatic determinations with rotating bioreaction and continuous flow stopped flow processing for choline esters in pharmaceuticals. The initial rate of response was measured and used for quantitation of succinylcholine and acetylcholine. The calibration graph was linear from 0.1 to 1.6 mM for acetylcholine.	[119]

Acetylcholineesterase	The surface photovoltage technique was applied to the fabrication of a biosensor based on immobilized cholineesterase. The type of the cholineesterase used depends on the type of substrate. On the surface of the silicon wafer the cholineesterase layers were immobilized by using 3-aminopropyl triethoxysilane and glutaraldehyde. Characteristics of the sensor were studied in phosphate buffered saline containing 15 mM NaCl and 1 mM phosphate buffer (pH 7).	The detection limits of the substrates were 9×10^{-7} M, 2.7×10^{-6} and 4.1×10^{-6} for butyrylcholine iodide, acetylcholine iodide and acetylcholine chloride, respectively. The activity of the cholinesterase was inhibited by the presence of alkaloids such as physostigmine and neostigmine.	[120]

(continued)

Table 1 (continued)

Biological system	Biosensor design	Results and remarks	Reference
Acetylcholineesterase, choline oxidase and butyrylcholine esterase	Different nonconducting polymers were synthesized on the surface of a Platinum electrode to assemble fast-response and sensitive amperometric biosensor for acetyl-choline choline and butyrylcholine based on ChO, AChE or butyryl-choline esterase co-immobilized by crosslinking with BSA and glutaraldehyde. Also, the electropolymerization conditions such as scan rate and time were optimized for each monomer in order to obtain permselectivity, the highest rejection of interference and stability.	The procedure of immobilization was fast, simple with better long-term stability and higher loading of the enzyme. Studies on reproducibility, life time of immobilized enzyme, pH, temperature, interference, buffer, response time, storage and operational time of the biosensor were carried out.	[121]

Acetylcholineesterase and choline oxidase	The detector fabrication of glutaraldehyde co-crosslinking of AChE and ChO with bovine serum albumin on the Pt working electrode of a conventional thin layer electrochemical flow cell. A mobile phase of phosphate buffer 0.1 M; pH 6.5 containing 5 mM-sodium hexane sulfonate and 10 mM-tetramethylammonium phosphate was used in the ion-pair, reversed phase liquid chromatography.	A liquid chromatography detector based on a fast response and sensitive bienzyme amperometric biosensor for ACh and Ch was described. Linear responses were observed over at least four decades and the absolute detection limit were 12 and 27 fmol injected for choline and acetylcholine, respectively.	[122]

(*continued*)

Table 1 (continued)

Biological system	Biosensor design	Results and remarks	Reference
Acetylcholineesterase and choline oxidase	The electrochemical sensor for acetylcholine/acetylthio choline and choline/thiocholine are developed using two enzyme reactors: AChE encapsulated organically modified sol-gel glass and choline oxidase immobilized within mediator [tetracyanoquinodimethane, tetrathiafulvalene and dimethylferrocene]-modified graphite paste electrode. The AChE immobilized into organically modified sol-gel glass behaves as the reactor for enzymatic hydrolysis of ACh into choline whereas mediator and choline oxidase modified paste electrode are used for the detection of choline through mediated mechanism.	The typical response curves for the detection of choline using mediators and choline oxidase-modified electrodes below 0.24V versus silver/silver chloride in 0.1 M Tris–HCl buffer pH 8 are reported. Comparative analytical performance on the mediated electrochemical responses of the biosensors is discussed.	[123]

Horseradish peroxidase (HPP), choline oxidase and acetylcholineesterase	A three enzyme layered assembly on Au electrodes or Au quartz crystal, consisting of HRP, ChO and AChE is used to sense ACh by the HRP-mediated oxidation of 3,3′,5,5′-tetramethyl benzidine (1) by H_2O_2 and the formation of the insoluble product (2) on the respective transducers. Acetylcholine is hydrolyzed by AChE to choline that is oxidized by ChO and O_2 to yield the respective betaine and H_2O_2. The amount of generated H_2O_2 and the resulting insoluble product on the transducer correlates with the concentration of acetylcholine in the samples.	Acetylcholine was sensed by the tri-component enzyme layered electrode using Faradic independence spectroscopy, Cyclic Voltammetry and Microgravimetric quartz crystal microbalance transduction methods. Acetylcholine concentrations correspond to 1×10^{-5} M are easily sensed by the different transducers.	[124]

monitoring of the pH. The end point was detected using a derivative potential time response curve, and differentiation and smoothing were carried out using the Stavitzky-Golay algorithm. The experiment was carried out in a triangle-programmed coulometric flow titration system. Calibration graphs were established and the detection limit was 80 μM for acetylcholine.

3.5 Flow injection method

Yao *et al.* reported a flow injection analytical system for the simultaneous determination of acetylcholine and choline that made use of immobilized enzyme reactors and enzyme electrodes [25]. Acetylcholineesterase–choline oxidase and choline oxidase were separately immobilized by reaction with glutaraldehyde onto alkylamino-bonded silica, and incorporated in parallel as the enzyme reactors in a flow injection system. The sample containing acetylcholine and choline in 0.1 M phosphate buffer (pH 8.3) carrier solution was injected into the system. The flow was split to pass through the two reactors, recombined, and mixed with 0.3 mM $K_4Fe(CN)_6$ reagent solution before reaching a peroxidase immobilized electrode. Because each channel had a different residence time, two peaks were obtained for choline and total acetylcholine and choline. Response was linear for 5 μM–0.5 mM choline, and for 5 μM–1 mM acetylcholine plus choline. The detection limits were 0.4 μM for choline and 2 μM for acetylcholine.

3.6 Amperometric methods

Gunaratna and Wilson reported the optimization of the multi-enzyme flow reactors for the determination of acetylcholine [26]. After separation of acetylcholine and choline by HPLC (Nucleosil C_{18}), the analytes were eluted into a flow reactor containing immobilized acetylcholineesterase (which hydrolyzed acetylcholine to choline). Choline oxidase then is used to convert choline into H_2O_2, which is then determined amperometrically. Of several immobilization techniques investigated, immobilization through the avidin–biotin linkage gave the highest activity of the immobilized enzyme, leading to improved detection limits. Better sensitivity and detection limits were obtained with both enzymes immobilized together on the same support through the avidin–biotin linkage than with separate immobilization in two columns. The post column system was applied to the determination of acetylcholine and choline in brain tissue extracts, and linear calibration graphs were obtained from 0.1 to 500 pmol.

Yao *et al.* described an online amperometric assay method for the determination of acetylcholine using microdialysis probes and immobilized enzyme reactors [27]. The probe consisted of a fused silica tube (350 μm i.d.), in which another two fused-silica tubes (75 μm i.d.) were inserted to serve as inlet and outlet. A regenerated cellulose dialysis fiber (200 μm i.d.) was fixed to the interior of the first tube. A stream (2 μL/min) containing 147 mM sodium chloride, 4 mM potassium chloride, 1.2 mM calcium chloride, and 1 mM magnesium chloride was pumped through the microdialysis probe, which was immersed in the analyte solution. The dialysate was merged with the buffer stream (0.1 M phosphate buffer of pH 7 or 0.1 M borate buffer pH 8) and the mixed stream was pumped through a PTFE reactor coil (20 mm × 0.5 mm i.d.) packed with controlled pore glass immobilized with a mixture of acetylcholineesterase and choline oxidase, for the determination of acetylcholine. Choline was removed from the dialysate. The resulting H_2O_2 was detected amperometrically using Pt disc electrode modified by coating with a film of poly(1,2-diaminobenzene). The detection limit was 1.5 μM for acetylcholine.

4. SPECTROPHOTOMETRIC METHODS OF ANALYSIS

4.1 Infrared spectrophotometric method

Mynka *et al.* reported a method for the infrared spectrophotometric identification and the determination of acetylcholine in the presence of other esters in pharmaceuticals [3]. Acetylcholine was determined at the maximum absorbance of $1755\,cm^{-1}$. Beer's law was obeyed in the concentration range 1–10 μg/mL, and the relative standard deviation was less than 2.7%. Results obtained were reported to be reliable and reproducible.

4.2 Photometric methods

Eksborg and Persson reported a photometric method for the determination of acetylcholine in rat brain after selective isolation by ion-pair extraction and microcolumn separation [28]. The same authors also reported a photometric method for the determination of acetylcholine and choline in brain and urine samples after selective isolation by ion-pair column chromatography [29].

Barcuchova *et al.* reported an extractive photometric method for the determination of choline and its derivatives that made use of bromothymol blue [30]. The sample, containing the iodide of the choline derivative, was treated with a solution of bromothymol blue in pH 6.5 phosphate buffer medium. The aqueous solution was extracted with chloroform, and the absorbance of the extract was measured at 416 nm against a reagent blank.

4.3 Mass spectrometric method

Suzuki and Haug compared the use of magnetic switching selected ion monitoring with conventional voltage switching selected ion monitoring for applicability to acetylcholine and choline analyses in brain samples [31]. Deuterated and nondeuterated acetylcholine and choline derivatives were analyzed by mass spectroscopy with magnetic and voltage switching selected ion monitoring of dimethylmethyleneimmonium ions, m/z 58, 60 and 64. With both monitoring methods, a maxima peak was obtained at 400 ms/ion, but the smallest variation was obtained at 100 ms/ion. The method is reported to be an alternative method to voltage switching selected ion monitoring.

Lehmann *et al.* determined choline and acetylcholine levels in distinct rat-brain regions using a stable-isotope dilution and field-desorption mass spectrometry method [32]. Tissue samples (20 ng) were homogenized at 0°C for 30 min with 4% $HClO_4$ (1 mL), including internal standards of $^2H_2$ choline and $^2H_9$ acetylcholine (10–11 mL). Each homogenate was centrifuged for 20 min, and 350 μL of the supernatant solution was transferred to a Reacti-vial with 1.5 mL of aqueous 0.5 M–Na_2HPO_4 and 1 mL of dipicrylamine in CH_2Cl_2. After stirring the mixture for 20 min, the upper layer was discarded and 1 mL of carbonate buffer solution (pH 9) was added. The lower phase (700 μL portion) was transferred to another vial, the solvent removed by a stream of nitrogen, and the residue dissolved in 50 μL of methanol. Aliquots (5 μL) were subjected to cyclic scans at $m/z = 104$ and 106 for choline and at $m/z = 148$ and 155 acetylcholine (and the deuteron-analogues). The signal was accumulated through the use of multi-channel analyses.

4.4 Colorimetric methods

Tsubouchi reported a colorimetric procedure for the analysis of acetylcholine [33]. About 2–10 mL of a test solution containing acetylcholine is extracted into 1,2-dichloroethane in the presence of the ethanolic reagent (2 mL at 5 mM) and pH 10 0.1 M $Na_2B_4O_7$–0.2 M Na_2HPO_4 buffer

(5 mL) in a total volume of 25 mL. The absorptivity of the extract was measured at 615 nm.

Barletta and Ward described a procedure for the spectrophotometric analysis of acetylcholine levels in plasma [34]. 0.3 g of K_2HPO_4 and 0.3 g of Na_2CO_3 are added to a sample of plasma (0.5 mL) containing 4–20 ng of acetylcholine. To this solution is added 0.08% bromophenol blue solution in 30% K_2HPO_4 solution (0.5 mL), and the resultant extracted with 1,2-dichloroethane containing 4% isoamyl alcohol. One then measures the absorptivity of the organic phase at 600 nm. The absorbance derived from a similar treatment of a sample of plasma taken before the administration of angiotensin II is measured. The change in the concentration of acetylcholine in turtle blood following the administration of angiotensin II is measured by a modification of the method of Mitchell and Clark [35].

Das and Abdulla reported a modified method for estimation of acetylcholine [36]. Acetylcholine reacts with hydroxylamine in alkaline medium to give hydroxamic acid, which produces a red–violet color when reacted with $FeCl_3$. Absorbance measurements are made at 540 nm. The method can be applied to saliva samples stored at 0°C for as long as 72 h.

Abernethy *et al.* reported an enzymatic method for erythrocyte acetylcholineesterase [37]. Haemolysate 50 μL was mixed with 67 mM phosphate buffer of pH 7.4 (2 mL) and 21.5 mM acetylcholine (50 μL), and the mixture incubated for 20 min at 37°C. A color reagent (500 μL) containing 4 mM phenol, 0.8 mM aminoantipyrine, peroxidase (5 iUL), choline oxidase (4 L iU/L), and 0.86 mM physostigmine plus the buffer was added and, after 20 min the absorbance was measured at 500 nm.

Sakai *et al.* reported a novel flow injection method for the selective spectrophotometric determination of acetylcholine using thermochromism of ion associates [38]. Samples (0.14 mL) containing acetylcholine were injected into a flow injection system with a buffered (pH 11) carrier stream and a reagent stream (10 mM tetrabromo-phenolphthalein ethyl ester in dichloroethane) at 0.8 mL/min. The temperature of the flow cell was 45°C which reduced interference and improved recovery, and the detection was at 610 nm.

Sakai *et al.* reported the use of batchwise and flow-injection methods for the thermo-spectrophotometric determination of acetylcholine and choline with tetrabromophenolphthalein ethyl ester [39]. An aqueous sample solution was injected into a carrier stream of 0.3 M KH_2PO_4/0.1 M sodium borate (adjusted to pH 11 with 1 M sodium hydroxide), which merged with the extracted solution of tetrabromophenolphthalein ethyl ester in dichloromethane. The extract was separated with the use of a PTFE porous membrane and passed through a micro-flow cell at 45°C before its absorbance was measured at 610 nm.

Takeuchi *et al.* published a mechanized assay of serum cholinesterase by specific colorimetric detection of the released acid [40]. The cholinesterase reaction was carried out on a thermostatted rack at 30°C with a reaction mixture of serum (10 μL), 50 mM barbitone–HCl assay buffer (pH 8.2; 140 μL), and 12.5 mM acetylcholine solution (50 μL). The solutions were prepared by programmed needle actions, and a sample blank was also prepared. The reaction was stopped after 9.7 min by injection of the mixture into a flow injection analysis system to determine the quantity of acetic acid formed. The carrier stream (water, at 0.5 mL/min) was merged with a stream (0.5 mL/min) of 20 mM 2-nitrophenylhydrazine hydrochloride in 0.2 M HCl and a stream (0.5 mL/min) of 50 mM 1-ethyl-3-(3-dimethyl-aminopropyl) carbodiimide hydrochloride in ethanol containing 4% of pyridine. The sample was injected into this mixture (pH 4.5), passed through a reaction coil (10 m × 0.5 mm) at 60°C, 1.5 M NaOH was added, and, after passing through a second reaction coil (1 m × 0.5 mm) at 60°C, the absorbance was measured at 540 nm.

4.5 Fluorimetric methods

O'Neill and Sakamoto reported an enzymatic fluorimetric method for the determination of acetylcholine in biological extracts [41]. Nanomolar amounts of acetylcholine were determined in perchloric acid extracts of biological materials (brain tissues) by use of a system containing acetylcholineesterase, acetyl CoA synthetase, maleate dehydrogenase, and citrate synthase. The production of $NADH_2$ was stoichiometrically related to the amount of acetylcholine in the system, and was followed fluorimetrically. Interfering fluorescent substances in the brain extracts were removed with acid-washed Florisil.

Sutoo *et al.* developed a high sensitivity and high linearity fluorescence microphotometry system for distribution analysis of neurotransmitter and related substances in the small brain regions [42]. The method makes use of the fluorescence intensity of not more than 10,000 points in animal brain slices, which are immunohistochemically and histochemically stained.

Ricny *et al.* determined both acetylcholine and choline by flow-injection with immobilized enzymes and fluorimetric and luminescence detection [43]. The analytes are injected into a continuous stream of a medium flowing through a sequence of enzyme reactors containing acetylcholineesterase, choline oxidase, and peroxidase. Additional reactors with choline oxidase and catalase are used to remove endogenous choline from tissue extracts in which the contents of acetylcholine is to be measured. Reaction products are detected fluorimetrically or luminometrically. The limits of sensitivity are about 10 pmol/sample with luminometric and 0.2 pmol/sample with fluorimetric detection.

A fluorimetric assay method for the determination of acetylcholine with picomole sensitivity was reported by MacDonald [44]. The method is based on the hydrolysis of acetylcholine to choline and acetate, catalyzed in the presence of acetylcholineesterase, oxidation of choline to betaine, and H_2O_2 in the presence of choline oxidase, and oxidation of 4-hydroxyphenylacetic acid by H_2O_2 to a fluorescent product, catalyzed by peroxidase. The interference in the analysis of brain homogenates was discussed.

Roulier *et al.* reported a sensitive and specific method for the measurement of choline and hydrogen peroxide in sea-water [45]. Choline was oxidized by choline oxidase to produce betaine and H_2O_2. The latter was used with horse-radish peroxidase to oxidize hydroxyphenyl-propionic acid to produce a fluorescent diphenol end product. The resulting fluorescence at 410 nm (excitation at 320 nm) was proportional to the amount of H_2O_2, and could thus be used to measure the amount of choline present in the sample. Only 2-dimethyl aminoethanol interfered. The method was optimized, and used to determine 0–45 nM choline in coastal sea-water.

Brennan *et al.* used a method to detect the reaction of acetylcholineesterase with acetylcholine [46]. The method was based on the use of a monolayer, consisting of fatty acids having C_{16} chain lengths, which were covalently attached to quartz wafers and which contained a small amount of nitrobenzoxadiazole dipalmitoyl phosphatidylethanolamine (NBD-PE) (partitioned from water into the membrane). The enzyme–substrate reaction produced a decrease in fluorescence intensity from the monolayer, and the detection system was sensitive to the changes in bulk concentration of as small as 0.1 μM, with a limit of detection of 2 μM of acetylcholine. The mechanism of transduction of the enzymatic reaction was investigated using spectrofluorimetric methods and fluorescence microscopy.

A kinetic-fluorimetric method for the determination of choline and acetylcholine by oxidation with cerium (IV) was reported by Lunar *et al.* [47]. To sample solutions containing 0.017–1.0 mM choline and/or acetylcholine were successively added 6 M H_2SO_4 (5 mL) and 7.1 mM Ce(IV) solution (0.35 mL), the mixture diluted to 10 mL with water, and the solution heated to 80°C for 2 min. A portion of the solution was transferred to a cell maintained at 20 ± 0.1°C, and after 1 min the Ce(III) fluorescence intensity was measured at 360 nm (excitation at 260 nm) as a function of time.

4.6 Chemiluminescence methods

Ternaux and Chamoin described an enhanced chemiluminescence assay method for the determination of acetylcholine [48]. Reaction medium was prepared by mixing 250 iu/mL of choline oxidase (100 μL), 2 mg/mL of horse-radish peroxidase (50 μL) and 10–120 μM luminol in 100 μL of 0.1 M Tris buffer (pH 8.6), or 100 μL of 10–100 μM 7-dimethylamino-naphthalene-1,2-dicarbonic acid hydrazide, for 10 min in 5 mL of 0.1 M sodium phosphate buffer (pH 8.6). Aqueous 0.325–80 pmol of acetylcho-lineesterase (50 μL) purified on a Sephadex G50 coarse column was mixed with 450 μL of reaction mixture, and the chemiluminescence was measured at 27°C.

Luterotti and Maysinger reported a chemiluminescence method for the determination of acetylcholine and choline-related substances in pharmaceutical preparation by dot-blot [49]. The sample (1 μL) was applied as a dot to a Hybond-N nylon membrane (Amersham), and 1 μL of a mixture comprising 50 μL of 0.1 M phosphate buffer at pH 8.6, 10 μL of choline oxidase (25 iu/mL), 2 μL of horse-radish peroxidase solution (5 mg/mL), and 50 μL of ECL western blotting reagent mixture (Amersham, 1:1) was added. Hyperfilm ECL chemiluminescence film (Amersham) was immediately exposed to the luminescing dots for 20 min, and then the film was developed, washed with water, fixed and washed again before densitometric evaluation. Acetylcholine was hydrolyzed by AChE before the analysis. This method can be used also to determine the sum total of choline and acetylcholine.

Fan and Zhang determined acetylcholine and choline in rat brain tissue by a fluorescence immunoassay method, making use of immobilized enzymes and chemiluminescence detection [50]. Tissue was homogenized with a 10 fold volume of 0.6 M $HClO_4$, the homogenates were kept on ice for 30 min, and then centrifuged at 2000 G for 20 min. The pellets were discarded and the supernatants neutralized with 2 M K_2CO_3 solution (pH 4.2). Portions of the neutralized supernatant were injected into a carrier stream, consisting of 0.05 M Tris hydrochloride buffer (0.6 mL/min) in a flow injection manifold, and the stream passed through two reactors containing acetylcholineesterase and choline oxidase immobilized on glass beads (30 × 2 mm i.d.). The resultant H_2O_2-containing analyte stream was merged with streams of 1 mM luminol in 0.1 M NaOH and 25 μM $CoCl_2$ (both at 2.5 mL/min) and the chemiluminescence intensity signal measured.

Hasebe *et al.* described a simultaneous flow-injection method for the determination of acetylcholine and choline based on luminol chemiluminescence in a micellar system with online dialysis [51]. The method was based on the determination of H_2O_2 produced from acetylcholine and choline by enzymatic reaction. A water carrier, a 0.02 M disodium hydrogen phosphate/NaOH buffer (pH 11), a 0.02 M phosphate buffer (pH 7), a 1 μM Co(II) solution in 0.01 M HCl (the catalyst for the chemiluminescence reaction), and a mixed solution of 0.5 μM luminol, 0.1 M sodium hydrogen carbonate, and 0.18% SDS in borate buffer solution (pH 11), were pumped by the three double-plunger pumps. An 800 μL sample was injected into the carrier, dialyzed to remove protein, and passed through an ion-exchange column to remove other interfering substances. Choline was converted to H_2O_2 at another immobilized enzyme column. The reactions were performed at 37°C. Hydrogen peroxide produced from choline was passed through a delay coil before the chemiluminescence of the acetylcholine and choline reaction products was measured from 350 to 650 nm. This method determines down to 1 μM at a sampling rate of six samples per hour.

Emteborg *et al.* developed a peroxyoxalate chemiluminescence method for the determination of acetylcholine in aqueous solution [52]. The method was used for the enzymatic detection of the generated H_2O_2 in FIA and LC separation modes. The analyte stream (at 0.5–0.6 mL/min) was merged with the reagent stream (1 mM 1,1′-oxalyl diimidazole) in acetonitrile, and the flow was passed through the detector cell where the chemiluminescence was detected using a photomultiplier tube. The chemiluminescence signal was independent of the composition of the analyte stream over the pH range 6–7. The detection method was applied to a FIA system for choline and acetylcholine determination, based on LC separation, and followed by a choline oxidase acetylcholineesterase enzyme reactor. Choline and acetylcholine were separated using a cation exchanger with 5 mM tetramethylammonium nitrate in 30 mM phosphate buffer at pH 7 as the mobile phase. The linear range for both analytes was 50 nM (detection limit) to 10 μM. The FIA system was employed to determine choline and acetylcholine in diluted urine.

Wetherell and French determined both acetylcholine and choline by a chemiluminescence method [53]. The optimum conditions for this determination were determined by using the method of Israel and Lesbats, in which acetylcholine was first hydrolyzed to choline by acetylcholineesterase and choline oxidase. The reaction products were then further reacted by the oxidation of choline to hydrogen peroxide and betaine, which in the presence of peroxidase oxidizes luminol to aminophthalate with the emission of light [54]. Linear standard curves were obtained for acetylcholine and choline in the picomole range. Since the enzymes differ in their pH optima, a compromise must be made, so the use of 0.1 M glycine buffer (pH 9.2) is recommended for light-sensitivity determinations (1–20 pmol) of acetylcholine and choline.

Israel and Lesbats reported a chemiluminescence method for the determination of acetylcholine, and continuous detection of its release from torpedo electric organ synapses and synaptosomes [55]. Birman described a new chemiluminescence assay method for the determination of acetylcholineesterase activity with the natural substrate [56]. The method involved monitoring the increase in light emission produced by accumulation of choline, or measuring the amount of choline generated.

A comparison between the chemiluminescence and the radioenzymatic assay methods that are used for the measurement of acetylcholine released from a rat phrenic nerve hemidiaphragm preparation was reported by Ehler *et al.* [57]. The comparison demonstrated quantitative equivalency and limits of detection for different analytes as 2 pmol.

5. BIOLOGICAL ASSAY METHODS OF ANALYSIS

A sensitive biological assay method for the assay of acetylcholine was used by Vapaatalo *et al.* [58]. Superfused hamster stomach strip was used in the bioassay of acetylcholine. The preparation proved very stable and had no spontaneous movement, and the sensitivity to acetylcholine was in the range of 10–12 g/mL. Satyanarayana *et al.* reported the use of a bioassay method for the analysis of acetylcholine using the arterial blood pressure of cats [59].

Karube and Yokoyama presented an overview on the developments in the biosensor technology [60]. The overview describes the use of micromachining fabrication techniques for the construction of detection units for FIA, electrochemical flow cells and chemiluminescence detectors. Acetylcholine microsensors using carbon fiber electrodes and glutamate microsensors for neuroscience were discussed.

Aizawa presented an overview on the principles and applications of the electrochemical and optical biosensors [61]. The current development in the biocatalytic and bioaffinity bensensor and the applications of these sensors were given. The optical enzyme sensor for acetylcholine was based on use of an optical pH fiber with thin polyaniline film.

Mascini and Guilbault presented a review covering the methods used for immobilization of the enzymes, enzyme electrodes, and sensors for choline and acetylcholine [62]. A number of reports have appeared in the literature [21, 22, 63–124] describing biosensors used for the detection and the analysis of acetylcholine. The important characteristics of some of these reports are summarized in Table 1.

6. CHROMATOGRAPHIC METHODS OF ANALYSIS

6.1 Thin layer chromatography (TLC)

Acetylcholine was separated by Paul and Ebrahimian from adrenaline and serotonin [125]. The thin layer chromatographic behavior of acetylcholine and other quaternary ammonium halides were studied on silica gel by Iorio *et al.* [126].

6.2 Gas chromatography (GC)

Acetylchloline has been separated and analyzed by many gas chromatographic systems. Table 2 summarizes some of these systems [127–130]. The important characteristics of other gas chromatography systems [131–140] will be outlined in the following paragraphs.

Budai *et al.* described a GC method for the isolation and determination of choline and choline esters from Krebs-Ringer solution [130]. Samples containing 5–2000 pmol of choline and choline esters were added to 2 mL of Krebs-Ringer solution containing 120 mM NaCl, 4.7 mM KCl, 25 mM $NaHCO_3$, 2.6 mM $CaCl_2$, 1.2 mM $MgSO_4$, 1.2 mM KH_2PO_4, and 10 mM glucose and gassed with 5% carbon dioxide in oxygen. Glucose was

Table 2

Gas Chromatographic Systems

Column	Detector	Temperature	Carrier gas and [flow rate]	Sample and remarks	Reference
6 ft × 0.25 inch of 1% phenyl diethanolamine succinate on Polypak 1 (80–120 mesh).	Flame ionization	180°	Nitrogen [45 mL/min]	Acetylcholine and related compounds micro estimation	[127]
1.8 × 2 mm of Porapak Q (80–100 mesh), Chromosorb 101 (80–100 mesh) or Chromosorb 105 (80–100 mesh).	^{63}Ni electron-capture	145–170°	Helium or nitrogen [20 mL/min]	Serum and urine, 10 fmol can be detected	[128]
2 m × 3 mm of 0.4% Carbowax 1500 on graphite (60–80 mesh).	Flame ionization	—	Nitrogen	Tests in urine by head-space GLC	[129]
A glass silane-treated column 1.8 m × 2 mm packed with Pennwalt 223 amine packing (80–120 mesh).	N sensitive	170°	Nitrogen [40 mL/min]	Krebs-Ringer solution	[130]

removed by chromatography on a column (2 cm × 6 mm) of Amberlite CG-50 (100–200 mesh, H^+ form), from which choline and its esters were eluted with 2 mL of 0.25 M HCl. The eluent was freeze-dried, acetonitrile (1 mL) was added, and the solution was centrifuged (15 min; 3000 G). The supernatant solution was evaporated, choline was converted into propionylcholine by reaction (10 min) with propionyl chloride (5 μL), and the dried sample containing butyrylcholine (internal standard) was demethylated (80°C, 30 min) with 200 μL of 50 mM sodium benzenethiolate and 25 mM benzenethiol in ethyl methyl ketone. After cooling, 0.5 M citric acid (30 μL) and pentane (300 μL) were added, the aqueous layer washed with pentane (2 × 300 μL), and then mixed with ethyl ether (200 μL) and 2.5 M ammonium citrate in aqueous 7.5 M ammonia (30 μL). The ether layer was evaporated (40°C) to 25% of its volume, and a portion (5 μL) was analyzed by gas chromatography (see Table 2).

Jenden *et al.* reported the use of a GC method for the estimation of choline esters in rat brains [131]. Their earlier gas liquid chromatographic method [132], which was used for determining acetylcholine as the volatile dimethylaminoethyl acetate, has been applied successfully to the study of acetylcholine in rat brain and integrated with a mass spectrometric system. More sensitivity and more efficient recovery were obtained. Green and Szilagyi measured acetylcholine by pyrolysis gas chromatographic method [133]. Acetylcholine was extracted from the tissue with acetonitrile containing propionylcholine as an internal standard, and after centrifugation the acetonitrile was removed by shaking with toluene–diethyl ether mixture. A solution of $KI–I_2$ was added to the aqueous solution to precipitate the quaternary ammonium compound. After being dissolved in aqueous acetonitrile, the precipitate was passed through an ion-exchange resin to convert the periodides of the quaternary compounds to chlorides, which were simultaneously pyrolyzed and analyzed by GC.

Karlen *et al.* analyzed acetylcholine and choline by an ion pair extraction and gas phase method [134]. The gas chromatographic estimation of the isolated acetylcholine and choline was carried out after demethylation either with benzenethiolate or by controlled pyrolysis. Acetylcholine was quantitated by flame ionization detection at the nmol level. Mass fragment analysis was employed for the determination of acetylcholine in pmol amounts.

Takahashi and Mouharo developed a pyrolysis apparatus which can be easily connected, and subsequently disconnected, from a conventional gas chromatograph for the simultaneous determination of acetylcholine and choline in brain tissue [135]. Rat brain samples were mixed with propionyl chloride and butyryl choline and heated in the pyrolyzing system. Thereafter, the propionylated sample was analyzed by its passage through the attached gas chromatograph. Three peaks with increasing retention time were observed, corresponding to acetyl-, propionyl-, and butyryl-choline.

Maruyama *et al.* reported the use of a simple pyrolysis gas chromatographic method for the determination of choline and acetylcholine in brain tissue [136]. Schmidt and Speth reported a simultaneous analysis of choline and acetylcholine levels in rat brain tissue by a pyrolysis gas chromatographic method [137]. Kosh *et al.* reported an improved gas chromatographic method for the analysis of acetylcholine and choline in rat brain tissue [138]. Mikes *et al.* used a syringe micro-pyrolyzer for the gas chromatographic determination of acetylcholine, choline and other quaternary ammonium salts [139].

Gilberstadt and Russell determined picomolar quantities of acetylcholine and choline in a physiological salt solution [140]. Carbon-14 labeled choline and acetylcholine standards in Krebs–Ringer solution were subject to chromatography on columns (12 mm × 8 mm) packed with Bio-Sil A (200–400 mesh), and it was found that 95–98% of choline and acetylcholine were retained. Of the bound choline, 84–97% was eluted in 1.5 mL of 0.075 M HCl, and then 95–98% of the bound acetylcholine was eluted in 1.5 mL of 0.03 M HCl in 10% butan-2-one.

Findeis and Farwell described a gas chromatographic method for the determination of acetylcholine and other biogenic quaternary ammonium compounds on fused-silica capillary columns with a splitless injector as a chemical reactor-pyrolysis chamber [141]. A Hewlett Packard model 5880 A chromatograph was used with a split/splitless injector, fused silica column (20 m × 0.32 mm) of DB-1 or (30 m × 0.25 mm) of DB-1 or DB-1701 and flame ionization detection. The column was heated at 35°C for 10 min, and then the temperature was programmed to 200°C at 7.5°C/min. N-Demethylation was achieved at a concentration of 4.3 μg/mL–1.5 mg/mL by pyrolysis, or by coinjection with methanolic 0.5% ethanolamine. Studies were also made using gas chromatography.

Cranmer used a gas chromatographic method for the estimation of the acetylcholine levels in brain tissue by gas chromatography [142]. Acetylcholine was extracted from frozen rat brain with ethanolic formic acid. The extract was evaporated to dryness at 50°C under reduced pressure, and the residue was heated at 95°C with 1 M NaOH. The resulting hydrolyzate was neutralized with 1 M HCl, diluted with water, and analyzed for its acetic acid content by gas chromatography. The glass column (6 ft × 0.25 in.) was packed with Porapak Q-S (40–50 mesh) and operated at 165°C, using nitrogen (45 mL/min) as the carrier gas and a hydrogen flame ionization detector. Recovery ranged from 93 to 100% and the method was reported to be simple and rapid.

6.3 High performance liquid chromatography (HPLC)

The application of immobilized enzyme reactors in HPLC was reported by Shimada *et al.* [9] and by Shi and Du [143]. The latter authors cited the reactors used for the determination of choline and acetylcholine in their review.

Duan *et al.* reported the use of a rapid and simple method for the determination of acetylcholine and choline in mouse brain by high performance liquid chromatography, making use of an enzyme-loaded post column and an electrochemical detector [144]. Perchloric acid extracts of small brain tissue were injected onto the HPLC system with no prior clean-up procedure. Detection limits for both compounds were 1 pmol, and this method was successfully applied to the measurement of acetylcholine in discrete brain areas of the mouse.

Tao *et al.* described an HPLC method for the determination of acetylcholine in a pharmaceutical preparation [145]. Utilizing reverse phase ion pairing, acetylcholine was determined in lyophilized ophthalmic preparations. Analysis of degraded commercial samples showed the utility of the method in quantitation, being stability indicating, and useful in separating acetylcholine from choline.

Rhodes *et al.* correlated steroid sulfatase inhibition, DU-14 [P-O-(sulfamoyl)-N-tetra-decanoyltyramine] with changes in blood levels of dihydroepiandrosterone sulfate and dihydroepiandrosterone, acetylcholine levels in the hippocampus, and the blockade of scopolamine amnesia in rats [146]. Rats were administered DU-14 by intraperitoneal injection (30 mg/kg, for 14 days before memory testing) using a passive avoidance paradigm. A separate group of animals, following pretreatment with DU-14, were anesthetized and a microdialysis probe surgically implanted into the hippocampus. Microdialysis samples were collected and analyzed for acetylcholine *via* HPLC with electrochemical detection. Hippocampal acetylcholine increased by almost a factor of three.

Other high performance liquid chromatography methods reported for the determination of acetylcholine are summarized (together with their conditions) in Table 3 [4, 117, 147–193].

6.4 Liquid chromatography (LC)

Niwa *et al.* determined acetylcholine and choline with platinum-black ultramicroarray electrodes using liquid chromatography with a post-column enzyme reactor [194]. An array of platinum-black microelectrodes was utilized for the indirect detection of acetylcholine and choline following microbore liquid chromatography with post-column enzymatic reaction. The liquid chromatography system consisted of a Unijet acetylcholine analytical column (53 cm × 1 mm i.d.) coupled with Unijet ACh/Ch immobilized enzyme column and an electrochemical detection cell equipped with the working electrode, a Ag/AgCl reference electrode, and a stainless steel auxiliary electrode. The working electrode was prepared by a plating method and consisted of Pt-black particles (0.1–0.3 μm) deposited onto an Au film electrode. Each Pt-black particle operated as an ultramicroelectrode. The detector was operated at +0.5 V, and the injection volume was 5 μL. The mobile phase (eluted at a rate of 120 μL/min) was 50 mM Na_2HSO_4 buffer at pH 8 containing 0.5 mM EDTA and 5 mL/L 1% Kathon CG reagent. Detection limits of 5.7 and 6 fmol were achieved for acetylcholine and choline respectively.

Mayer determined acetylcholine and choline by enzyme-mediated liquid chromatography with electrochemical detection [195]. The two compounds were separated by passing the eluted fractions through a post-column reactor containing immobilized Acetylcholineesterase and choline oxidase. In the presence of either compound, the dissolved oxygen was converted into hydrogen peroxide, which was detected amperometrically at a platinum electrode. This method was used to determine choline in rat brain homogenates.

The acetylcholine levels in rat brain regions were determined by Ikarashi *et al.* [196]. The method used was an application of FRIT fast-atom-bombardment liquid chromatography-mass spectrometry. This technique was used to determine acetylcholine levels in seven regions of rat brain by monitoring the intact molecular cations of

Table 3

High Performance Liquid Chromatography Systems

Column	Mobile phase and [flow rate]	Detection	Sample and remarks	Reference
Discovery C_{18} column (15 cm × 2.1 mm i.d.)	20 mM ammonium acetate/20 mM hepta-fluoro butyric acid in methanol/water (1:9; pH 3.2) [0.4 mL/min].	Tandem ion-trap MS	Rat brain	[4]
MF-8904 cation exchange (53 cm × 1 mm i.d.) (10 μm)	50 mM Na_2HPO_4 containing Kathon CG (5 mL/L) of pH 8 [140 μL/min].	Vitreous C electrode operated at + 100 mV versus Ag/AgCl.	Detection of basal acetyl-choline in rat brain microdialysate	[117]
15 cm of Bio-Sil ODS-5S fitted with an ODS-5 guard column	0.01 M sodium acetate-0.02 M citric acid (pH 5) containing 4.5 mg/L of sodium octyl sulfate and was 1.2 mM in tetramethyl ammonium chloride.	Electrochemical	Neuronal tissues. Reversed-phase HPLC.	[147]

(*continued*)

Table 3 (continued)

Column	Mobile phase and [flow rate]	Detection	Sample and remarks	Reference
Reversed-phase packing post column reactor 3 cm × 2.1 mm of Aquapore AX 300.	20 mM Tris acetate (pH 7)–1 mM-tetramethylammonium chloride-2 μM octanesulfonate containing 2% of acetonitrile.	Electrochemical	Tissue extracts, AChE and ChO are absorbed and immobilized on the post column	[148]
Hamilton PRP-1 polystyrene cartridge (10 cm × 4.6 mm) or column of Ultrasphere ODS.	20 mM Tris-acetate (pH 7) 200 mM in octanosulfonate and containing 2.5% of acetonitrile [2 mL/min].	Electrochemical	Rat or mouse brain or salivary gland tissues	[149]
Chemcosorb C-18 (3 μm)	—	Amperometric	Rat brain tissues; determines 10 pmol	[150]

25 cm × 4 mm of ODS-120 T.	10 mM Sodium acetate (pH 5)–1 mM tetra-methylammonium chloride–10 ppm sodium octyl sulfate [1 mL/min]	Amperometric	Use of reactor containing ChO and AChE on LiChrosorb-NH_2	[151]
10 cm × 4.6 mm of reversed phase material	10 mM Sodium acetate, sodium octyl sulfate (30 mg/L and 2.7% of acetonitrile, adjusted to pH 5 with citric acid [0.8 mL/min].	Electrochemical	Brain tissues, passed through a column containing ChO and AChE	[152]
Radial-Pak μBondapak C_{18} cartridge.	Butanol–methanol–acetic acid–water (4:2:1:43) containing 0.15 mM-1-phenethyl-2-picolinium bromide [3 mL/min].	UV at 254 nm	Soya bean leaves. Detection limit 1 nmol	[153]
PRP-1styrene-divinyl-benzene (10 μm)	Water–acetonitrile–propanol (90:7:3) containing 0.45 M-KH_2PO_4 and 0.1 mM-hexadecyltrimethyl ammonium bromide [2 mL/min].	UV at 254 nm	Limit of detection 2 μg	[154]

(continued)

Table 3 (continued)

Column	Mobile phase and [flow rate]	Detection	Sample and remarks	Reference
Develosil Ph-5 and Develsil Ph (15–30 μm) guard column	Potter *et al.*, (*Anal. Abstr.*, 1984, 46, 9D167).	Electrochemical	Rat brain limit of detection 5 pmol.	[155]
20 cm × 4 mm of styrene divinyl benzene copolymer	Aqueous 2 mM hexansulfonic acid-acetonitrile.	Electrical conductivity.	Choline in plants. RP ion-pair HPLC.	[156]
25 cm × 4.6 mm of Microsorb-silica	(A) Acetonitrile:water: ethanol:acetic acid: 0.83 M sodium acetate (800:127:68:2:3) at pH 3.6. (B) (400:400:68:53: 79).	Liquid scintillation counting.	Biological material, choline metabolite	[157]
7.5 cm × 4 mm of Nucleosil 5 SA and a guard (7.5 cm × 2.1 mm) cation exchange.	0.2 M phosphate buffer (pH 7.1) and 20 mM tetramethyl ammonium [0.8 mL/min].	Electrochemical	Brain tissue. Eluate passed through a column of ChO and AChE.	[158]

15 cm × 4.6 mm of Nucleosil C_{18} at 37°.	Sodium acetate (1.36 g/L) Na_2EDTA (3.72 g/L) sodium octyl sulfate (25 mg/L)–tetramethyl ammonium chloride (1.2 mM)–Water (pH 5) [0.8 mL/min].	Electrochemical	Rat brain tissues. Eluate passed through the AChE and ChO column.	[159]
7.5 cm × 4.6 mm of Chemosorb 70 DS-L 300 (7 μm)	Ikarashi *et al*, (*Anal. Abstr.*, 1985, **322**, 191).	Electrochemical	Brain tissue	[160]
15 cm × 4 mm Nucleosil 5 SA	100 mM sodium phosphate of pH 7.6, 7 mM in tetramethyl ammonium perchlorate [700 μL/min].	Electrochemical	Biological tissues	[161]
12.5 cm × 4 mm of Yanapak ODS 120 T.	10 mM phthalic acid, 1.2 mM triethylamine and 76 mM sodium octanesulfonate adjusted to pH 5 with KOH solution [0.4 mL/min].	Chemi-luminescence	Eluate was passed through a column of AChE and ChO.	[162]

(*continued*)

Table 3 (continued)

Column	Mobile phase and [flow rate]	Detection	Sample and remarks	Reference
15 cm × 4.6 mm of PLRP-S (7 μm) and an immobilized enzyme column (BAS Japan).	50 mM sodium phosphate (pH 8.3) containing 1 mM tetra-methyl ammonium chloride and 70 μM sodium octane-1-sulfonate [1 mL/min].	Electrochemical with a platinum electrode at +500 mV versus AgAgCl.	Brain tissue. Recoveries were 108–110%.	[163]
10 cm × 2 mm of Spherisorb S5 C6	Acetone–aqueous buffer solution (71:929) containing 50 mM NaH_2PO_4, 15 mM tetramethyl ammonium chloride and 0.3 mM sodium dodecyl sulfate adjusted to pH 7.7 with NaOH [200 μL/min].	Electrochemical	Spinal cord of rat. Eluate passed through reactor of AChE and ChO	[164]

25 cm × 4.6 mm of Gynkotek Hypersil ODS (5 μm)	0.1 M KH_2PO_4 containing 500–700 mg/L of tetramethyl ammonium chloride at pH 7.	Electrochemical	Brain tissues. Post column enzymatic reaction.	[165]
7.5 cm × 2.1 mm of Nucleosil 5 SA.	50 mM potassium phosphate buffer (pH 7.4) containing 20 mM tetramethyl ammonium nitrate [0.5 mL/min]	Chemi-luminescence	Serum and urine. Detection limit 0.7 and 0.5 pmol for ACh and Ch.	[166]
5 cm × 4.6 mm of RLRPS (5 μm)	10 mM Trizma acetate buffer (pH 7.8)–140 μM octanesulfonic acid–1 mM tetraethyl ammonium chloride [1.3 mL/min].	Electrochemical	Neuronal tissues. Detection limit 0.5 and 1 pmol for Ch and ACh.	[167]
10 cm × 3 mm of Chromspher 5 C_{18} loaded with sodium dodecyl sulfate and a pre-column and a guard column.	0.2 M Potassium phosphate (pH 8) [0.6 mL/min].	Electrochemical	Spinal cord of rats. Eluate passed through a reactor of AChE and ChO.	[168]

(*continued*)

Table 3 (continued)

Column	Mobile phase and [flow rate]	Detection	Sample and remarks	Reference
100 mm × 3.0 mm reversed phase chromsphere S C_{18} with lauryl sulfate	0.2 M KH_2PO_4 buffer and 0.005 M KCl at pH 8.	Electrochemical	Blood plasma and red cells in humans and in mice.	[169]
Chemosorb ACh-II reversed phase.	—	Coulometric	Rat brain. Detection limit was 10–20 pmol for Ch and ACh.	[170]
15 cm × 6 mm of Shodex RSpak DE613 medium polarity methacrylate gel.	0.1 M sodium phosphate buffer (pH 8.3) containing 1.2 mM tetramethyl ammonium chloride and 300 mg/L of sodium decan-1-sulfonate [1 mL/min].	Amperometric	Human cereprospinal fluid. Post-column enzymatic reaction with AChE and ChO.	[171]

Eicompak AC-GEL	0.07 M-Phosphate buffer containing 60 ppm of Na_2 EDTA, 0.065% of tetramethyl ammonium chloride and 0.3 of sodium octane-sulfonate	Electrochemical	Mouse brain tissues. An immobilized enzyme column with ChO and AChE was used.	[172]
10 cm × 3.1 mm of Nucleosil 5 C_{18} with guard column (1 cm × 2.1 mm)	0.2 M phosphate–0.1 mM EDTA (pH 7.9) containing 0.2 g/L of sodium of octane-1-sulfonate and 0.4 mM tetramethyl ammonium bromide.	Electrochemical	Human cerebrospinal fluid. An immobilized enzyme was used.	[173]
3 or 6 cm × 2 mm of Dynosphere PD-051-R (5 μm)	3:47 Acetone-aqueous buffer (pH 7.5)	Electrochemical	Both AChE and ChO were immobilized in post column reactor.	[174]
An ODS guard column (5 mm × 4 mm). An AC Gel (10 μm) separation column (15 cm × 6 mm), CA trap and AC-Enzympak column	0.1 phosphate buffer (pH 8) containing 65 ppm of tetramethyl ammonium chloride and 300 ppm of sodium octane-sulfonate [1 mL/min].	Electrochemical	Cerebrospinal fluids of rats	[175]

(continued)

Table 3 (continued)

Column	Mobile phase and [flow rate]	Detection	Sample and remarks	Reference
10 cm × 7 mm of Spherisorb ODS 2 (3 μm)	0.05 M KH_2PO_4 containing tetramethyl ammonium chloride and sodium octanesulfonate (pH 7) [1.2 mL/min]	Electrochemical	Mice brain tissues. Post column reactor containing ChO and AChE was used.	[176]
15 cm × 5 mm of Eicom-pak AC-GEL	0.1 M phosphate buffer (pH 8.3) containing 1.2 mM tetramethyl ammonium chloride and 300 ppm of sodium decansulfonate [1 mL/min]	Electrochemical	Human plasma. Eluent passed through an immobilized enzyme column.	[177]
3 to 25 cm × 4 to 4.6 mm of cation exchange columns. Hamilton PRP-X 200 column.	100 mM phosphate buffer pH 7.6 containing 7 mM tetramethyl ammonium perchlorate [0.5 to 1.5 mL/min]	—	Best results were obtained with Hamilton column	[178]

10 cm × 3.2 mm of C_{18} (3 μm)	Tris buffer solution (pH 7.5) containing acetic acid, tetramethyl ammonium chloride, sodium octylsulfate, NaN_3, EDTA, and acetonitrile [0.9 mL/min].	Electrochemical	Serum. Detection limit was 3 pmol for choline.	[179]
6 cm × 4 mm of polymeric styrene-based packing	0.05 M phosphate–1.0 mM Na_2 EDTA-0.4 mM sodium 1-octansulfonate of pH 8.4 [0.8 mL/min].	Electrochemical	Rat brain	[180]
15 cm × 3 mm of Li Chrosorb RP18 (25 μm) 25 × 4.1 mm of Hamilton PRPX 200 (10 μm). 25 cm × 4 mm of Supelcosil LC8 (5 μm) and a LiChroCART cartridge 12.5 cm × 2 mm of Superspher 60 RP-Select B).	0.1 M phosphate buffer (pH 7.4) containing 0.1 mM EDTA and 4 mM tetramethyl ammonium perchlorate with the use of a reactor cartridge (3 cm × 2.1) of immobilized Cho and AChE [0.5 mL/min].	Electrochemical	Erythrocytes extract from heart or brain tissue. Reversed phase column provided better pH stability and longer working life.	[181]

(*continued*)

Table 3 (continued)

Column	Mobile phase and [flow rate]	Detection	Sample and remarks	Reference
10 cm × 1 mm of Aminex A-9	0.2 M phosphate buffer solution (pH 8) containing 5 mM NaCl and 0.1% Kathon CG [60 μL/min].	Electrochemical	Brain tissues microdialysis or cerebrospinal fluids.	[182]
15 cm × 1 mm of Spheron Micro-DEAE 300.	0.1 M sodium phosphate buffer (pH 7.4) containing 0.1 mM acetylcholine and 0.1 mM EDTA [60 μL/min].	Electrochemical	Plasma. Detects anti AChE compound with post column reaction	[183]

Polymeric reversed phase column (15 cm × 3 mm i.d.) of ACH-3 (5 μm) with an ACH-3-G guard cartridge.	100 mM sodium phosphate of pH 8 containing 0.5 mM tetramethyl ammonium chloride, 0.005% Reagent MB (ESA) and 2 mM octanesulfonic acid [0.35 mL/min].	Electrochemical using an ESA model 5040 solid state cell with a Kel-F or PEEK Pt-target working electrode at 300 mV versus a Pd reference electrode.	Microdialysis in rat striatum. Analysis in the presence and absence of esterase inhibitors.	[184]
Analytical column and post column immobilized enzyme reactor supplied as part of a ACh/Ch assay kit.	Sodium phosphate buffer pH 8.5 containing Kathon CG (Rohm and Haas, PA, USA), [1 mL/min].	Electrochemical at + 0.5 V versus Ag/AgCl.	Human plasma and peritoneal dialysis effluent.	[185]
(15 cm × 1.7 mm i.d.) of SFPAC-ODSO5-S15 (5 μm).	0.2% trifluoroacetic acid in water containing 1% glycerol, [40 μL/min for 5 min. then 11 μL/min].	Flow fast atom bombardment mass spectrometry	Brain tissues of rodents. Detection limit was 5 and 2 for Ch and ACh.	[186]

(continued)

Table 3 (continued)

Column	Mobile phase and [flow rate]	Detection	Sample and remarks	Reference
10 μL loop injector on a 10 μm bioanalytical system ACh column (53 cm × 1 mm i.d.)	28 mM Na_2HPO_4–0.5% Kathon (pH 8.5 [1 mL/min].	Electrochemical at 0.5 V versus AgCl.	Brain dialysis of rats. An immobilized-enzyme reactor was used.	[187]
A BASJ ion-exchange microbore column (45 cm × 1 mm i.d.)	0.05 Sodium phosphate buffer of pH 8.5 containing 0.1 mM EDTA [60 μL/min].	Electrochemical at horseradish peroxidase osmium redox polymer-modified vitreous C electrodes at 0 mV versus Ag/AgCl.	Rat frontal cortex dialysate samples	[188]
Allteck C-18 (25 cm × 1 mm i.d.) (5 μm).	Aqueous 10 mM ammonium acetate–2 mM pyridine–230 μM sodium octanesulfonate of pH 5 [10 μL/min].	Electrospray ionisation MS detection at *m/z* 146.	Rat cell culture system. Detection limit 10 pM of acetylcholine.	[189]

(1) A Stipstik column (53 cm × 1 mm i.d.) with ChO/AChE reactor and guard columns.	29 mM NaH_2PO_4–22 mM sodium acetate–5 mM Kathon of pH 8.5.	Electrochemical detection	ACh in microdialysis samples.	[190]
(2) A polymeric reversed phase column (12 cm × 2 mm i.d.)	50 mM Na_2HPO_4–0.3 mM sodium octane-1-sulfonate–1 mM Na_2-EDTA of pH 8.5.	Amperometric	ACh in microdialysis samples.	[190]
A polymer gel column (10 cm × 4 mm i.d.)	50 mM Na_2HPO_4 of pH 8 containing 0.5% Kathon CG [0.8 mL/min].	Electrochemically using Pt electrode at +0.5 V versus Ag/AgCl.	ACh and Ch in micro-dialysates from rat brain.	[191]
Bioanalytical systems acetylcholine separation column (6 cm × 4 mm i.d.) with post column (5 cm × 4 mm i.d.) containing immobilized AChE and ChO.	50 mM phosphate buffer (pH 8.4) containing 1 mM EDTA and 0.4 mM sodium-1-octanesulfonate [0.8 mL/min].	Amperometric	Acetylcholine and related metabolite, detection limit are 2–5 pM.	[192]

(*continued*)

Table 3 (continued)

Column	Mobile phase and [flow rate]	Detection	Sample and remarks	Reference
Crosslinked polystyrene column (3 μm) (6 cm × 4 mm i.d.) with a precolumn (1 cm × 4 mm i.d.) by a slurry packing with acetone.	0.05 M phosphate buffer (pH 8.4) containing 1 mM EDTA and 0.4 mM sodium 1-octanesulfonate [0.7 mL/min].	Amperometric at +0.5 V (versus Ag/AgCl)	Determines ACh and Ch in biological samples.	[193]

acetylcholine and an internal deuterated standard. Results were compared with other methods, such as pyrolysis GC–MS and LC with electrochemical detection.

Ikarashi *et al.* reported the development of a liquid chromatography multiple electrochemical detector (LCMC) and its application in neuroscience [197]. The system developed consisted of four parallel liquid chromatographic systems equipped with multiple electrochemical detectors, and the fourth was used for the assay of acetylcholine with a detection limit of 0.1–0.4 pmol.

6.5 Cation-exchange liquid chromatography

Stein reported the separation of choline and acetylcholine by a cation-exchange chromatographic method [198]. Four columns were compared by Salamoun *et al.* for the separation of choline and acetylcholine by cation exchange liquid chromatography [178]. The mobile phase used was 0.1 M phosphate buffer (pH 7.4), containing 0.1 mM EDTA and 4 mM tetramethylammonium perchlorate, and was eluted at a rate of 0.5 mL/min. The system made use of a reactor cartridge (3 cm × 2.1 mm) containing immobilized choline oxidase and acetylcholineesterase, and electrochemical detection at 0.45 V. The columns were (15 cm × 3 mm i.d.) LiChrosorb RP-18 (5 μm), (25 cm × 4.1 mm i.d.) Hamilton PRP-X 200 (10 μm), (25 cm × 4 mm i.d.) Supelcosil LC-8 (5 μm), and a LiChroCART cartridge (12.5 cm × 2 mm i.d.) containing Supersphere 60 RP-Select B. Appropriate guard columns were used as well. The use of reversed phase columns provided better pH stability and a longer working life than the normal silica-based cation exchangers.

6.6 High performance liquid chromatography–mass spectrometry (HPLC/MS)

Ishimaru *et al.* reported the use of high performance liquid chromatography combined with continuous flow fast atom bombardment mass spectrometry for the simultaneous determination of choline and acetylcholine in rodent brain regions [186]. Brain tissues were homogenized in 2 mL of acetonitrile containing 50 mg magnesium sulfate and 10 nmol each of the d_4 analogs of choline and acetylcholine (as internal standards). The mixture was centrifuged, and the solution extracted with two 2 mL aliquots of hexane. The acetonitrile was evaporated under nitrogen, and the residue was dissolved in 0.2 mL of the mobile phase. A 20 μL portion of the resulting solution was analyzed by HPLC (see Table 3 for more details).

Acevedo *et al.* quantified acetylcholine in a cell culture system by semi-micro high performance liquid chromatography and electrospray ionization mass spectrometry [189]. Rat cells were washed with Dullecco's PBS solution (containing 10 μM Physostigmine), and incubated in 2.5 mL 10 μM Physostigmine/50 mM-KCl in Dullecco's PBS for 10 min. Portions (2.5 mL) were mixed with 500 pM N,N,N-trimethyl-d_9-acetylcholine (internal standard), and were applied to a C-18 SPE column. After washing with water, acetylcholine and choline were eluted with aqueous 50% acetic acid. The first three fractions (500 μL) were combined, frozen on dry ice, and lyophilized. After reconstitution in 50 μL of aqueous 10 mM ammonium acetate–2 mM pyridine–230 μM sodium octanesulfonate (pH 5), 5 μL portions were analyzed on a 5 μm Alltech C-18 column (25 cm × 1 mm i.d.). Electrospray ionization with MS detection was used at *m/z* 104 and 146 for choline and acetylcholine, respectively.

6.7 Gas chromatography–mass spectrometry (GC/MS)

Jenden reported the use of gas chromatography–mass spectrometry method to measure tissue levels and the turnover of acetylcholine [199]. The systems that catalyze and hydrolyze acetylcholine were inactivated rapidly by freezing the brain tissue in nitrogen, pulverizing, and extracting in 15% 1 N formic acid in acetone. After addition of the internal standard, the homogenate was centrifuged to obtain a supernatant containing the bound forms of acetylcholine. Acetylcholine was esterified with propionyl chloride, dried in a stream of nitrogen, demethylated, heated to 80°C for 45 min, and then partitioned between aqueous acid and pentane. The tertiary amines were removed by extraction with CH_3Cl, and the acetylcholine injected into a gas chromatograph. Quantitative determination was achieved by mass spectrometry using ^{2}H-labeled compounds as tracers and internal standards.

Polak and Molenaar described a method for the determination of acetylcholine from brain tissue by pyrolysis-gas chromatography-mass spectrometry [200]. The deuterium-labeled acetyl-choline is pyrolytically demethylated with sodium benzenethiolate, followed by quantitative GC–MS analysis. In this method, care must be taken so that the samples do not contain appreciable amounts of choline since exchange of deuterium-labeled groups between acetylcholine and choline during pyrolysis may yield erroneous results. The same authors have also reported a method for the determination of acetylcholine by slow pyrolysis combined with mass fragment analysis on a packed capillary column [201].

Mikes reported the use of a new injection micropyrolysis apparatus for acetylcholine and choline determination by gas chromatography–mass spectrometry [202].

Singh and Drewes described an improved method for the analysis of acetylcholine and choline in canine brain and blood samples by capillary gas chromatography–mass spectrometry [203]. Frozen samples were mixed with butyryl choline (internal standard) and extracted. The GC–MS system was equipped with an electron-impact ionizer and a 15 m capillary column coated with a bonded methyl-silicone phase (HP-SE 54). The column was operated at 60°C for 3 min, and then temperature programmed (at 25°C/min) to 200°C. The ion at *m/z* 58 was selected for monitoring, and the separation of the three compounds was achieved in less than 5 min.

Liberato *et al.* reported the separation and quantification of acetycholine and choline by thermospray liquid chromatography–mass spectrometry [204]. Choline and acetylcholine were extracted from mouse brain homogenates, in the presence of $^{2}H_{9}$-choline and $^{2}H_{4}$-acetylcholine, and the extract was analyzed directly by HPLC–MS on a column (15 cm × 4.6 mm) of Ultrasphere-I.P. ODS. The mobile phase was 0.1 M ammonium acetate–2 mM pyridine containing octanesulfonic acid (50 mg/L, pH 5), and was eluted at a flow rate of 1.25 mL/min. The column was coupled to a quadrupole mass spectrometer by the thermospray interface, and selected ions were monitored at *m/z* 104, 113, 146 and 150 for choline, [$^{2}H_{9}$] choline, acetylcholine and [$^{2}H_{4}$] acetylcholine, respectively.

Watanabe *et al.* used a pyrolysis gas chromatography–mass spectrometric method for the determination of acetylcholine in human blood [205]. Acetylcholine was extracted by an $HClO_4$ treatment of the whole blood in the presence of eserine to inhibit cholineesterase activity, and was then determined by pyrolysis GC–MS with butyrylcholine as the internal standard. Pyrolysis was carried out for 7.5 s at 300°C, and the product was analyzed on a glass column (2 m × 2 mm) of 5% of DDTS and 5% of OV-101 on Gas Chrom Q (100–200 mesh) at 115°C. Helium was used as the carrier gas (25 mL/min), and detection was effected using MS at 28 eV with an ion source temperature of 205°C. The ion at *m/z* = 58 was monitored for both acetylcholine and the internal standard.

Pomfret *et al.* measured both choline and choline metabolite concentrations using HPLC and GC–MS [206]. Tissues were extracted with chloroform/methanol/water, and the separated phases were evaporated. The organic phase residue was dissolved in chloroform/methanol (1:1), and a portion was applied on a silica gel for TLC analysis with chloroform/methanol–water (65:30:4) as the mobile phase. Zones corresponding to phosphatidylcholine and lysophosphatidylcholine were removed and hydrolyzed in methanolic HCl to liberate choline. The aqueous phase residue was dissolved in methanol/water (10:1), and the hydrophilic metabolites were separated by HPLC on a column (8.3 cm × 4.6 mm) of Pecosphere-3 CSi (5 μm) with a Supelguard LCSi guard column (2 cm × 4.6 mm) and acetonitrile–ethanol–acetic acid–1 M ammonium acetate–water–0.1 M sodium phosphate (800:68:2:3:127:10 changing to 400:68:44:88:400:10) as the mobile phase (flow rate of 1.5 mL/min). Fractions corresponding to betaine, acetylcholine, phosphocholine, glycerophosphocholine, and choline were separated and hydrolyzed as before. Choline was determined by GC–MS of each fraction, and metabolites could be detected down to the 200 pmol level in 10 mg samples.

Terry *et al.* reported the synthesis of novel selenium-containing choline and acetylcholine analogues and their quantitation in biological tissues, using a pyrolysis-gas chromatography–mass spectrometry assay method [207]. Biological tissues were homogenized in formic acid-acetonitrile, the homogenate was centrifuged, and the supernatant solution was extracted with 2 mM dipicrylamine in CH_2Cl_2 and TAPS buffer (pH 9.2). After centrifugation, the CH_2Cl_2 layer was evaporated under nitrogen, silver *p*-toluenesufonate in acetonitrile and propionyl chloride were added, and the solution was set aside for 10 min at room temperature. The mixture was evaporated to dryness under nitrogen, the residue was dissolved in sodium acetate buffer (pH 4), and the solution was washed with butan-2-one before evaporation to dryness at 50°C under nitrogen. The residue was dissolved in aqueous 90% acetonitrile, and this was pyrolyzed for 10 s at 325°C in a quartz tube. The pyrolysis products were trapped in a cold trap for three min before analysis by GC on a column (30 m × 0.25 mm) coated with Stabilwax (0.5 μm) with helium as the carrier gas and temperature programming from 30°C to 200°C (held for 2.5 min) or 30°C/min and 68 eV EIMS with selected ion monitoring at *m/z* 122 and 125. Calibration graphs were linear from 20 pmol to 20 nmol of selenium choline propionyl ester and acetyl-selenium choline.

Patterson and Kosh described a method for the simultaneous quantitation of arecoline, acetylcholine, and choline in mouse brain tissues using gas chromatography–electron-impact mass spectrometry [208]. Mouse brain tissues were homogenized in 1 M formic acid – acetonitrile (17:3), washed with ethyl ether, and centrifuged for two min. Arecoline, acetylcholine, and choline in the supernatant were extracted as ion pairs with 5 mM sodium tetraphenylboron in heptan-3-one. After centrifugation, the organic layer was evaporated to dryness under nitrogen at 60°C, and the dried residue was mixed with 5 mM silver *p*-toluene sulfonate in acetonitrile. Propionyl chloride was added, and after 5 min the mixture was evaporated to dryness. The residue was mixed with sodium thiophenoxide solution and incubated at 80°C for 30 min. The cooled solution was mixed with 0.5 mL citric acid and washed with ethyl ether and pentane. The solvents were evaporated, chloroform was added to the solution in an ice bath, which was then followed by 2 M ammonium citrate–7.5 M ammonium hydroxide buffer. Following centrifugation, the chloroform layer was analyzed by gas chromatography–mass spectrometry. Hexadeuterated arecoline and non-deuterated acetylcholine and choline were used as internal standards. The analysis was performed on a column (30 m × 0.25 mm) of Stabilwax with ramped temperature programming for 50°C (held for 0.75 min) to 200°C (held for 3.5 min) at 35°C/min and then to 220°C at 25°C/min. Electron impact MS was carried at 68 eV with SIMS monitoring of $m/z = 140$ and 146 for arecoline and $m/z = 58$ and 64 for acetylcholine and choline. The detection limit was 25 pmol for all the three compounds, and the calibration curves were linear from 0.025 to 50 nmol.

6.8 Capillary electrophoresis

Kappes *et al.* evaluated the potentiometric detection of acetylcholine and other neurotransmitters through capillary electrophoresis [209]. Experiments were performed on an in-house capillary electrophoresis instrument that made use of detection at a platinum wire, dip-coated in 3.4% potassium tetrakis (4-chlorophenyl) borate/64.4% *o*-nitrohenyl octyl ether/32.2% PVC in THF. The results were compared to those obtained using capillary electrophoresis with amperometric detection at a graphite electrode. Samples prepared in the capillary electrophoresis buffer were electrokinetically injected (7 s at 5 kV) into an untreated fused silica capillary (88 cm × 25 μm i.d.) and separated with 20 mM tartaric acid adjusted to pH 3 with MgO as the running buffer. The system used an applied potential of 30 kV, and detection versus the capillary electrophoresis ground electrode.

6.9 Radioassay methods

The enzymatic radioassay method for the analysis of acetylcholine and choline in brain tissue has been reported by Reid *et al.* [210]. The method describes the determination of nanogram amounts of acetylcholine and choline in as little as 10 mg of brain tissue, involves isolation of acetylcholine by high-voltage paper electrophoresis, alkaline hydrolysis of acetylcholine to choline, and conversion of this into [^{32}P]-phosphoryl choline in the presence of choline kinase and [γ^{32}P] ATP. The labeled derivative is isolated by column chromatography on Bio-Rad AG1-X8 resin, using Tris buffer solution as the eluent. Cerenkov radiation from ^{32}P is counted (at 33% efficiency) in a liquid scintillation spectrometer. The amount of phosphorylcholine is proportional to the amount of choline over the range of 0.08–8.25 nmol.

Bluth *et al.* described a simplified radio-enzymatic assay method for the determination of acetylcholine and choline in discrete structures of rat brain [211].

Gilberstadt and Russell reported the determination of picomole quantities of acetylcholine and choline in physiologic salt solutions [140]. Carbon-14 labeled choline and acetylcholine standards in Krebs–Ringer solution were subjected to chromatography. The unlabeled choline and acetylcholine were assayed radioenzymatically by using [γ^{32}P] ATP, choline kinase and acetylcholineesterase. The calibration graphs were rectilinear in the range 10–3000 pmol acetylcholine and 30–3000 pmol of choline.

Dick *et al.* used a radio gas chromatography method for monitoring the *in-vivo* labeling of postmortem [^{3}H] choline production [212]. After homogenization in 1 M formic acid–tetrahydrofuran (3:17) and derivatization with butyryl chloride, brain tissues from rats injected with [^{3}H] choline were assayed for choline, acetylcholine, and their ^{3}H-analogues. The analysis was performed using GC on a column of Chromosorb 750, coated with 10% OV-17 and 10% Triton X-100. The method used temperature programming for 130–150°C, helium as the carrier gas (flow rate of 40 mL/min), and N-P detection. For subsequent scintillation counting, 20% of the column effluent was trapped on *p*-terphenyl, coated as described above. Propionylcholine and [^{3}H]-hexanoylcholine were used as internal standards. Detection limits for [^{3}H]-choline and [^{3}H]-acetylcholine were 150 and 730 fmol, respectively.

A multiple enzyme and multi-substrate cycling system was described by O'Neill and Hurschak for the assay of choline acetyltransferase [213]. Assay tubes containing sodium [^{14}C] acetate (13 mCi/mmol) and (final concentration) 0.2 M Tris–HCl, 5 mM choline chloride, 0.1 mM EDTA, 0.05 mM dithiothreitol, 7 mM potassium acetate, 20 mM $MgCl_2$, 0.2 mM eserine sulfate, 0.36 mM ADP, 300 mM KCl, 20 mM-phosphoenol pyruvate, 0.1 mM-coenzyme A, 0.20% of bovine serum albumin, pyruvate kinase, phosphate acetyltransferase, and acetate kinase were centrifuged (1000 G at 3–4°C for 10 min), and kept at 37°C for 20 min. Brain homogenate was added, and the mixtures were centrifuged and incubated at 37°C for 1–60 min. The reaction was stopped by placing the tubes in ice-water. A 10 mM-acetylcholine solution was added, and the mixture was applied to a column of Ag1-X8. Elution was with water, and the eluates were collected in phials containing Chaikoff scintillation solution and assayed by liquid scintillation counting.

Cooper reported a simple method to determine the released acetylcholine in the presence of choline [214]. The use of this method obviates the necessity both for preventing the re-uptake of choline by synaptosomes during determination of released acetylcholine, and for subsequently separating radiolabeled acetylcholine from radiolabeled choline. To the sample or standard solution were added [^{3}H]-choline (5 nmol), choline oxidase (10–150 min), and Krebs-Ringer bicarbonate buffer solution (pH 7.4) to yield a final volume of 0.2 mL. The mixture was incubated at 37°C for 5 miU, and then 1 mL of 1% tetraphenylborate solution in butyronitrile was added. The mixture was vortex-mixed for 30 s, and then centrifuged. A 0.5 mL portion of organic phase was added to 5 mL of Opti-Fluor scintillator for the measurement of radioactivity.

Eckernas and Aquilonius described a simple radioenzymatic procedure for the determination of choline and acetylcholine in brain regions of rats sacrificed by microwave irradiation [215]. The methods are based on acetylation (in phosphate buffer) of free choline with [^{14}C]-acetylcholine using purified choline acetyltransferase. After ion-pair extraction of [^{14}C]-acetylcholine with tertiary phenyl borate contained in toluene-based scintillation cocktail, the radioactivity was measured.

Spector *et al.* developed a specific radioimmunoassay method for the determination of acetylcholine [216]. The described method is suitable for determining 2–137 pmol of acetylcholine; the radioligand used is

^{3}H-acetylcholine, and the hapten is 2-(trimethyl-amino)ethyl-6-amino-hexanoate hydrobromide toluene-*p*-sulfonate, which is conjugated to bovine serum albumin to yield the antigen. The antiserum is raised in rabbits and is used at a dilution of 1:1000 in phosphate-buffered saline containing bovine γ-globulin and eserine (as a cholinesterase inhibitor). After incubation of the mixture for 1 or 16 h at 4°C, the bound antigen is precipitated by adding ammonium sulfate solution, centrifuged off, and dissolved in water for scintillation counting. Cross-reactivity of the antiserum with choline is only 0.1%.

Saelens *et al.* determined acetylcholine and choline by an enzymatic assay method [217]. This assay method has been made more practical, primarily by preparing choline acetyltransferase in large quantities in advance. The preparation of the enzyme from mouse brain is described. The acetylcholine and choline were separated from brain-tissue extracts by electrophoresis at 500 V for one hour in a formic acid-acetic acid-water buffer solution, with tetraethylammonium bromide being used as a marker. The analyte bands were located through staining with iodine vapor. The acetylcholine was eluted and hydrolyzed, and the resulting choline was reacetylated with [acetyl-^{14}C]-coenzyme A in the presence of choline acetyltransferase. The [acetyl-^{14}C]-acetylcholine was then counted using a liquid scintillation system. The choline was also separately eluted from the electropherogram and determined by the same technique. Higher values for acetylcholine and choline were obtained after sacrificing mice with liquid nitrogen than after other methods of sacrifice. The brain tissue was extracted with formic acid-acetone.

Ladinsky and Consolo used an enzymatic radioassay method for the determination of acetylcholine and choline [218]. The method was based on the electrophoretic separation of acetylcholine and choline, hydrolysis of acetylcholine to form choline, and acetylation of the choline with labeled AcCoA and choline acetyltransferase. The labeled acetylcholine formed was isolated and quantitated. The method was sensitive and specific, and permitted the routine handling of a large number of samples in a single experiment. The standard curves were linear up to at least 42.5 ng (0.4 nmol) choline and 45 ng (0.3 nmol) acetylcholine. The lower limit of sensitivity was 2 ng, and the recovery of acetylcholine was 95% when carried through the entire procedure.

Saelens *et al.* measured choline, acetylcholine, and their metabolites by combined enzymatic and radiometric techniques [219]. Choline and acetylcholine were isolated from tissue extracts by a low voltage electrophoresis and assayed by transfer of the Ac-^{14}C group of acetyl-^{14}C-coenzyme A to choline by choline-acetyltransferase. The lower limit of sensitivity for this method was 10^{-11} mol of choline or acetylcholine. When choline-methyl-^{3}H was infused intravenously into rats, the specific activity of choline and acetylcholine increased curvilinearly in the cortex, midbrain, and brainstem. The metabolism of choline involved in acetylcholine synthesis and the metabolism of acetylcholine itself were extremely rapid.

7. OTHER METHODS OF ANALYSIS

7.1 Ion-selective method

Jaramillo *et al.* published an article regarding some ion-selective microelectrodes used in the analysis of acetylcholine and choline [220].

7.2 Microwave methods

Szepesy *et al.* used a microwave apparatus in the determination of acetylcholine and choline in the central nervous system samples [221]. Schmidt *et al.* reported the use of a microwave radiation method for the determination of acetylcholine in the rat brain [222].

7.3 Microdialysis

Kehr *et al.* described microsurgical methods for the simultaneous implantation of a microdialysis probe and intraventricular injection cannula *via* their respective guide cannulas into the mouse brain [223]. Basal and stimulated release of acetylcholine was determined in the ventral hippocampus of freely moving mice. No significant differences in basal potassium-stimulated or scopolamine-induced extracellular acetylcholine levels were observed in 4-month-old wide-type (WT) mice, or the mice with a loss of function mutation of the galanin gene (KO). In the aged, 10-month-old animals, the basal extracellular acetylcholine levels were significantly reduced in both WT and KO groups.

8. ACKNOWLEDGMENT

The authors wish to thank Mr. Tanvir A. Butt, Department of Pharmaceutical Chemistry, College of Pharmacy, King Saud University, Riyadh, Saudi Arabia, for typing the manuscript of this profile.

9. REFERENCES

1. E.S. Vizi, L.G. Harsing, Jr., D. Duncalf, H. Nagashima, P. Potter, and F.F. Foldes, *J. Pharmacol. Methods*, ***13***, 201 (1985).
2. W. Wiegrebe and M. Vibig, *Pharm. Zig.*, ***126***, 381 (1981).
3. A.F. Mynka, V.V. Ogurtsov, and M.L. Lyuta, *Khim.-Farm. Zh.*, ***22***, 491 (1988).
4. X.Y. Zhu, P.S.H. Wong, M. Gregor, G.F. Gitzen, L.A. Coury, and P.T. Kissinger, *Rapid Commun. Mass Spectrom.*, ***14***, 1695 (2000).
5. L.M. Lewin and N. Marcus, *Anal. Biochem.*, ***10***, 96 (1965).
6. S.E. Unger and T.M. Ryan, *SIA, Surf. Interface Anal.*, ***3***, 12 (1981).
7. Y. Maruyama, Y. Ikarashi, and G.L. Blank, *Trend Pharmacol. Sci.*, ***8***, 84 (1987).
8. I. Hanin, *Mod. Methods Pharmacol.*, 129 (1982).
9. K. Shimada, t. Oe, and T. Nambora, *J. Chromatogr., Biomed. Appl.*, ***84***, 345 (1989).
10. T.H. Tsai, *J. Chromatogr. Biomed. Appl.*, ***747***, 111 (2000).
11. ***The United States Pharmacopoeia (USPXX)***, United State Pharmacopoeial Convention, Inc., Rockville, MD, 20852, p. 1045 (1980).
12. M. Szakacs-Pinter and I. Perl-Molnar, *Acta Pharm. Hung.*, ***54***, 224 (1984).
13. T. Kakutani, T. Ohkouchi, T. Osakai, T. Kakiuchi, and M. Senda, *Anal. Sci.*, ***1***, 219 (1985).
14. T. Osakai, T. Kakutani, and M. Senda, *Bunseki Nagaku*, ***33***, E371 (1984).
15. G. Baum, *Analyt. Lett.*, ***3***, 105 (1970).
16. S. Nakazawa and K. Tanaka, *Bunseki Kagaku*, ***27***, 100 (1978).
17. O.E.S. Godinho, T.C. Rodrigues, M. Tubino, G. de-Oliveira-Neto, and L.M. Aleixo, *Anal. Proc.*, ***32***, 333 (1995).
18. A.F. Maslova, ***Choline, Acetylcholine; Handbook Chemical Assay Methods***, I. Hanin, ed., Raven Press, New York, p. 215 (1974).
19. T. Matsue, A. Aoki, I. Uchida, and T. Osa, *Chem. Lett.*, ***5***, 957 (1987).
20. T. Osakai, T. Nuno, Y. Yamamoto, A. Saito, and M. Senda, *Bunseki Kagaku*, ***38***, 479 (1989).
21. P.D. Hale, L.F. Liu, and T.A. Skotheim, *Electroanalysis*, ***3***, 751 (1991).

22. J.L. Kawagoe, D.E. Niehaus, and R.M. Wightman, *Anal. Chem.*, ***63***, 2961 (1991).
23. V. Marecek and Z. Samec, *Anal. Lett.*, ***14***, 1241 (1981).
24. R.E. Gyurcsanyi, Z. Feher, and G. Nagy, *Talanta*, ***47***, 1021 (1998).
25. T. Yao, Y. Matsumoto, and T. Wasa, *J. Flow Injection Anal.*, ***4***, 112 (1987).
26. P.C. Gunaratna and G.S. Wilson, *Anal. Chem.*, ***62***, 402 (1990).
27. T. Yao, S. Suzuki, H. Nishino, and T. Nakahara, *Electrolysis*, *7*, 1114 (1995).
28. S. Eksborg and B.A. Persson, *Acta Pharm. Suec.*, ***8***, 205 (1971).
29. S. Eksborg and B.A. Persson, ***Choline, Acetylcholine; Handbook Chemical Assay Methods***, I. Hanin, ed., Raven Press, New York, p. 181 (1974).
30. A. Barcuchova, J. Gasparic, and A. Oulehlova, *Cesk. Farm.*, ***33***, 343 (1984).
31. M. Suzuki and P. Haug, *Koenshu-Iyo Masu Kenkyukai*, ***5***, 153 (1980).
32. W.D. Lehmann, H.R. Schulten, and N. Schroeder, *Biomed. Mass Spectrom.*, ***5***, 591 (1978).
33. M. Tsubouchi, *J. Chem. Soc. Japan, Pure Chem. Sect.*, ***91***, 1190 (1970).
34. M.A. Barletta and C.O. Ward, *J. Pharm. Sci.*, ***59***, 879 (1970).
35. Mitchell and Clark, *Proc. Soc. Exp. Biol. Med.*, ***81***, 105 (1952), through reference 34.
36. D.G. Das and K.M. Abdulla, *Indian Drugs*, ***22***, 221 (1985).
37. M.H. Abernethy, H.P. Fitzgerald, and K.M. Ahern, *Clin. Chem.*, ***34***, 1055 (1988).
38. T. Sakai, Y. Gao, N. Ohno, and N. Ura, *Chem. Lett.*, ***1***, 163 (1991).
39. T. Sakai, Y.H. Gao, N. Ohno, and N. Ura, *Anal. Chim. Acta*, ***255***, 135 (1991).
40. T. Takeuchi, Y. Kabasawa, R. Horikawa, and T. Tanimura, *Clin. Chim. Acta*, ***205***, 117 (1992).
41. J.J. O'Neill and T. Sakamoto, *J. Neurochem.*, ***17***, 1451 (1970).
42. D. Sutoo, K. Akiyama, and I. Maeda, *Nippon Yakurigaku Zasshi*, ***91***, 173 (1988).
43. J. Ricny, J. Coupek, and S. Tucek, *Anal. Biochem.*, ***176***, 221 (1989).
44. R.C. MacDonald, *J. Neurosci. Methods*, ***29***, 73 (1989).
45. M.A. Roulier, B. Palenik, and F.M.M. Morel, *Mar. Chem.*, ***30***, 409 (1990).

46. J.D. Brennan, R.S. Brown, D. Foster, R.K. Kallury, and U.J. Krull, *Anal. Chim., Acta*, ***255***, 73 (1991).
47. M.L. Lunar, S. Rubio, and D. Perez-Bendito, *Anal. Lett.*, ***24***, 979 (1991).
48. J.P. Ternaux and M.C. Chamoin, *J. Biolumin. Chemilumin.*, ***9***, 65 (1994).
49. S. Luterotti and D. Maysinger, *J. Pharm. Biomed. Anal.*, ***12***, 1083 (1994).
50. W.Z. Fan and Z.J. Zhang, *Microchem. J.*, ***53***, 290 (1996).
51. T. Hasebe, J. Nagao, and T. Kawashima, *Anal. Sci.*, ***13***, 93 (1997).
52. M. Emteborg, K. Irgum, C. Gooijer, and U.A.T. Brinkman, *Anal. Chim. Acta*, ***357***, 111 (1997).
53. J.R. Wetherell and M.C. French, *Biochem. Soc. Trans.*, ***14***, 1148 (1986).
54. M. Israel and B. Lesbats, *J. Neurochem.*, ***39***, 248 (1982).
55. M. Israel and B. Lesbats, *Neurochem. Int.*, ***3***, 81 (1981).
56. S. Birman, *Biochem. J.*, ***225***, 825 (1985).
57. K.W. Ehler, E.A. Hoops, R.J. Storella, and G.G. Beirkamper, *J. Pharmacol. Methods*, ***16***, 271 (1986).
58. H. Vapaatalo, I.B. Linden, and J. Parantainen, *J. Pharm. Pharmacol.*, ***28***, 188 (1976).
59. M. Satyanarayana, K.R. Rajeswari, and P.B. Sastry, *Indian J. Pharm. Sci.*, ***46***, 205 (1984).
60. I. Karube and K. Yokoyama, *Sens. Actuators B*, ***B13***, 12 (1993).
61. M. Aizawa, *Anal. Chim. Acta*, ***250***, 249 (1991).
62. M. Mascini and G. Guilbault, *G. Ital. Chem. Clin.*, ***11***, 241 (1986).
63. M.F. Suaud-Chagny and J.F. Pujol, *Analusis*, ***13***, 25 (1985).
64. M. Mascini and D. Moscone, *Anal. Chim. Acta*, ***179***, 439 (1986).
65. M. Gotoh, E. Tamiya, M. Momoi, Y. Kagawa, and I. Karube, *Anal. Lett.*, ***20***, 857 (1987).
66. M.E. Eldefrawi, S.M. Sherby, A.G. Andreou, N.A. Mansour, Z. Annau, N.A. Blum, and J.J. Valdes, *Anal. Lett.*, ***21***, 1665 (1988).
67. L. Campanella, M.P. Sammartino, and M. Tomassetti, *Anal. Lett.*, ***22***, 1389 (1989).
68. U. Loeffler, U. Wollenberger, F. Scheller, and W. Goepel, *Fresenius' Z. Anal. Chem.*, ***335***, 295 (1989).
69. J.L. Marty, K. Sode, and I. Karube, *Anal. Chim. Acta*, ***228***, 49 (1990).

70. A.V. Bratov, YuG. Vlasov, L.P. Kuznetsova, S.S. Levichev, E.B. Nikol'Skaya, and YuA. Tarantov, *Zh. Anal. Khim.*, ***45***, 1416 (1990).
71. R.M. Morelis and P.R. Coulet, *Anal. Chim. Acta*, ***231***, 27 (1990).
72. H. Taguchi, N. Ishihara, K. Okumura, and Y. Shimabayashi, *Anal. Chim. Acta*, ***236***, 441 (1990).
73. M. Hoshi, E. Matsumura, K. Mogami, T. Hayashi, and E. Wantanabe, *Nippon Suisan Gakkaishi*, ***57***, 79 (1991).
74. E. Tamiya, Y. Sugiura, E.N. Navera, S. Mizoshita, K. Nakajima, A. Akiyama, and I. Karube, *Anal. Chim. Acta*, ***251***, 129 (1991).
75. U. Wollenberger, K. Setz, F.W. Scheller, U. Loeffler, W. Goepel, and R. Gruss, *Sens. Actuators B*, ***B4***, 257 (1991).
76. E.N. Navera, K. Sode, E. Tamiya, and I. Karube, *Biosens Bioelectron.*, ***6***, 675 (1991).
77. R. Rouillon, N. Mionetto, and J.L. Marty, *Anal. Chim. Acta*, ***268***, 347 (1992).
78. M.G. Garguilo, N. Huynh, A. Proctor, and A.C. Michael, *Anal. Chem.*, ***65***, 523 (1993).
79. E.N. Navera, M. Suzuki, E. Tamiya, T. Takeuchi, and I. Karube, *Electroanalysis*, ***5***, 17 (1993).
80. Y.K. Zhou, X.D. Yuan, Q.L. Hao, and S. Ren, *Sens. Actuators B*, ***B12***, 37 (1993).
81. Lopez-Ruiz, E. Dempsey, C. Hua, M.R. Smyth, and J. Wang, *Anal. Chim. Acta*, ***273***, 425 (1993).
82. E.N. Navera, M. Suzuki, K. Yokoyama, E. Tamiya, T. Takeuchi, I. Karube, and J. Yamashita, *Anal. Chim. Acta*, ***281***, 673 (1993).
83. I. Karube, K. Yokoyama, and E. Tamiya, *Biosens. Bioelectron.*, ***8***, 219 (1993).
84. M. Goto, K. Morikage, Y. Esaka, B. Uno, and K. Kano, *Bunseki Kagaku*, ***43***, 403 (1994).
85. A.M. Nyamsi-Hendji, N. Jaffrezic-Renault, C. Martelet, A.A. Shul'ga, S.V. Dzydevich, A.P. Soldatkin, and A.V. El'sakya, *Sens. Actuators B*, ***B21***, 123 (1994).
86. C. Eppelsheim and N. Hampp, *Analyst*, ***119***, 2167 (1994).
87. K. Kano, K. Morikage, B. Uno, Y. Esaka, and M. Goto, *Anal. Chim. Acta*, ***229***, 69 (1994).
88. M.G. Garguilo and A.C. Michael, *Trends Anal. Chem.*, ***14***, 164 (1995).
89. M. Stoytcheva, *Electroanalysis*, ***7***, 560 (1995).

90. K.S. Bronk, K.L. Michael, P. Pantano, and D.R. Walt, *Anal. Chem.*, ***67***, 2750 (1995).
91. A. Guerrieri, G.E. De-Benedetto, F. Palmisano, and P.G. Zambonin, *Analyst*, ***120***, 2731 (1995).
92. A. Riklin and I. Willner, *Anal. Chem.*, ***67***, 4118 (1995).
93. R.F. Taylor, I.G. Marenchic, and E.J. Cook, *Anal. Chim. Acta*, ***213***, 131 (1988).
94. H.Z. Bu, *Fenxi-Huaxue*, ***20***, 841 (1992).
95. D.M. Ivintskii and J. Rishpon, *Anal. Chim. Acta*, ***282***, 517 (1993).
96. Z.X. Huang, R.L. Villarta-Snow, G.J. Lubrano, and G.G. Guilbault, *Anal. Biochem.*, ***215***, 31 (1993).
97. Q. Xin and R.M. Wightman, *Anal. Chim. Acta*, ***341***, 43 (1997).
98. A. Cagnini, I. Palchetti, M. Mascini, and A.P.F. Turner, *Mikrochim. Acta*, ***121***, 155 (1995).
99. R.T. Andres and R. Narayanaswamy, *Talanta*, ***44***, 1335 (1997).
100. T. Horiuchi, K. Torimitsu, K. Yamamoto, and O. Niwa, *Electroanalysis*, ***9***, 912 (1997).
101. O. Niwa, T. Horiuchi, R. Kurita, and K. Torimitsu, *Anal. Chem.*, ***70***, 1126 (1998).
102. Q.A. Chen, Y. Kobayashi, H. Takeshita, T. Hoshi, and J.I. Anzai, *Electroanalysis*, ***10***, 94 (1998).
103. R.E. Gyurcsanyi, Z. Vagfoldi, K. Toth, and G. Nagy, *Electroanalysis*, ***11***, 712 (1999).
104. M. Masoom, *Anal. Chim. Acta*, ***214***, 173 (1988).
105. M. Goto, E. Tamiya, and I. Karube, *J. Membr. Sci.*, ***41***, 291 (1989).
106. D.P. Nikolelis and C.G. Siontorou, *Anal. Chem.*, ***67***, 936 (1995).
107. A.N. Reshetilov, O.V. Fedoseeva, T.P. Eliseeva, V. YuSergeev, E.P. Medyantseva, and G.K. Budnikov, *Zh. Anal. Khim*, ***50***, 453 (1995).
108. R. Kataky and D. Parker, *Analyst*, ***121***, 1829 (1996).
109. J. Chen, X.Q. Lin, Z.H. Chen, S.G. Wu, and S.Q. Wang, *Anal. Lett.*, ***34***, 491 (2001).
110. C. Tran Minh, P.C. Pandy, and S. Kumaran, *Biosens. Bioelectron*, ***5***, 461 (1990).
111. G. Shen and J. Li, *Yaowu Fenxi Zazhi*, ***11***, 4 (1991).
112. K.R. Roger, J.C. Fernando, R.G. Thompson, J.J. Valdes, and M.E. Eldefrawi, *Anal. Biochem.*, ***202***, 111 (1992).
113. D.P. Nikolelis, M.G. Tzanelis, and U.J. Krull, *Anal. Chim. Acta*, ***281***, 569 (1993).

114. L. Ilcheva and T. Kolusheva, *Izv. Khim.*, ***23***, 39 (1990).

115. L. Campanella, M.P. Sammaritno, and M. Tomassetti, *Ann. Chim.*, ***81***, 639 (1991).

116. R.M. Morelis, P.R. Coulet, a. Simplot, C. Boisson, and G. Guibaud, *Clin. Chim. Acta*, ***203***, 295 (1991).

117. T.H. Huang, L. Yang, J. Gitzen, P.T. Kissinger, M. Vreeke, and A. Heller, *J. Chromatogr. Biomed. Appl.*, ***670***, 323 (1995).

118. H. Cui, Z.H. Lin, L.B. Zhang, and G.L. Sen, *Fenxi Kexue Xuebao*, ***11***, 7 (1995).

119. B. Lopez Ruiz, M. Sanchez Cabezudo, and H.A. Mottola, *Analyst*, ***121***, 1695 (1996).

120. O.V. Fedosseeva, H. Vchida, T. Katsube, Y. Ishimaru, and T. Iida, *Sens. Actuators B*, ***B65***, 55 (2000).

121. A. Curulli, S. Dragulescu, C. Cremisini, and g. Palleschi, *Electroanalysis*, ***13***, 236 (2001).

122. A. Guerrieri and F. Palmisano, *Anal. Chem.*, ***73***, 2875 (2001).

123. P.C. Pandy, S. Upadhyay, H.C. Pthak, C.M.D. Pandy, and I. Tiwari, *Sens. Actuators, B*, ***B62***, 109 (2000).

124. L. Alfonta, E. Katz, and I. Willner, *Anal. Chem.*, ***72***, 927 (2000).

125. J. Paul and S. Ebrahimian, *Microchem J.*, ***27***, 425 (1982).

126. M.A. Iorio, M. Molinari, and A. Laurenzi, *Farmaco, Ed. Prat.*, ***38***, 126 (1983).

127. D.J. Jenden, I. Hanin, and S.I. Lamb, *Anal. Chem.*, ***40***, 125 (1968).

128. J.L.W. Pohlmann and S.L. Cohan, *J. Chromatogr.*, ***131***, 297 (1977).

129. T.J. Davies and N.J. Hayward, *J. Chromatogr. Biomed. Appl.*, ***32***, 11 (1984).

130. D. Budai, P. Szerdahelyi, and P. Kasa, *Anal. Biochem.*, ***159***, 260 (1986).

131. D.J. Jenden, B. Campbell, and M. Roch, *Anat. Biochem.*, ***35***, 209 (1970).

132. D.J. Jenden et al., *Anal. Abstr.*, ***16***, 2069 (1969).

133. J.P. Green and P.I.A. Szilagyi, ***Choline, Acetylcholine; Handbook Chemical Assay Methods***, I. Hanin, ed., Raven Press, New York, N.Y., p. 151 (1974).

134. R. Karlen, G. Lundgren, I. Nordgren, and B. Holmstedt, ***Choline, Acetylcholine; Handbook Chemical Assay Methods***, I. Hanin, ed., Raven Press, New York, N.Y., p. 163 (1974).

135. K. Takahashi and T. Mouharo, *A & R*, ***16***, 69 (1978).

136. Y. Maruyama, M. Kusaka, J. Mori, A. Horikawa, and Y. Hasegawa, *J. Chromatogr. Biomed. Appl.*, **6**, 121 (1979).
137. D.E. Schmidt and R.C. Speth, *Anal. Biochem.*, **67**, 353 (1975).
138. J.W. Kosh, M.B. Smith, J.W. Sowell, and J.J. Freeman, *J. Chromatogr. Biomed. Appl.*, **5**, 206 (1979).
139. F. Mikes, G. Boshart, K. Wuethrich, and P.G. Waser, *Anal. Chem.*, **52**, 1001 (1980).
140. M.L. Gilberstadt and J.A. Russell, *Anal. Biochem.*, **138**, 78 (1984).
141. P.M. Findeis and S.O. Farwell, *J. High Resolut. Chromatogr. Chromatogr. Commun.*, **7**, 19 (1984).
142. M.F. Cranmer, *Life Sci.*, **7**, 995 (1968).
143. J. Shi and H. Du, *Sepu.*, **10**, 204 (1992).
144. W.Z. Duan, Z.W. Qu, and J.T. Zhang, *Yao Xue Xue Bao*, **32**, 920 (1997).
145. F.T. Tao, J.S. Thurber, and D.M. Dye, *J. Pharm. Sci.*, **73**, 1311 (1984).
146. M.E. Rhodes, P.K. Li, A.M. Burke and D.A. Johnson, *A.A.C.P.*, Annual Meeting American Association of Colleges of Pharmacy Annual Meeting 1998; 99 (July); 90.
147. P.E. Potter, J.L. Meek, and N.H. Neff, *J. Neurochem.*, **41**, 188 (1983).
148. J.L. Meek and C. Eva, *J. Chromatogr.*, **317**, 343 (1984).
149. C. Eva, M. Hadjiconstantinou, N.H. Neff, and J.L. Meek, *Anal. Biochem.*, **143**, 320 (1984).
150. Y. Ikarashi, T. Sasahara, and Y. Maruyama, *J. Chromatogr.*, **322**, 191 (1985).
151. T. Yao and M. Sato, *Anal. Chim. Acta*, **172**, 371 (1985).
152. F.P. Bymaster, K.W. Perry, and D.T. Wong, *Life Sci.*, **37**, 1775 (1985).
153. R.S. Jones and C.A. Stutte, *J. Chromatogr.*, **319**, 454 (1985).
154. C.D. Raghuveeran, *J. Liquid Chromatogr.*, **8**, 537 (1985).
155. N. Kaneda and T. Nagatsu, *J. Chromatogr. Biomed. Appl.*, **42**, 23 (1985).
156. J. Gorham, *J. Chromatogr.*, **362**, 243 (1986).
157. M. Liscovitch, A. Freese, J.K. Blusztajn, and R.J. Wurtman, *Anal. Biochem.*, **151**, 182 (1985).
158. G. Damsma, B.H.C. Westerink, and A.S. Horn, *J. Neurochem.*, **45**, 1649 (1985).
159. M. Asano, T. Miyauchi, T. Kato, K. Fujimori, and K. Yamamoto, *J. Liquid Chromatogr.*, **9**, 199 (1986).

160. N. Kaneda, M. Asano, and T. Nagatsu, *J. Chromatogr.*, ***360***, 211 (1986).
161. H. Stadler and T. Nesselhut, *Neurochem. Int.*, ***9***, 127 (1986).
162. K. Honda, K. Miyagushi, H. Nishino, H. Tanaka, T. Yao, and K. Imai, *Anal. Biochem.*, ***153***, 50 (1986).
163. K. Fujimori and K. Yamamoto, *J. Chromatogr., Biomed. Appl.*, ***58***, 167 (1987).
164. N. Tyrefors and P.G. Gillberg, *J. Chromatogr., Biomed. Appl.*, ***67***, 85 (1987).
165. A. Beley, A. Zekhnini, S. Lartilliot, D. Fage, and J. Bralet, *J. Liquid Chromatogr.*, ***10***, 2977 (1987).
166. P. Van-Zoonen, C. Gooijer, N.H. Velthorst, R.W. Frei, J.H. Wolf, J. Gerrits, and F. Flentge, *J. Pharm., Biomed. Anal.*, ***5***, 485 (1987).
167. N.M. Barnes, B. Costall, A.F. Fell, and R.J. Naylor, *J. Pharm. Pharmacol.*, ***39***, 727 (1987).
168. G. Damsma, D. Lammerts-van-Bueren, B.H.C. Westerink, and A.S. Horn, *Chromatographia*, ***24***, 827 (1987).
169. G. Damsma and F. Flentge, *J. Chromatogr.*, ***428***, 1 (1988).
170. H. Takeda, *Showa Igakkai Zasshi*, ***48***, 553 (1988).
171. S. Okuyama and Y. Ikeda, *J. Chromatogr., Biomed. Appl.*, ***75***, 389 (1988).
172. S. Murai, H. Miyate, H. Saito, N. Nagahama, Y. Masuda, and T. Itoh, *J. Pharmacol. Methods*, ***21***, 255 (1989).
173. A.W. Teelken, H.F. Schuring, W.B. Trieling, and G. Damsma, *J. Chromatogr., Biomed. Appl.*, ***94***, 408 (1990).
174. N. Tyrefors and A. Carlsson, *J. Chromatogr.*, ***502***, 337 (1990).
175. M. Matsumoto, H. Togashi, M. Yoshioka, M. Hirokami, K. Morii, and H. Saito, *J. Chromatogr., Biomed. Appl.*, ***91***, 1 (1990).
176. N. Bertrand, J. Bralet, and A. Beley, *J. Neurochem.*, ***55***, 27 (1990).
177. Y. Fujiki, Y. Ikeda, S. Okuyama, K. Tomoda, K. Ooshiro, H. Matsumura, T. Itoh, and T. Yamauchi, *J. Liquid Chromatogr.*, ***13***, 239 (1990).
178. E. Haen, H. Hagenmaier, and J. Remien, *J. Chromatogr.*, ***537***, 514 (1991).
179. R.B. Miller and C.L. Blank, *Anal. Biochem.*, ***196***, 377 (1991).
180. Y. Ikarashi, H. Iwatsuki, L.C. Blank, and Y. Maruyama, *J. Chromatogr., Biomed. Appl.*, ***113***, 29 (1992).
181. J. Salamoun, Phuc Trung Nguyen, and J. Remien, *J. Chromatogr.*, ***596***, 43 (1992).

182. F. Flentge, K. Venema, T. Koch, and J. Korf, *Anal. Biochem.*, ***204***, 305 (1992).
183. J. Salamoun and J. Remien, *J. Pharm., Biomed. Anal.*, ***10***, 931 (1992).
184. M.D. Greaney, D.L. Marshall, B.A. Bailey, and I.N. Acworth, *J. Chromatogr., Biomed. Anal.*, ***133***, 125 (1993).
185. L.E. Webb, T.M. Pavlina, and J.T. Hjelle, *Clin. Biochem.*, ***26***, 173 (1993).
186. H. Ishimaru, Y. Ikarashi, and Y. Maruyama, *Biol. Mass. Spectrom.*, ***22***, 681 (1993).
187. T.R. Tsai, T.M. Cham, K.C. Chen, C.F. Chen, and T.H. Tsai, *J. Chromatogr., Biomed. Appl.*, ***678***, 151 (1996).
188. T. Kato, J.K. Lui, K. Yamamoto, P.G. Osborne, and O. Niwa, *J. Chromatogr., Biomed. Appl.*, ***682***, 62 (1996).
189. L.D. Acevedo, Y.D. Xu, X. Zhang, R.J. Pearce, and A. Yergey, *J. Mass. Spectrom.*, ***31***, 1399 (1996).
190. A.J. Carter and J. Kehr, *J. Chromatogr., Biomed. Appl.*, ***692***, 207 (1997).
191. W.L. Ye, X.F. Ma, and Z.T. Mei, *Sepu.*, ***16***, 375 (1998).
192. Y. Ikarashi, C.L. Blank, T. Kawakubo, Y. Suda, and Y. Maruyama, *J. Liquid Chromatogr.*, ***17***, 287 (1994).
193. Y. Ikarashi, C.L. Blank, Y. Suda, T. Kawakubo, and Y. Maruyama, *J. Chromatogr.*, ***718***, 267 (1995).
194. O. Niwa, T. Horiuchi, M. Morita, T. Huang, and P.T. Kissinger, *Anal. Chim. Acta*, ***318***, 167 (1996).
195. G.S. Mayer, *Curr. Sep.*, *7*, 47 (1986).
196. Y. Ikarashi, K. Itoh, and Y. Maruyama, *Biol. Mass. Spectrom.*, ***20***, 21 (1991).
197. Y. Ikarashi, C.L. Blank, K. Itoh, H. Satoh, H.K. Inoue, and Y. Maruyama, *Nippon Yakurigaku Zasshi*, ***97***, 51 (1991).
198. R.L. Stein, *J. Chromatogr.*, ***214***, 148 (1981).
199. D.J. Jenden, *Advan. Biochem. Psychopharmacol.*, *7*, 69 (1973).
200. R.L. Polak and P.C. Molenaar, *J. Neurochem.*, ***23***, 1295 (1974).
201. R.L. Polak and P.C. Molenaar, *J. Neurochem.*, ***32***, 407 (1979).
202. F. Mikes, *Chem. Rundsch.*, ***33***, 1 (1980).
203. A.K. Singh and L.R. Drewes, *J. Chromatogr., Biomed. Appl.*, ***40***, 170 (1985).
204. D.J. Liberato, A.L. Yergey, and S.T. Weintraub, *Biomed. Environ.-Mass-Spectrom*, ***13***, 171 (1986).
205. M. Watanabe, A. Kimura, K. Akasaka, and S. Hayashi, *Biochem. Med. Metab. Biol.*, ***36***, 355 (1986).

206. E.A. Pomfret, K.A. DaCosta, L.L. Schurman, and S.H. Zeisel, *Anal. Biochem.*, ***180***, 85 (1989).

207. A.V. Terry Jun, L.A. Silks, R.B. Dunlap, J.D. Odom, and J.W. Kosh, *J. Chromatogr.*, ***585***, 101 (1991).

208. T.A. Patterson and J.W. Kosh, *Biol. Mass Spectrom*, ***21***, 299 (1992).

209. T. Kappes, P. Schnierle, and P.C. Hauser, *Electrophoresis*, ***21***, 1390 (2000).

210. W.D. Reid, D.R. Haubrich, and G. Krishna, *Analyt. Biochem.*, ***42***, 390 (1971).

211. R. Bluth, R. Langnickel, K.H. Raubach, and H. Fink, *Acta Biol. Med. Ger.*, ***39***, 881 (1980).

212. R.M. Dick, J.J. Freeman, and J.W. Kosh, *J. Chromatogr.*, ***347***, 387 (1985).

213. J.J. O'Neill and K.A. Hruschak, *Neurochem. Res.*, ***12***, 515 (1987).

214. J.R. Cooper, *Life Sci.*, ***45***, 2041 (1989).

215. S.A. Eckernas and S.M. Aquilonius, *Acta Physiol. Scand.*, ***100***, 446 (1977).

216. S. Spector, A. Felix, G. Semenuk, and J.P.M. Finberg, *J. Neurochem.*, ***30***, 685 (1978).

217. J.K. Saelens, M.P. Allen, and J.P. Simke, *Archs. Int. Pharmacodyn. Ther.*, ***186***, 279 (1970).

218. H. Ladinsky and S. Consolo, ***Choline, Acetylcholine; Handbook Chemical Assay Methods***, I. Hanin, ed., Raven Press, New York, pp. 1–17 (1974).

219. J.K. Saelens, J.P. Simke, M.P. Allen, and C.A. Conroy, *Methods Neurochem.*, ***4***, 69 (1973).

220. A. Jamarillo, S. Lapez, J.B. Justice, J.D. Jun Salamone, and D.B. Neill, *Anal. Chim. Acta*, ***146***, 149 (1983).

221. G. Szepesy, K. Bansaghy, and P. Kasa, *Kiserl Orvostud*, ***33***, 174 (1981).

222. D.E. Schmidt, R.C. Speth, F. Welsch, and M.J. Schmidt, *Brain Res.*, ***38***, 377 (1972).

223. J. Kehr, T. Yoshitaka, F.H. Wang, D. Wynick, K. Holmberg, U. Lendahl, T. Bartfai, M. Yamaguchi, T. Hokfelt, and S.O. Orgen, *J. Neurosci. Methods*, ***109***, 71 (2001).

Benazepril Hydrochloride: Comprehensive Profile

F. Belal, H.H. Abdine and Abdullah A. Al-Badr

Department of Pharmaceutical Chemistry
College of Pharmacy, King Saud University
P.O. Box 2457, Riyadh-11451
Kingdom of Saudi Arabia

PROFILES OF DRUG SUBSTANCES,
EXCIPIENTS, AND RELATED
METHODOLOGY – VOLUME 31
DOI: 10.1016/S0000-0000(00)00000-0

CONTENTS

1. **Description** 119
   1.1 Nomenclature 119
      1.1.1 Systematic chemical name 119
      1.1.2 Nonproprietary names 120
      1.1.3 Proprietary names 120
   1.2 Formulae 120
      1.2.1 Empirical formula, molecular weight, CAS number 120
      1.2.2 Structural formula 120
   1.3 Elemental analysis 120
   1.4 Appearance 121
   1.5 Uses and application 121
2. **Method of preparation** 121
3. **Physical characteristics** 123
   3.1 Solubility characteristics 123
   3.2 Optical activity 123
   3.3 X-Ray powder diffraction pattern 123
   3.4 Thermal methods of analysis 123
      3.4.1 Melting behavior 123
      3.4.2 Differential scanning calorimetry 123
   3.5 Spectroscopy 124
      3.5.1 UV/VIS spectroscopy 124
      3.5.2 Vibrational spectroscopy 124
      3.5.3 Nuclear magnetic resonance spectrometry . . 124
   3.6 Mass spectrometry 128
4. **Methods of Analysis** 129
   4.1 Compendial methods 129
   4.2 Statistically compared methods 130
   4.3 Potentiometric methods 131
   4.4 Spectrophotometric methods 132
   4.5 Colorimetric method 143
   4.6 Voltammetric methods 147
   4.7 Chromatographic methods of analysis 147
      4.7.1 Thin-Layer chromatographic methods 147
      4.7.2 Gas chromatographic–mass spectrometric methods (GC–MS) 148
      4.7.3 High performance liquid chromatography 150
      4.7.4 Capillary electrophoresis methods 156

5. **Stability** 158
6. **Absorption and Pharmacokinetics** 158
   6.1 Absorption, metabolism and fate 158
   6.2 Excretion 158
7. **Acknowledgments** 159
8. **References** 159

1. DESCRIPTION

1.1 Nomenclature

1.1.1 Systematic chemical names [1–3]

(3*S*)-3-[[(1*S*)-1-Carbethoxy-3-phenylpropyl]amino]-2,3,4,5-tetrahydro-2-oxo-1*H*-1-benzazepine-1-acetic acid, hydrochloride.

[*S*-(*R**,*R**)]-3-[[1-(Ethoxycarbonyl)-3-phenylpropyl]amino]-2,3,4,5-tetrahydro-2-oxo-1*H*-1-benzazepine-1-acetic acid, hydrochloride.

{(3*S*)-3-[(1*S*)-1-Ethoxycarbonyl-3-phenylpropylamino]-2,3,4,5-tetrahydro-2-oxo-1*H*-1-benzazepin-1-yl}acetic acid, hydrochloride.

(3*S*)-1-(Carboxymethyl)-[[(1*S*)-1-(ethoxycarbonyl)-3-phenylpropyl]-amino]-2,3,4,5-tetrahydro-1*H*-[1]benzazepin-2-one, hydrochloride.

1-Carboxymethyl-3-[1-ethoxycarbonyl-3-phenyl-(1*S*)-propylamino]-2,3,4,5-tetrahydro-1*H*-1(3*S*)-benzazepin-2-one, hydrochloride.

(3*S*)-3-[[(1*S*)-1-Carboxy-3-phenylpropyl]amino]-2,3,4,5-tetrahydro-2-oxo-1*H*-1-benzazepin-1-acetic acid-3-ethyl ester, hydrochloride.

3-{[1-Ethoxycarbonyl-3-phenyl-(1*S*)-propyl]amino}-2,3,4,5-tetrahydro-2-oxo-1*H*-(3*S*)-benzazepine-1-acetic acid, hydrochloride.

3-[1-(Ethoxycarbonyl)-3-phenyl-(1*S*)-propylamino]-2,3,4,5-tetrahydro-2-oxo-1*H*-1-(3*S*)-benzazepine-1-acetic acid, hydrochloride.

1*H*-1-Benzazepine-1-acetic acid, [*S*-(*R**,*R**)]-3-[[1-(ethoxycarbonyl)-3-phenyl-propyl]amino]-2,3,4,5-tetrahydro-2-oxo-, monohydrochloride.

1*H*-1-Benzazepine-1-acetic acid, 3-[[1-(ethoxycarbonyl)-3-phenylpropyl]-amino]-2,3,4,5-tetrahydro-2-oxo-, [*S*-(*R**,*R**)].

1.1.2 Nonproprietary names [1–4]
Benazepril, Benazepril hydrochloride, CGS-14824 A.

Benazeprilat, CGS-14831.

1.1.3 Proprietary names [1–4]
Briem, Cibacen, Cibacene, Fortekor, Labopal, Lotensin, Tensanil and Zinadril, Cibace.

1.2 Formulae

1.2.1 Empirical formula, molecular weight, CAS number [1]

Benazepril	$C_{24}H_{28}N_2O_5$	424.00	[86541–75–5]
Benazepril HCl	$C_{24}H_{28}N_2O_5 \cdot HCl$	460.50	[86541–74–4]
Benazeprilat	$C_{22}H_{24}N_2O_5$	396.00	[86541–78–8]

1.2.2 Structural formula

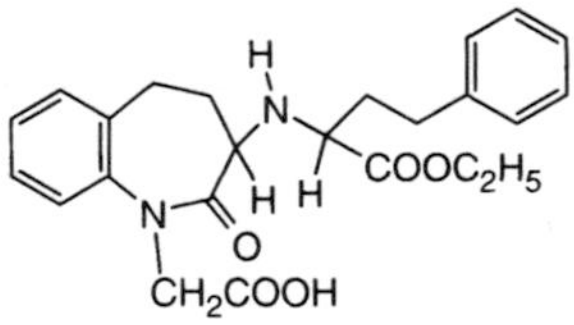

Benazepril 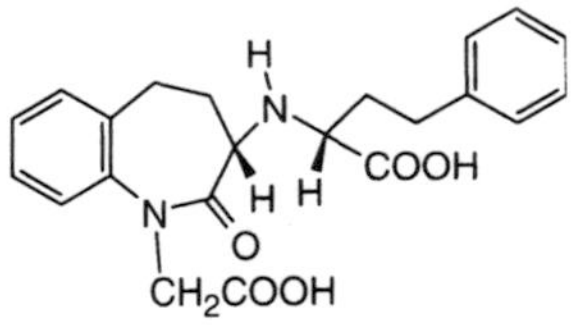 Benazeprilat

1.3 Elemental analysis [1]

The calculated elemental compositions are as follows:

	Benazepril	Benazepril HCl	Benazeprilat
Carbon	67.91%	62.54%	66.67%
Hydrogen	06.65%	06.34%	06.06%
Chlorine	—	07.68%	—
Nitrogen	06.60%	06.07%	07.07%
Oxygen	18.84%	17.37%	20.20%

1.4 Appearance

Benazepril hydrochloride is obtained as a white to off-white crystalline powder [3].

1.5 Uses and application

Benazepril hydrochloride, an angiotensin-converting enzyme (ACE) inhibitor, is used in the treatment of hypertension and heart failure. The family of ACE inhibitors inhibits the angiotensin-converting enzyme, which is involved in the conversion of angiotensin I to angiotensin II. Angiotensin II stimulates the synthesis and secretion of aldosterone and raises blood pressure *via* a potent direct vasoconstrictor effect. ACE is identical to bradykininase or kininase II, and ACE inhibitors may reduce the degradation of bradykinin [2].

Benazepril hydrochloride, being a prodrug of the diacid benazeprilat, is rapidly absorbed and converted to the active ACE inhibitor, benazeprilat. The drug has a half-life of 10–11 h. It is approved for treating essential hypertension and may be effective for treating congestive heart failure [2].

Benazepril owes its activity to benazeprilat, to which it is converted after oral administration. The drug is given by mouth as the hydrochloride salt, but doses are usually expressed in terms of the free base. In the treatment of hypertension, the usual initial dose is 10 mg once daily. An initial dose of 5 mg once daily is suggested for patients with renal impairment, or who are receiving a diuretic. The usual maintenance dose is 20–40 mg daily, which may be given in two divided doses if control is inadequate with a single dose. In the treatment of heart failure, the usual initial dose is 2.5 mg once daily, adjusted according to response to a maximum dose of 20 mg daily [2].

2. METHOD OF PREPARATION

Watthey and Watthey *et al.* [5–7] reported the synthesis of benazepril using the following procedure (which is illustrated in Scheme 1).

Chlorination of benzazepin-2-one (**1**) with phosphorous pentachloride in xylene gave dichlorolactam (**2**). Compound (**2**) was reduced to the monochlorolactam (**3**) by catalytic dehydrogenation (over Pd/C). Treatment of (**3**) with sodium azide in dimethyl sulfoxide at 80°C gave azide (**4**).

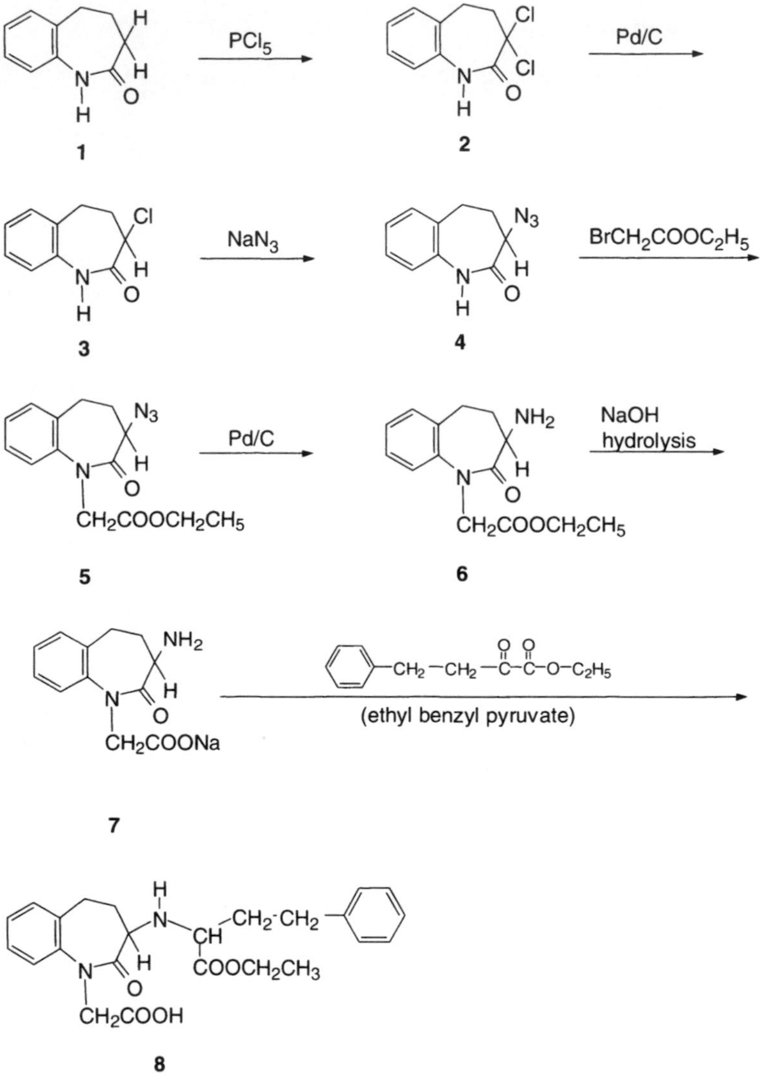

Scheme 1. Synthesis of benazepril, according to the method of Watthey and Watthey *et al.* [5–7].

Alkylation of (**4**) with ethyl bromoacetate was effected under phase-transfer conditions with powdered potassium hydroxide and tetrabutyl ammonium bromide in tetrahydrofuran to yield compound (**5**), and the azide function in (**5**) was then reduced by catalytic hydrogenation (over Pd/C) to give (**6**). The aminoester (**6**) was resolved into its enantiomers *via* the tartrate salt. The negatively rotating enantiomer was hydrolyzed to the sodium salt of the carboxylic acid (**7**), and this material was condensed with ethyl benzyl pyruvate in the presence of sodium cyanoborohydride. Two isomers were obtained in a ratio of 70:30, and the major isomer (**8**, benazepril) was obtained in its pure form by recrystallization.

3. PHYSICAL CHARACTERISTICS

3.1 Solubility characteristics [3]

The solubility of benazepril hydrochloride is greater than 1 g in 10 mL of water, ethanol, and methanol.

3.2 Optical activity [1]

The angle observed for the optical rotation of benazepril (C = 1.2% in ethanol) is – 159°, and the corresponding angle of rotation for benazepril hydrochloride (C = 0.9 in ethanol) is –141.0° and for benazeprilat (C = 1 in 3% aqueous ammonium hydroxide) is 200.5°.

3.3 X-Ray powder diffraction pattern

The X-ray powder diffraction pattern of benazepril hydrochloride was obtained using a Simons XRD-5000 diffractometer, and is shown in Figure 1. A summary of the observed scattering angles (degrees 2-θ), the interplanar d-spacings (Å), and the relative intensities (%) of the observed peaks of benazepril hydrochloride is found in Table 1.

3.4 Thermal methods of analysis

3.4.1 Melting behavior [1]

Benazepril melts at about 148–149°C, benazepril hydrochloride melts at about 189–190°C, and benazeprilat melts at about 270–272°C.

3.4.2 Differential scanning calorimetry

The differential scanning calorimetry (DSC) thermogram of benazepril hydrochloride was obtained using a DuPont TA-9900 thermal analyzer attached to a DuPont Data unit. The thermogram shown in Figure 2 was obtained at a heating rate of 10°C/min, and was run from 100 to 220°C. The compound was found to melt at 190.0°C.

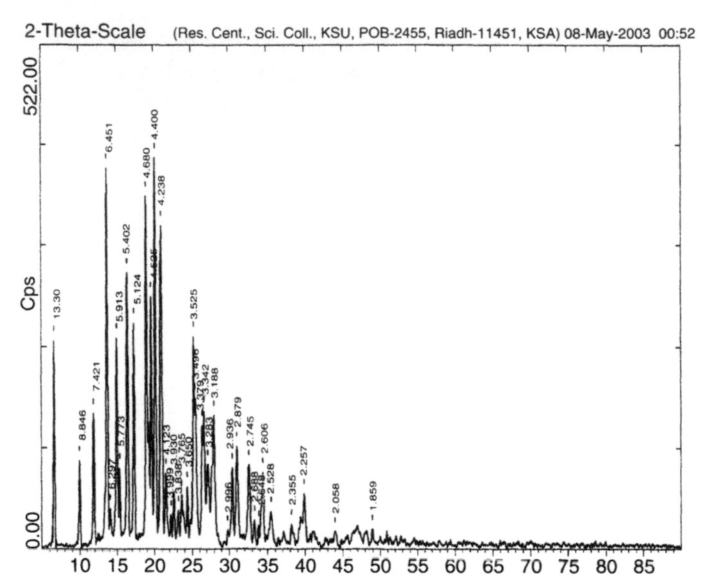

Figure 1. X-ray powder diffraction pattern of benazepril hydrochloride.

3.5 Spectroscopy

3.5.1 UV/VIS spectroscopy

The ultraviolet spectrum of benazepril hydrochloride (obtained at a concentration of 0.1 mg/mL in water), which is shown in Figure 3, was recorded using a Shimadzu model 1601 PC ultraviolet–visible spectrometer. The drug exhibited only one maximum at 238.5 nm, A[1%, 1 cm] = 198.7, and molar absorptivity = 9160.

3.5.2 Vibrational spectroscopy

The infrared absorption spectrum of benazepril hydrochloride was obtained in a KBr pellet using a Perkin-Elmer infrared spectrophotometer. The infrared spectrum is shown in Figure 4, and the principal peaks were detected at 3435, 2980, 2680, 2467, 1739, 1674, 1524, 1390, 1211, 1003, 754, 702 cm^{-1}. The assignments of the major infrared absorption bands of benazepril hydrochloride are shown in Table 2.

3.5.3 Nuclear magnetic resonance spectrometry

3.5.3.1 ^{1}H-NMR spectrum

The proton NMR spectra of benazepril hydrochloride were obtained using a Bruker Advance System operating at 300, 400, or 500 MHz.

Table 1

Scattering Angles, *d*-Spacings, and Peak Relative Intensities from the X-Ray Powder Diffraction Pattern of Benazepril Hydrochloride

Scattering Angle (degrees 2-θ)	***d*-spacing (Å)**	**Relative Intensity (%)**	**Scattering Angle (degrees 2-θ)**	***d*-spacing (Å)**	**Relative Intensity (%)**
6.641	13.2986	53.15	25.244	3.5250	54.16
9.991	8.8460	22.53	25.466	3.4947	38.30
11.916	7.4206	34.58	26.400	3.3732	31.50
13.715	6.4512	97.34	26.651	3.3420	35.07
14.053	6.2967	10.04	27.137	3.2833	21.61
14.970	5.9133	53.74	27.966	3.1878	34.13
15.335	5.7733	20.66	29.796	2.9960	4.70
16.397	5.4015	70.50	30.425	2.9355	20.68
17.291	5.1244	57.67	31.038	2.8789	26.22

(*continued*)

Table 1 (continued)

Scattering Angle (degrees 2-θ)	***d*-spacing (Å)**	**Relative Intensity (%)**	**Scattering Angle (degrees 2-θ)**	***d*-spacing (Å)**	**Relative Intensity (%)**
18.949	4.6795	90.20	32.593	2.7451	21.52
19.602	4.5250	64.26	33.299	2.6885	7.38
20.163	4.4005	100.00	33.818	2.6484	6.00
20.945	4.2378	82.76	34.393	2.6054	19.49
21.535	4.1231	19.30	35.474	2.5284	9.50
22.210	3.9993	8.60	38.175	2.3555	6.29
22.604	3.9304	14.46	39.905	2.2573	14.04
23.156	3.8380	9.86	43.967	2.05577	4.29
23.608	3.7655	13.80	48.951	1.8592	4.86
24.369	3.6496	15.88			

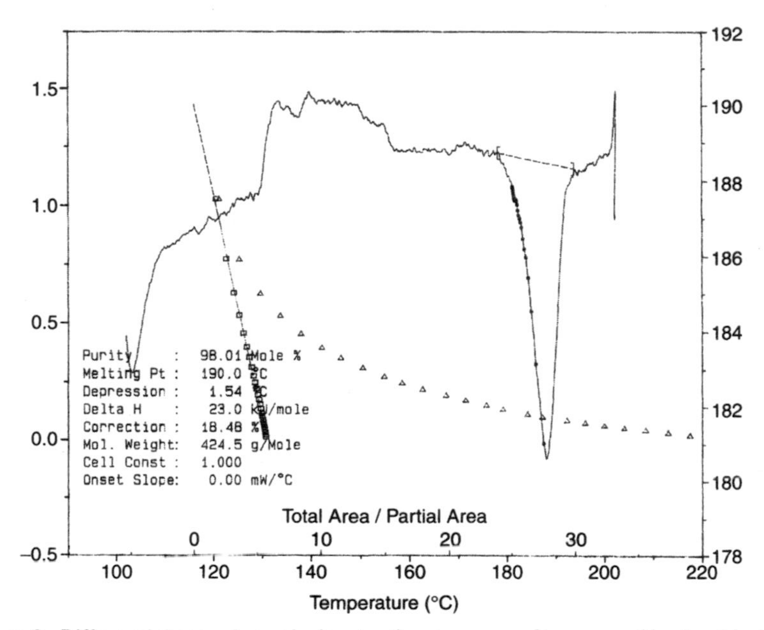

Figure 2. Differential scanning calorimetry thermogram of benazepril hydrochloride.

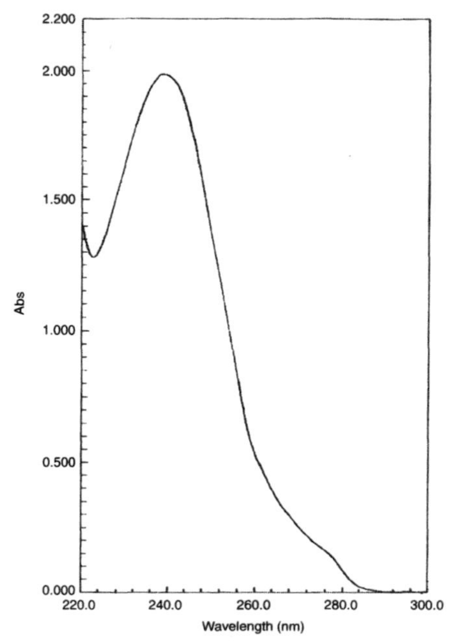

Figure 3. Ultraviolet absorption spectrum of benazepril hydrochloride (0.1 mg/mL, aqueous solution).

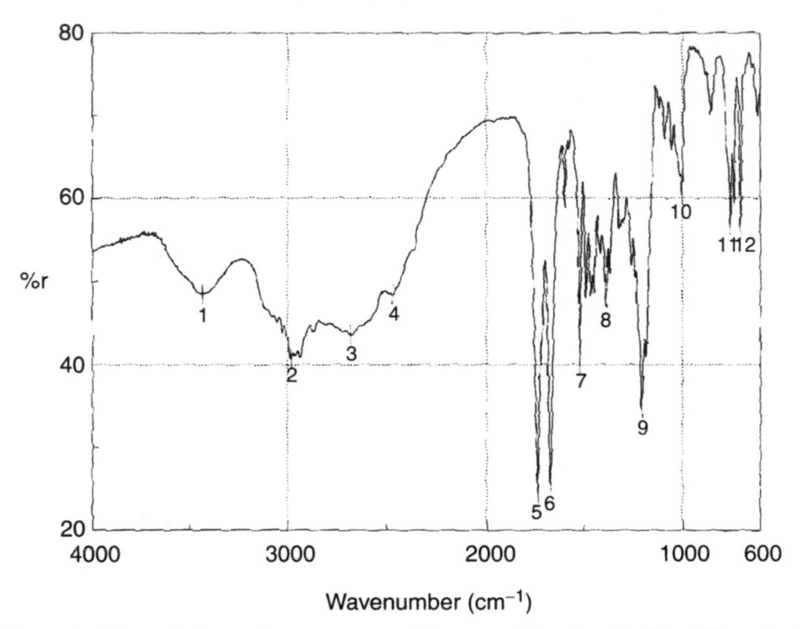

Figure 4. Infrared absorption spectrum of benazepril hydrochloride (KBr pellet).

Standard Bruker Software was used to execute the recording of DEPT, COSY, and HETCOR spectra. The sample was dissolved in D_2O, and all resonance bands were referenced to the tetramethylsilane (TMS) internal standard. The ^{1}H-NMR spectra of benazepril hydrochloride are shown in Figures 5–9. The ^{1}H-NMR spectral assignments for benazepril hydrochloride are provided in Table 3.

3.5.3.2 ^{13}C-NMR spectrum

The carbon-13 NMR spectra of benazepril hydrochloride were obtained using a Bruker Advance Instrument operating at 75, 100, or 125 MHz. The sample was dissolved in D_2O, and tetramethylsilane (TMS) was added to function as the internal standard. The ^{13}C-NMR spectra are shown in Figures 10–13, and the gradient HMQC NMR spectra are shown in Figures 14 and 15. The assignments for the observed resonance bands associated with the various carbons are provided in Table 4.

3.6 Mass spectrometry

The mass spectrum of benazepril hydrochloride was obtained using a Shimadzu PQ-5000 mass spectrometer. The parent ion was collided with

Table 2

Assignments for the Infrared Absorption Bands of Benazepril Hydrochloride

Frequency (cm^{-1})	Assignments
3435	Acid O–H stretching
2980	N–H stretching
2680	C–H stretching
2467	C–N stretching
1739	Acid C=O stretching
1674	C=C stretching
1524	$-CH_2$ bending
1390	$-CH_3$ bending
1211	C–C stretching
1003	C–O stretching
754	Di-substituted benzene
702	Mono-substituted benzene

helium as the carrier gas. Figure 16 shows the detailed mass fragmentation pattern, and Table 5 shows the mass fragmentation pattern of the substance.

4. METHODS OF ANALYSIS

4.1 Compendial methods

Benazepril hydrochloride is not yet a compendial article in any of the major pharmacopoeias, so no official tests have been certified as of yet.

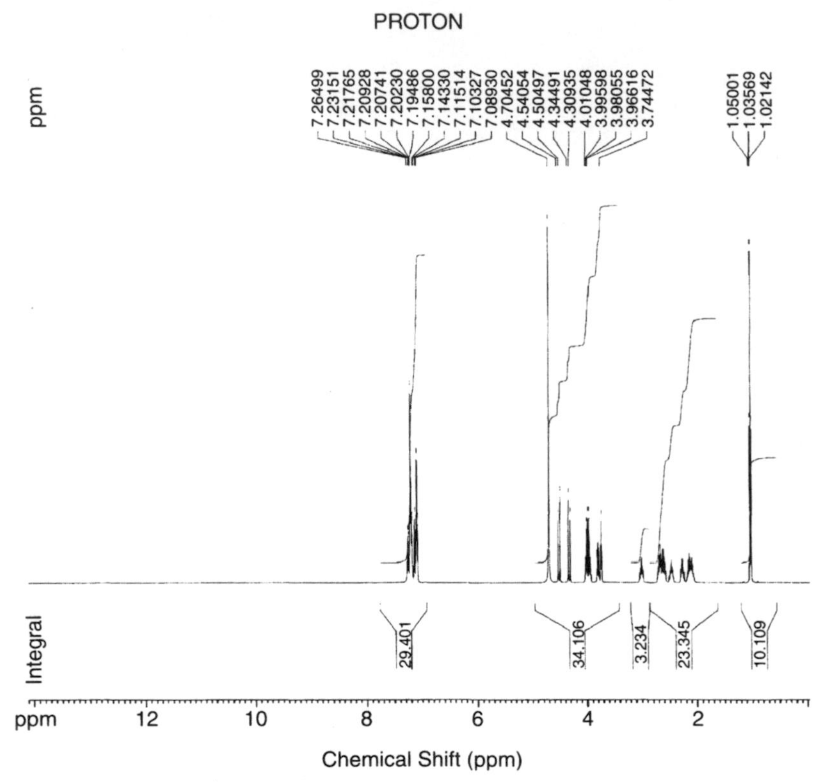

Figure 5. The ^{1}H-NMR spectrum of benazepril hydrochloride in D_2O.

4.2 Statistically compared methods

Wade *et al.* reported the use of a novel statistical approach for the comparison of analytical methods to measure angiotensin converting enzyme [peptidyl-dipeptidase A] activity, and to measure enalaprilat and benazeprilat [8]. Two methods were used to measure peptidyl-dipeptidase A, namely hippuryl histidyl leucine (HHL-method) [9], and inhibitor binding assay (IBA method) [10]. Three methods were used to measure enalaprilat, namely a radioimmunoassay (RIA) method [11], the HHL method, and the IBA method. Three methods were used to measure benazeprilat (then active metabolite of benazepril) in human plasma, namely gas chromatography–mass spectrometry (GC–MS method) [12], the HHL method, and the IBA method, and were statistically compared. First, the methods were compared by the paired *t* test or analysis of variance, depending on whether two or three different methods were under comparison. Secondly, the squared coefficients of variation of the

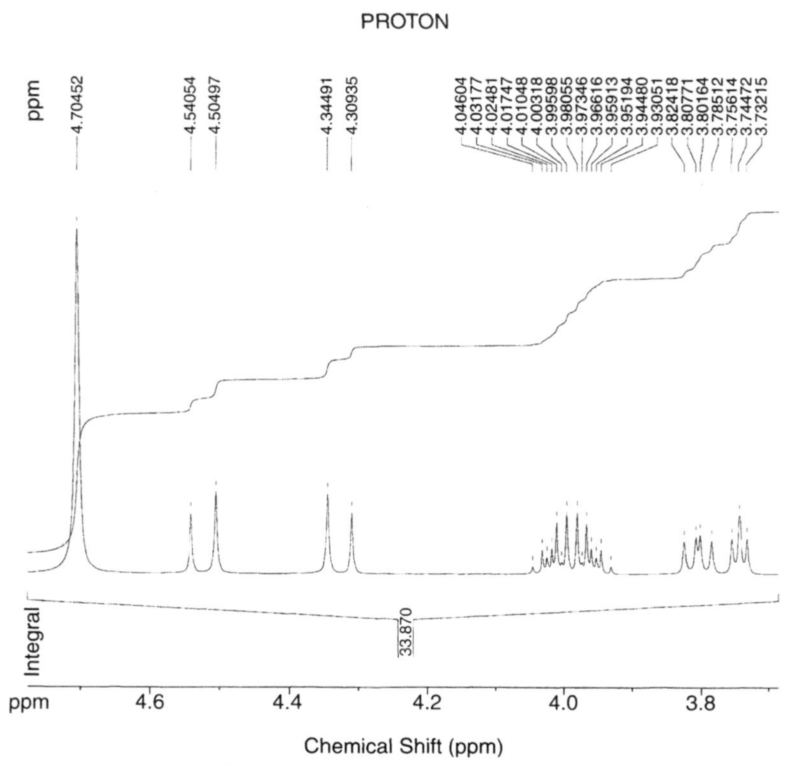

Figure 6. Expanded ^{1}H-NMR spectrum of benazepril hydrochloride in D_2O.

methods were compared by the Jackknife technique. No statistically significant difference was found between the two assays used to measure peptidyl-dipeptidase A. Differences were found between the three methods used to measure enalaprilat, although no obvious reason was determined for this. Significant differences were also found between the techniques used to determine benazeprilat; with these being attributed to the presence of metabolites interfering in the nonspecific assay methods.

4.3 Potentiometric methods

Khalil and El-Aliem constructed a coated-wire benazepril-selective electrode based on incorporation of the benazepril-tetraphenyl borate ion pair in a poly (vinyl chloride) coating membrane [13]. The influence of membrane composition, temperature, pH of the test solution, and foreign ions on the electrode performance was investigated. At 25°C, the electrode showed a Nernstian response over a benazepril concentration

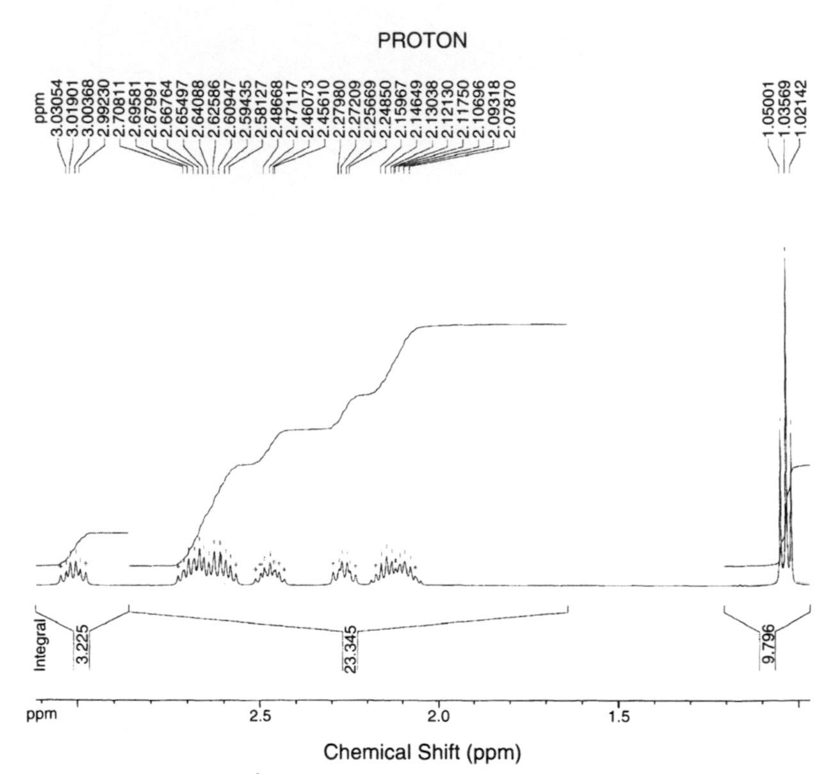

Figure 7. Expanded ^{1}H-NMR spectrum of benazepril hydrochloride in D_2O.

range of 1.26×10^{-5} to 0.58×10^{-2} M, at 25°C and was found to be very selective, precise, and usable over the pH range of 2.5–9.2. The standard electrode potentials were determined at 20, 25, 30, 35, 40 and 45°C, and used to calculate the isothermal temperature coefficient of the electrode. Temperatures higher than 45°C were found to seriously affect the electrode performance. The electrode was successfully used for the potentiometric determination of benazepril hydrochloride both in pure solutions and in pharmaceutical preparations.

4.4 Spectrophotometric methods

Erk and Onur used two new spectrophotometric methods for the simultaneous determination of benazepril hydrochloride and hydrochlorothiazide in their binary mixtures [14]. In a derivative spectrophotometric method, $dA/d\lambda$ values were measured at 260.7 and 239.8 nm for benazepril hydrochloride and hydrochlorothiazide, respectively, for

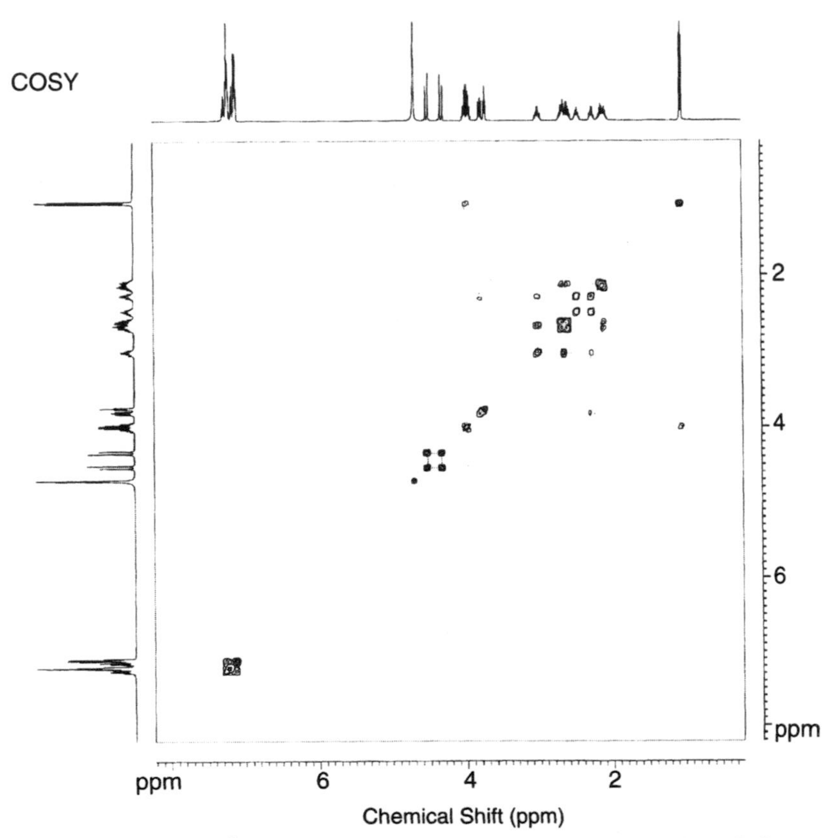

Figure 8. COSY ^{1}H-NMR spectrum of benazepril hydrochloride in D_2O.

their combination solutions. The relative standard deviation of the method was found to be 0.77% for benazepril hydrochloride and 0.28% for hydrochlorothiazide. In an absorbance ratio method, quantification of benazepril hydrochloride and hydrochlorothiazide was performed using the absorbances measured at 264, 258.3, and 236.4 nm in their combination solutions. The relative standard deviations of the method were found to be 1.26% and 0.36% for benazepril hydrochloride and hydrochlorothiazide, respectively. These two methods have been successfully applied to tablets containing only these drugs.

Panderi used a second-order derivative spectrophotometric method for the simultaneous determination of benazepril hydrochloride and hydrochlorothiazide in tablets [15]. The determination of benazepril hydrochloride in the presence of hydrochlorothiazide was achieved by

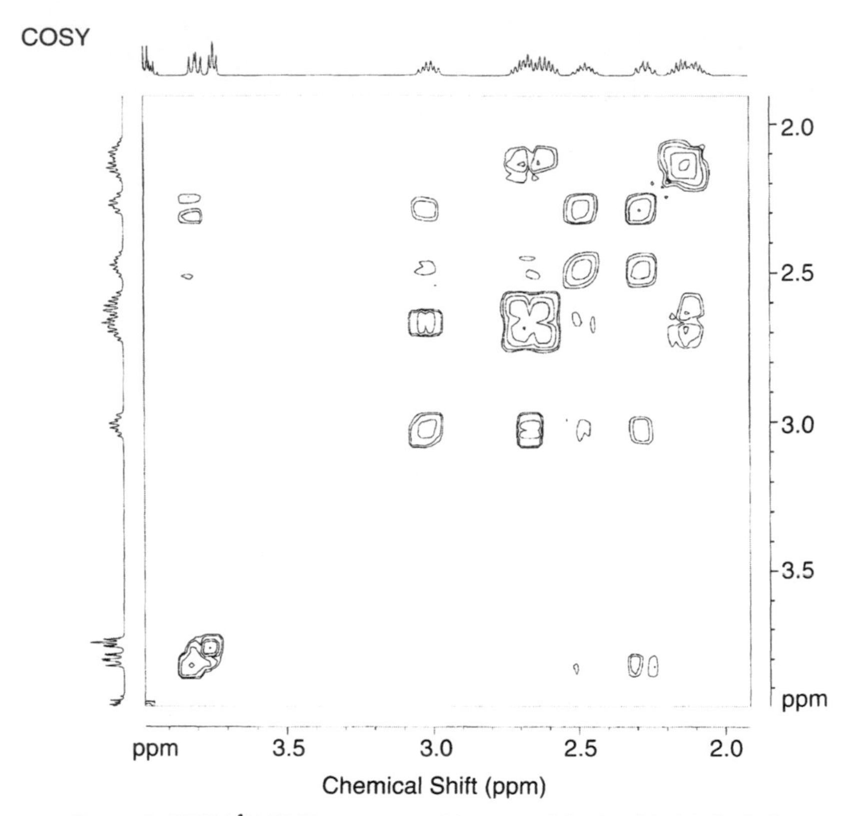

Figure 9. COSY ^{1}H-NMR spectrum of benazepril hydrochloride in D_2O.

measuring the second order derivative signals at 253.6 and 282.6 nm, while the second-order derivative signal at 282.6 nm was measured for the determination of hydrochlorothiazide. The linear dynamic ranges were 14.80–33.80 μg/mL for benazepril hydrochloride and 18.5–42.2 μg/mL for hydrochlorothiazide. The correlation coefficients for the calibration graphs exceeded 0.9998, the precision (% RSD; $n = 5$) was better than 1.43%, and the accuracy was satisfactory (error less than 0.99%). The detection limits were found to be 2.46 and 1.57 μg/ml for benazepril HCl and hydrochlorothiazide, respectively. The method was applied to the quality control of commercial tablets, and proved to be suitable for rapid and reliable analysis.

El-Yazbi *et al.* presented three spectrophotometric methods for determination of benazepril hydrochloride in its single and multi-component dosage forms [16]. The methods are (i) derivative

Table 3

Assignments for the Proton Nuclear Magnetic Resonance Spectrum of Benazepril Hydrochloride

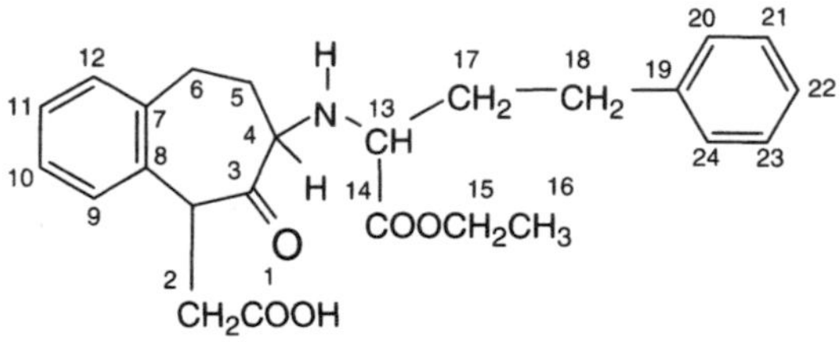

Chemical Shift (ppm, relative to TMS)	Number of Protons	Multiplicity*	Assignment (proton at carbon number)
1.04	3	s (7 Hz)	16
2.08–3.03	8	m	(5, 6, 17, 18)
3.73–3.76	1	t (6.5 Hz)	13
3.79–3.82	1	q (8.5 Hz)	4
3.93–4.04	2	m	15
4.31–4.54	2	dd	2
7.09–7.26	9	m	(9–12, 20–24)

*s: singlet; t: triplet; q: quartet; m: multiplet; dd: double doublet.

spectrophotometry to resolve matrix interferences, (ii) reaction with bromocresol green (BCG), and (iii) reaction with 3-methylbenzothialozone hydrazone (MBTH) after alkaline hydrolysis. Powdered tablets (coating removed for the MBTH method) equivalent to 10 mg of benazepril hydrochloride were mixed with 25 mL of water (10 mL MBTH method), shaken for 45 min, filtered, treated and analyzed. For method (i), derivative spectra were measured at 240 and 244 nm. For method (ii),

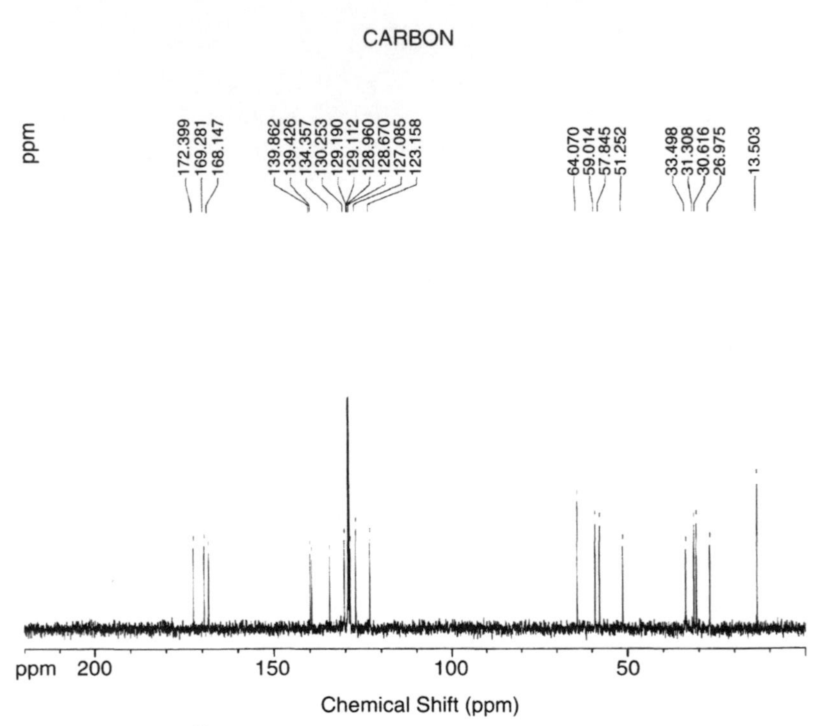

Figure 10. ^{13}C-NMR spectrum of benazepril hydrochloride in D_2O.

the color produced upon reaction with 4 mL 0.1% BCG at pH 2.2 was measured at 412 nm ($\varepsilon = 18523$). For method (iii), after alkaline hydrolysis, the color produced on reaction with 1 mL each of MBTH (0.2% in 0.1 M HCl), $FeCl_3$ (1% in 0.1 M HCl) and ethanol was measured at 593 nm ($\varepsilon = 11059$). The color of the BCG and MBTH products was found to be stable for 20 min. Beer's law was obeyed from 8–24, 4–24 and 4–40 µg/mL using methods (i), (ii) and (iii), respectively, with intraday RSD values ($n = 5$) being less than 2%. Recoveries were in the range of 98.1–101.8%. The BCG approach was found to be a general method, derivative spectrophotometry was used for benazepril hydrochloride in tablet forms (single or in combination with hydrochlorothiazide), and the MBTH method was applied to specific determinations of the degradation product of benazepril hydrochloride.

A derivative UV spectrophotometric method was used by Bonazzi *et al.* for the determination of benazepril hydrochloride and other angiotensin converting enzyme inhibitors in their pharmaceutical dosage forms [17].

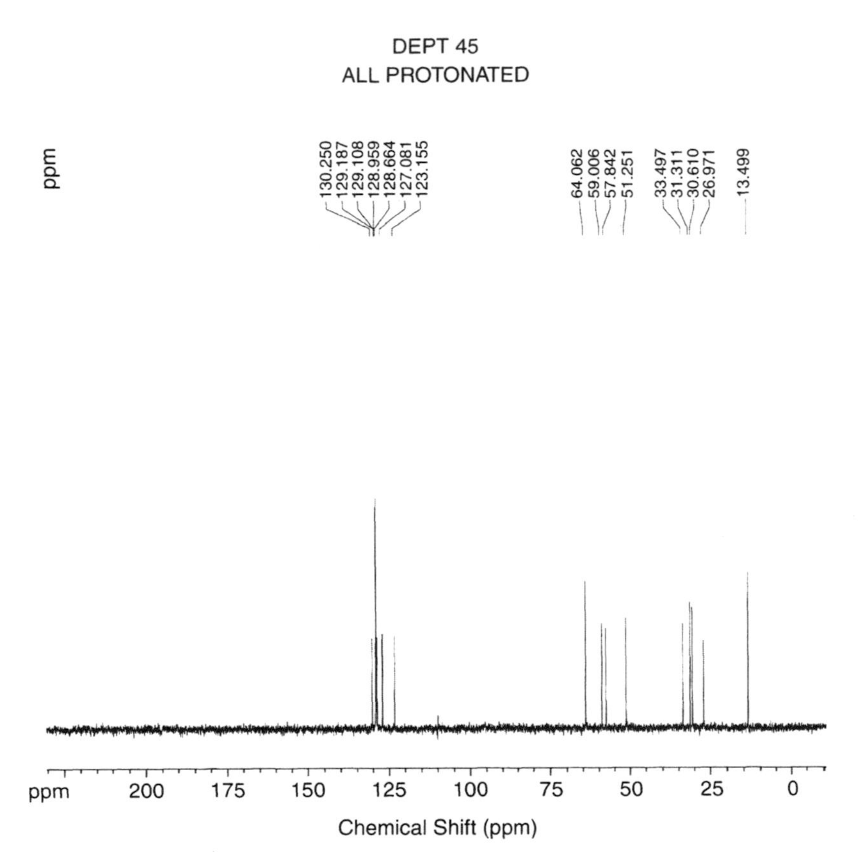

Figure 11. DEPT 45 ^{13}C-NMR spectrum of benazepril hydrochloride in D_2O.

The drug was determined by second derivative spectrophotometry using a Uvidec 610 double beam spectrophotometer, and the absorbance was measured at 259.2 nm. Beer's law was obeyed in range of 10–30 μg/mL.

Banoglu *et al.* reported the use of rapid, simple, and reliable spectroscopic methods (absorbance ratio method and Vierordt's method) for the quantitative determination of benazepril hydrochloride and hydrochlorothiazide in the dissolution testing of these two drugs in commercial tablets [18]. A 249 nm wavelength was chosen as the isobestic point in the absorbance ratio method, and the absorbance ratios A236/A249 nm for benazepril and A269/A249 nm for hydrochlorothiazide were used for calculation of the regression equation. For the Vierordt' method, $A_{[1\%,1\,cm]}$ values were obtained at 236 and 269 nm for both drugs and were used for their quantitative analyses.

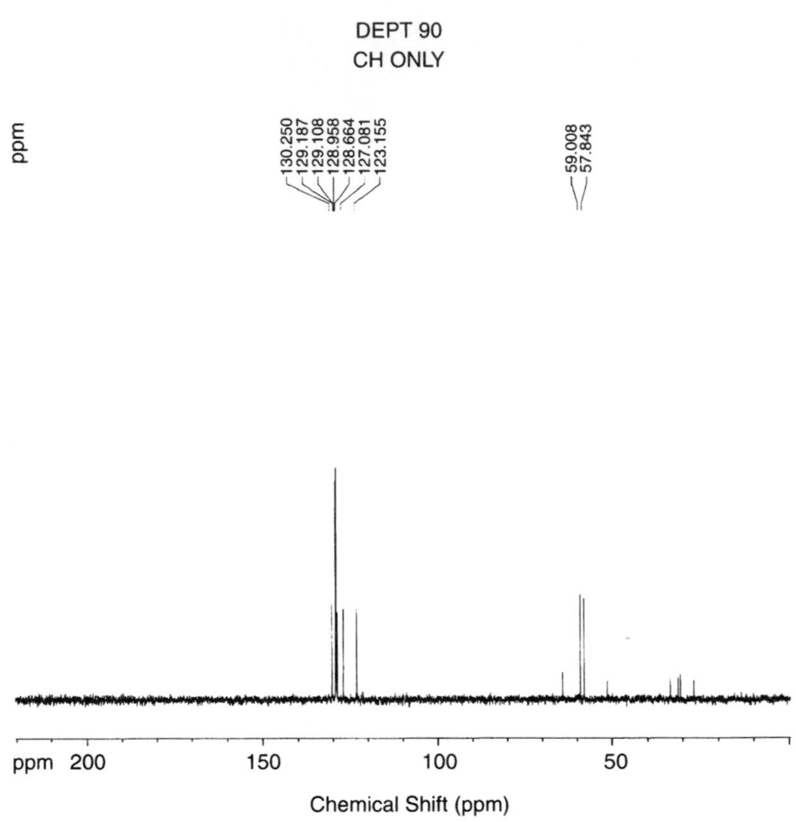

Figure 12. DEPT 90 ^{13}C-NMR spectrum of benazepril hydrochloride in D_2O.

Erk presented a ratio spectra derivative spectrophotometric method and a Vierordt's method for the simultaneous determination of hydrochlorothiazide in binary mixtures with benazepril hydrochloride, triametrene, and cilazapril [19]. A mass of powder equivalent to one tablet was dissolved in 1:1 methanol/0.1 M Hydrochloric acid (solvent A), shaken for 30 min, filtered and the residue was washed with 3 × 10 mL of solvent A. Combined solutions were diluted to 100 mL with solvent A, and then diluted with methanol for analysis by (i) Vierordt's method, and (ii) ratio spectra derivative spectrophotometry. For method (i), the absorbance of the solution was measured at 271.7, 238.1, 234.1, and 210.7 nm for hydrochlorothiazide, benazepril, triametrine, and cilazapril, respectively, and the corresponding concentrations were calculated using simultaneous equations. For method (ii), the absorption spectra of solutions containing different amounts of hydrochlorothiazide in a

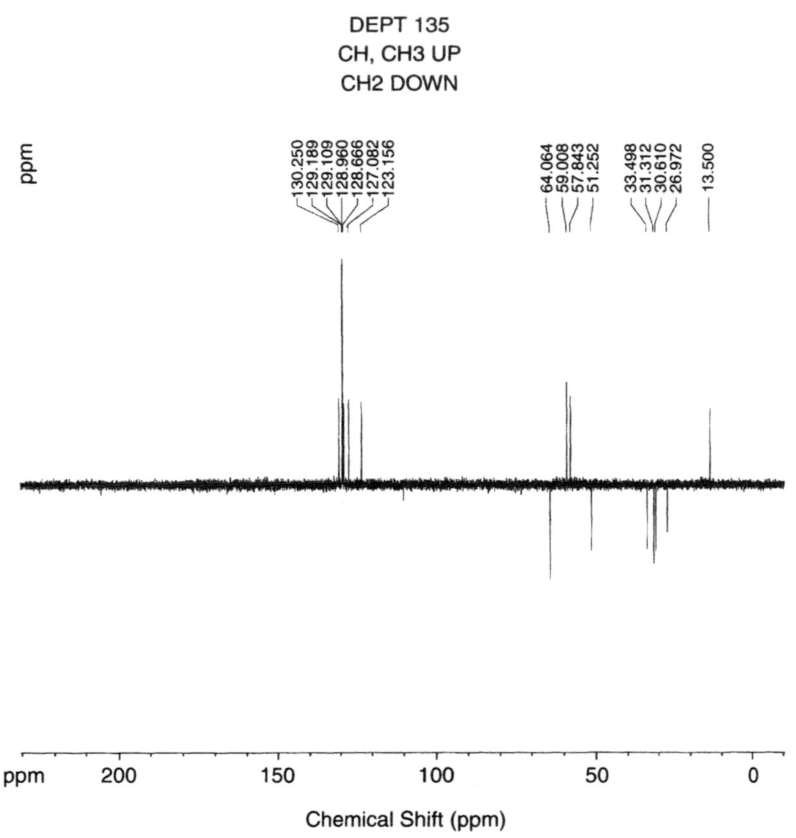

Figure 13. DEPT 135 ^{13}C-NMR spectrum of benazepril hydrochloride in D_2O.

binary mixture with benazepril hydrochloride, triametrine, or cilazapril were recorded within 200–350 nm ($\Delta\lambda = 1$ nm), and the first derivative ratio spectra obtained. The contents of hydrochlorothiazide with benazepril hydrochloride, with triametrine and with cilazapril were determined from signals at 269.5 and 238.8 nm, 260.7 and 284.4 nm, and 273.7 and 236.8 nm, respectively. For both methods, Beer's law was obeyed over the range of 8–36 μg/mL of benazepril hydrochloride. Relative standard deviation ($n = 5$) were 0.3–1.8%, and recoveries were 98.6–99.9%.

El-Gindy *et al.* presented different spectrophotometric methods for the simultaneous determination of benazepril hydrochloride and hydrochlorothiazide in pharmaceutical tablets [20]. The first method depends

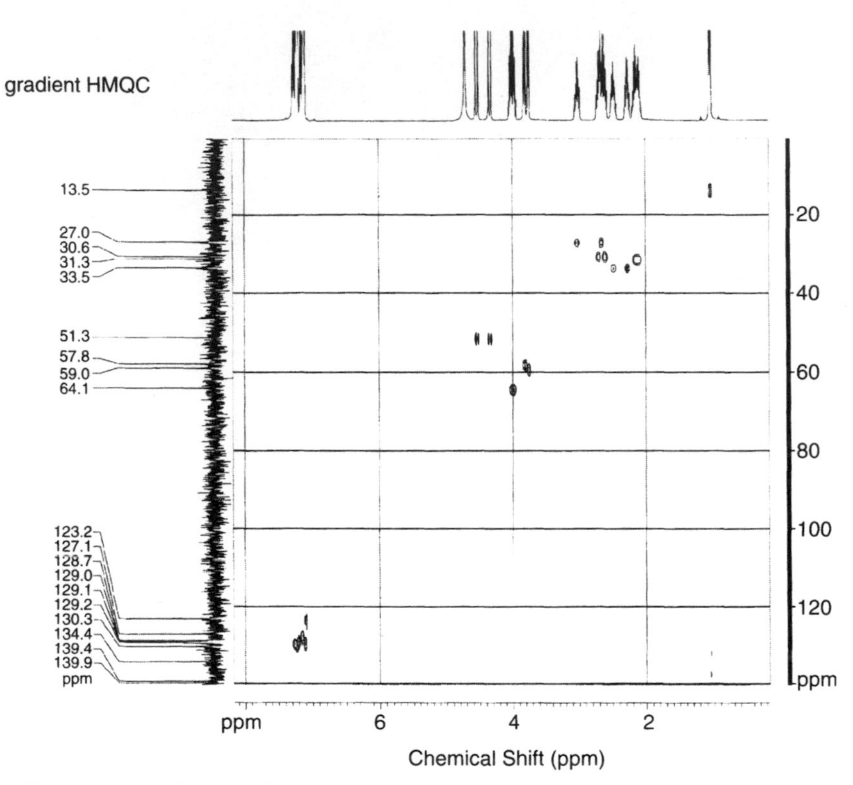

Figure 14. Gradient HMQC NMR spectrum of benazepril hydrochloride in D_2O.

on second-derivative ultraviolet spectrophotometry with zero crossing and peak-to-base measurements. The second derivative amplitude at 214.8 and 227.4 nm were selected for the assay of benazepril hydrochloride and hydrochlorothiazide, respectively. The second method depends on the second derivative of the ratio spectra by measurement of the amplitude at 241.2 and 273.2 nm for benazepril hydrochloride and hydrochlorothiazide, respectively. Chemometric methods, classical least squares, and principal component regression were applied to analyze the mixture. Both the chemometric methods were applied to the zero and first order spectra of the mixture. These methods were successfully applied to the determination of the two drugs in laboratory prepared mixtures and in commercial tablets.

Dinc reported that three chemometric techniques, classical least squares (CLS), inverse least square (ILS) and principal component

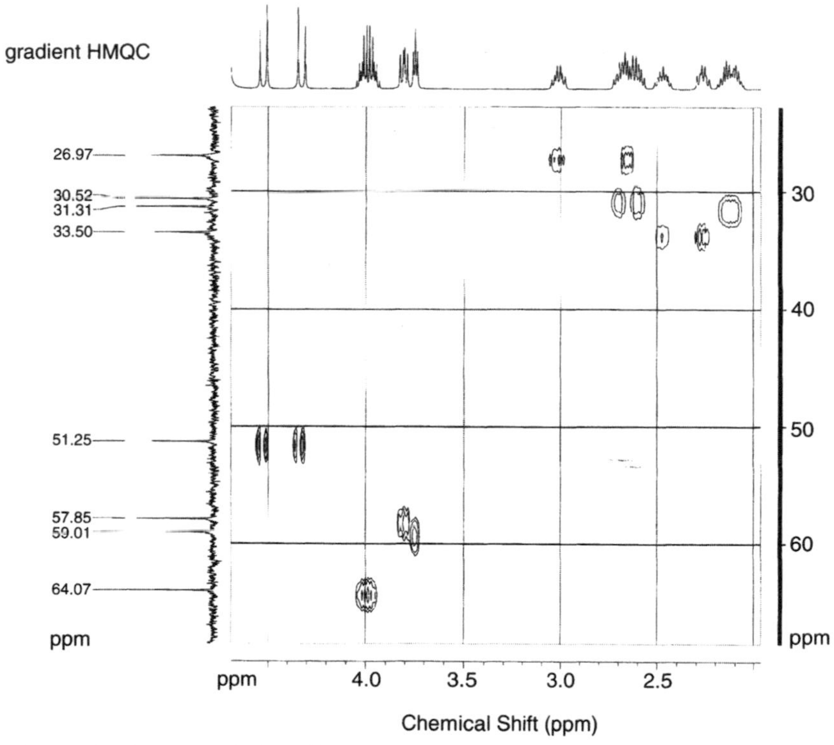

Figure 15. Gradient HMQC NMR spectrum of benazepril hydrochloride in D_2O.

regression (PCR), could be applied to the spectrophotometric analysis of two pharmaceutical tablet formulations containing hydrochlorothiazide and benazepril hydrochloride in the presence of their overlapping spectra [21]. A training set was randomly prepared by using the different composition mixtures containing 0–22 μg/mL of hydrochlorothiazide and 0–36μg/mL of benazepril hydrochloride in 0.1 M sodium hydroxide. The chemometric calibrations were constructed by using the prepared training set and its measurement of the absorbance values at 24 points in the spectrum from 220 to 350 nm. The obtained chemometric calibrations were used for the prediction of hydrochlorothiazide and benazepril hydrochloride levels in the samples. The validation of the three techniques was realized by analyzing the synthetic mixtures of the two drugs. The CLS, ILS, and PCR techniques are suitable for the chemometric determination of benazepril

Table 4

Assignments for the Carbon-13 Nuclear Magnetic Resonance Spectrum of Benazepril Hydrochloride

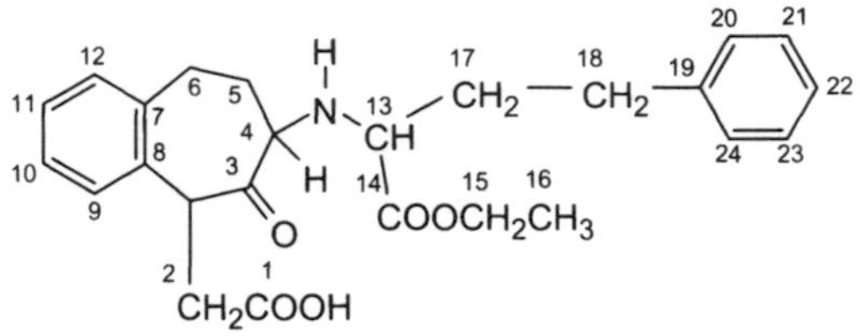

Chemical Shift (ppm relative to TMS)	Assignment (Carbon number)	Chemical Shift (ppm relative to TMS)	Assignment (Carbon number)
172.4	1	127.9	12
169.3	3	123.2	22
168.2	14	64.1	2
139.9	7	59.0	4
139.4	8	57.8	13
134.4	19	51.3	5
130.3	9	33.5	6
129.2	20,24	31.3	17
129.1	10	30.6	18
129.0	21,23	27.0	15
128.7	11	13.5	16

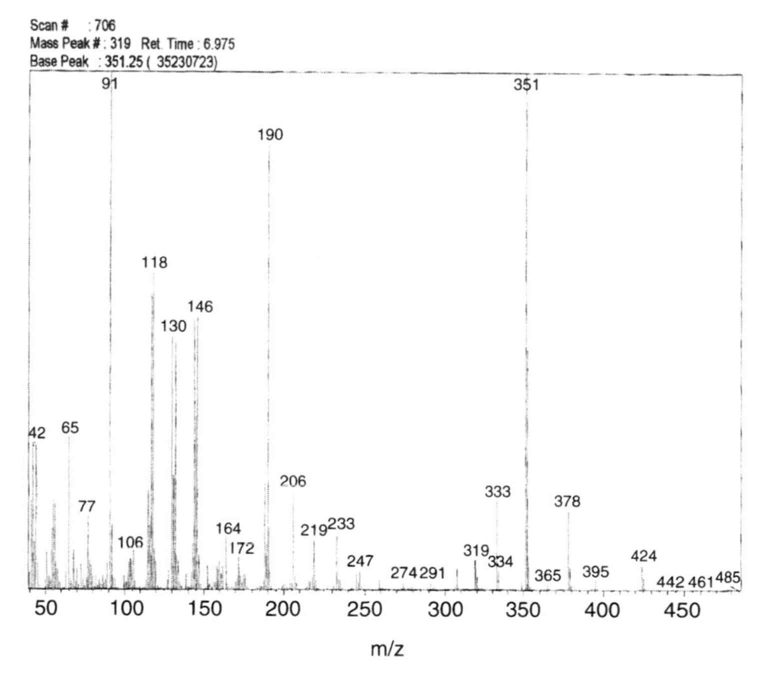

Figure 16. Mass spectrum of benazepril hydrochloride.

hydrochloride and hydrochlorothiazide in synthetic mixtures and in tablet formulations.

4.5 Colorimetric method

Belal *et al.* developed a simple and sensitive spectrophotometric method for the determination of benazepril hydrochloride in tablets [22]. The method is based on the reaction of the drug with potassium permanganate in the presence of sodium hydroxide to produce a bluish-green colored species measurable at 609.4 nm. The absorbance-concentration plot is linear over the range 1–8 μg/mL, with a limit of detection of 0.1 μg/mL (2.17×10^{-7} M). The molar absorptivity was 4.07×10^{4} L/mol/cm, and the Beer's Law correlation coefficient ($n = 6$) was 0.9991. The different experimental parameters affecting the development and stability of the color were studied carefully, and optimized. This method was applied successfully to the determination of benazepril hydrochloride in its dosage forms. The percentage recoveries ± SD ($n = 9$) were 99.79% ± 1.4 and 100.5 ± 1.48 for tablets containing 10 and 20 mg,

Table 5

Mass Spectral Fragmentation Pattern for Benazepril HCl

m/z	**Relative Intensity**	**Fragment**	
		Formula	**Structure**
424	2.5%	$C_{24}H_{28}N_2O_5$	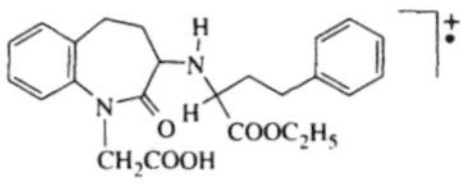
395	1.1%	$C_{22}H_{23}N_2O_5$	M^+–CH_2CH_3
378	7.7%	$C_{22}H_{22}N_2O_4$	M^+–CH_3CH_2OH
351	100%	$C_{21}H_{23}N_2O_3$	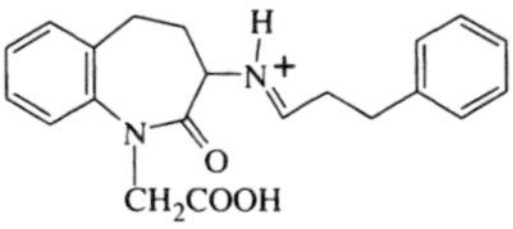
333	9.0%	$C_{21}H_{21}N_2O_2$	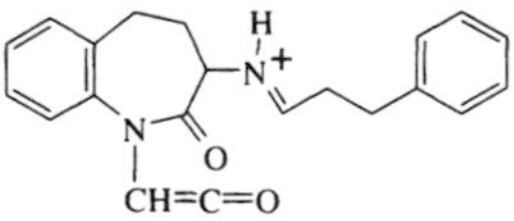
319	3.0%	$C_{16}H_{19}N_2O_5$	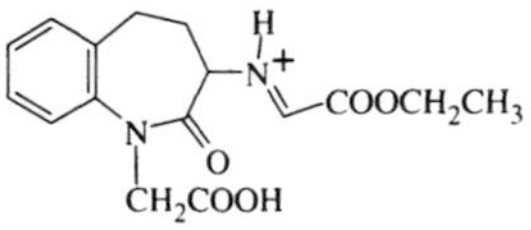
291	1.0%	$C_{14}H_{15}N_2O_5$	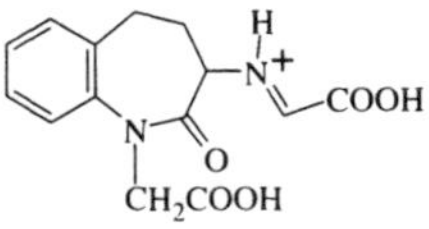

(*continued*)

Table 5 (continued)

m/z	Relative Intensity	Fragment	
		Formula	**Structure**
274	1.0%	$C_{15}H_{18}N_2O_3$	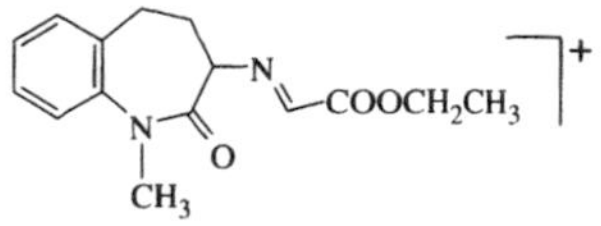
247	2.0%	$C_{13}H_{15}N_2O_3$	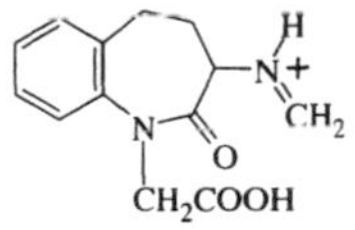
219	4.81%	$C_{12}H_{13}NO_3$	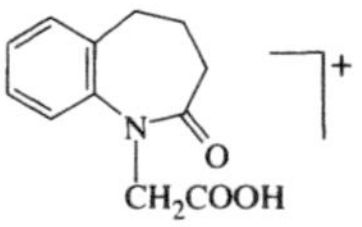
206	9.6%	$C_{12}H_{16}NO_2$	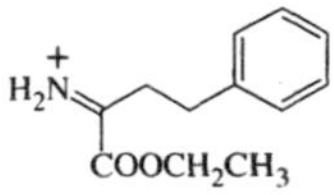
190	43%	$C_{12}H_{14}O_2$	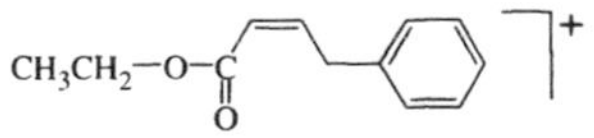
146	26.5%	$C_{10}H_{10}O$	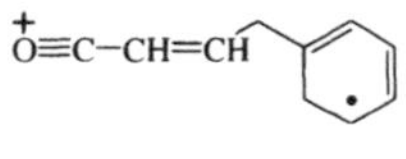
144	26.2%	$C_{10}H_{10}N$	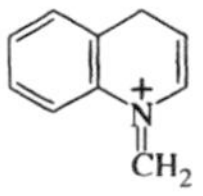

(continued)

Table 5 (continued)

m/z	**Relative Intensity**	**Fragment**	
		Formula	**Structure**
132	24.1%	$C_9H_{10}N$	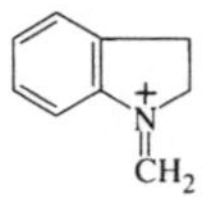 $\overset{+}{N}$=CH_2
130	24.6%	C_9H_8N	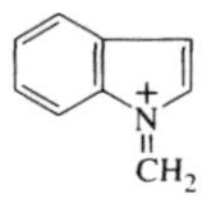 $\overset{+}{N}$=CH_2
118	30.8%	C_8H_8N	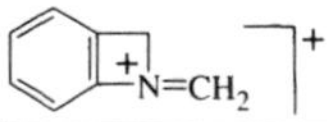 $\overset{+}{N}=CH_2$ $]^{+}$
117	28.5%	C_9H_9	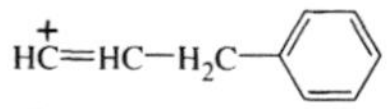 $H\overset{+}{C}=HC-H_2C-$
91	54.5%	C_7H_7	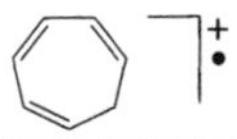 $]^{+\bullet}$
77	7.2%	C_6H_5	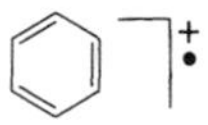 $]^{+\bullet}$
65	14.9%	C_5H_5	 $]^{+\bullet}$
42	14.31%	C_2H_2O	$H_2C=C=O$ $]^{+}$

respectively. The results obtained were in good agreement with those obtained using a reference spectrophotometric method. The method could also be applied to the determination of benazepril in the presence of co-formulated hydrochlorothiazide..

4.6 Voltammetric methods

Belal *et al.* developed a simple and highly sensitive voltammetric method for the determination of benazepril and ramipril [23]. The compounds were treated with nitrous acid, and the cathodic current produced by the resulting nitroso derivatives was measured. The voltammetric behavior was studied by adopting direct current (DCt), differential pulse (DPP), and alternating current (ACt) polarography. Both compounds produced well-defined, diffusion-controlled cathodic waves over the whole pH range when studied in Britton-Robinson buffer (BPb). At pH 3 and 5, the values of diffusion-current constants (Id), were 5.9 ± 0.4 and 6.66 ± 0.61 for benazepril and ramipril, respectively. The current concentration plots for benazepril were linear over the range of 1.5–40 and 0.1–30 μg/mL in the DCt and DPP modes, respectively. The limits of detection (S/N = 2) were 0.015 μg/mL (about 3.25×10^{-8} M) and 0.012 μg/mL (about 2.88×10^{-8} M) for benazepril and ramipril, respectively (when using the DPP mode). When applied to the determination of both compounds in dosage forms, the results obtained using this method were in good agreement with those obtained using reference methods. The method was applied to the determination of benazepril in spiked human urine and plasma. When using the DPP mode, the percentage recoveries were 96.2 ± 1.21 and 95.7 ± 1.61, respectively.

4.7 Chromatographic methods of analysis

4.7.1 Thin-Layer chromatographic methods

Hassib *et al.* developed a new, simple, sensitive, and fast thin layer chromatographic scanning densitometric method for the simultaneous determination of benazepril hydrochloride and hydrochlorothiazide using ethyl acetate–methanol–ammonia (85:20:10 v/v) as the developing system [24]. The R_f values were 0.33 and 0.68 for benazepril hydrochloride and hydrochlorothiazide, respectively. The minimum detection limit obtained was 0.12 μg/spot and 0.24 μg/spot for benazepril hydrochloride and hydrochlorothiazide, respectively. The mean percentage recoveries were 100.04 ± 1.02 and 99.31 ± 1.009 for benazepril hydrochloride and hydrochlorothiazide, respectively. The method could be successfully used for the determination of pure, laboratory made mixtures, and pharmaceutical dosage forms. The results obtained were compared with those obtained by spectrophotometric methods.

El-Gindy *et al.* described a high performance thin layer chromatographic (HPTLC) method for the simultaneous determination of benazepril hydrochloride and hydrochlorothiazide in binary mixtures [25]. The method is based on the HPTLC separation of the two drugs, followed by densitometric measurement of their spots at 238 and 275 nm for benazepril hydrochloride and hydrochlorothiazide, respectively. The separation was carried out on Merck HPTLC aluminum sheets coated with silica gel 60 F_{254}, using ethyl acetate–methanol–chloroform (10:3:2 v/v) as mobile phase. A second order polynomial equation was used for the regression line in the range 2–20 and 2.5–25 μg/spot for benazepril hydrochloride and hydrochlorothiazide, respectively. The method was successfully applied to the determination of the two drugs in laboratory prepared mixtures and in commercial tablets. No chromatographic interference from tablet excipients was found.

4.7.2 Gas chromatographic–mass spectrometric methods (GC–MS)

Sioufi *et al.* determined benazepril and its active metabolite, benazeprilat, in plasma and urine by capillary gas chromatography with mass-selective detection [12]. Plasma or urine was extracted with Amberlite XAD-2 or Extrelut 1, respectively, and then derivatized with diazomethane. A portion (2 μL) of the final derivative solution in toluene was analyzed on a fused silica column (12.5 m × 0.2 mm) coated with cross-linked methylsilicone, operated at 210°C (held for 0.5 min) to 290°C at 50°C/min, with helium as the carrier gas. Detection was by selected-ion monitoring at $m/z = 365$ for the methyl ester derivatives of benazepril and its free acid analogue, benazeprilat and at $m/z = 370$ for the methyl ester derivatives of deuterated benazepril and benazeprilat (internal standards). Detection limits were 4.5 and 6.5 nM for benazepril and benazeprilat, respectively, in plasma. The corresponding values in urine were 11 and 53 nM. The calibration graphs were linear from 5.4 to 217 and 10.9 to 542.5 nM of benazepril in plasma and urine, respectively, and the corresponding values for benazeprilat were 6.6–259.6 and 52.9–1310.4 nM

Sereda *et al.* developed an electron-capture gas chromatographic detection method for the detection of benazepril and other angiotensin-converting enzyme (peptidyl-dipeptidase) inhibitors [26]. Sample solutions (0.001–0.1 mL) containing the inhibitor or metabolite were vertex-mixed at 70°C for two hours with 10 mg of potassium acetate, 3 mL of acetone, 0.1 mL of 0.1% ramiprilat (the internal standard) solution, and 0.1 mL of alpha-bromopentafluorotoluene. A 0.25 mL portion of the cooled solution was evaporated to dryness at 100°C, the residue was dissolved in 0.25 mL of acetone, and a 0.5 μL portion of the resulting solution was injected onto a glass column (30 m × 0.252 mm) coated with methyl silicone and operated at 295°C. Helium was used as the carrier gas (25 psi), and the method employed a ^{63}Ni electron capture detector. Benazepril and eight ACE inhibitors were well separated, and calibration graphs were linear over the range of 0.3–15.8 ng of drug on the column.

Pommier *et al.* determined benazepril and its active metabolite (benazeprilat) in human plasma using capillary gas chromatography-mass-selective detection, with their respective labeled internal standards [27]. The method was developed and validated according to international regulatory requirements. After addition of the internal standards, the compounds were extracted from plasma by solid-phase extraction using an automated 96-well plate technology. After elution, the compounds were converted into their methyl ester derivatives by means of a safe and stable diazomethane derivative. The methyl ester derivatives were determined by gas chromatography using a mass-selective detector at $m/z = 365$ for benazepril and benazeprilat, and at $m/z = 370$ for the internal standards. Intra- and interday accuracy and precision were found to be suitable over a range of concentrations between 2.5 and 1000 ng/mL.

Kaiser *et al.* developed and validated a specific and sensitive gas chromatographic–mass spectrometric method for the simultaneous quantification of unchanged benazepril and its active metabolite (the dicarboxylic acid) in plasma and urine [28]. The $^{2}H_5$-labeled analogues of benazepril and its metabolite were used as internal standards. The compounds were isolated from plasma and urine under acidic conditions using XAD-2 resin or Extrelut 1 columns. Following derivatization with diazomethane, the samples were analyzed by packed-column gas chromatography–electron impact mass spectrometry with selected ion monitoring. The analysis of spiked plasma and urine samples demonstrated the good accuracy and precision for the method, which was found to be suitable for use in pharmacokinetic and bioavailability studies with benazepril hydrochloride in humans.

Maurer *et al.* developed a gas chromatographic–mass spectrometric screening procedure for the detection of benazepril and other angiotensin-converting enzyme inhibitors, their metabolites, and angiotensin II receptor antagonists in urine as part of systematic toxicologic analysis procedure for acidic drugs and poisons after extractive methylation [29]. The part of the phase-transfer catalyst remaining in the organic phase was removed by solid-phase extraction in a diol phase. The compounds were separated by capillary gas chromatography, and identified by computerized mass spectrum in the full scan mode. Using mass chromatography with ion detection at m/z of 157, 160, 172, 192, 204, 220, 234, 248, 249 and 262, the possible presence of the ACE inhibitors, their metabolites, and angiotensin II antagonist could be indicated. The identity of positive signals in these mass chromatograms was confirmed by comparison of the peak underlying full mass spectra with reference spectra recorded during this study. This method allowed detection of therapeutic concentrations of benazepril and five other ACE inhibitors, their metabolites (or both), as well as the therapeutic concentration of the angiotensin II antagonist (valsartan) in human urine samples. The overall recoveries ranged between 80 and 88%, with coefficient of variation of less than 10%, and the limit of detection of at least 10 ng/mL (signal-to-noise ratio 3) in the full-scan mode.

4.7.3 High performance liquid chromatography

Panderi and Parissi-Poulou developed a microbore liquid chromatographic method for the simultaneous determination of benazepril hydrochloride and hydrochlorothiazide in pharmaceutical dosage forms [30]. The use of a BDS C-18 microbore analytical column was found to result in substantial reduction in solvent consumption and in increased sensitivity. The mobile phase consisted of a mixture of 25 mM sodium dihydrogen phosphate buffer (pH 4.8) and acetonitrile (11:9 v/v), pumped at a flow rate of 0.4 mL/min. Detection was effected at 250 nm using an ultraviolet absorbance detector. The intra- and inter-day relative standard deviation values were less than 1.25% ($n = 5$), while the relative percentage error was less than 0.9% ($n = 5$). The detection limits obtained according to the IUPAC definition were 0.88 and 0.58 μg/mL for benazepril hydrochloride and hydrochlorothiazide, respectively. The method was applied to the quality control of commercial tablets and content uniformity test, and proved to be suitable for rapid and reliable analysis.

Radhakrishna *et al.* developed a gradient liquid chromatographic method for the determination and purity evaluation of benazepril hydrochloride in bulk and pharmaceutical dosage forms [31]. The method is simple, rapid, and selective. 5-Methyl-2-nitrophenol was used as the internal standard. The method is linear over the range of 50–800 μg. The precision of the inter- and intra-day assay variation of benazepril hydrochloride is below 1.6% relative standard deviation. The accuracy, determined as relative mean-error (RME) for the intra-day assay, is within ± 2%. The method is stability indicating, and is useful in the quality control of bulk manufacturing and also in pharmaceutical formulations.

Manna *et al.* described a rapid and accurate reversed-phase ion-paring liquid chromatographic method, with ultraviolet detection, for the simultaneous assay of benazepril hydrochloride, fosinopril sodium, ramipril and hydrochlorothiazide [32]. The separation was achieved on a LC-8 (12.5 cm × 4 mm, 5 μm particle size) column. The mobile phase consisted of 20 mM sodium heptansulfonate (pH 2.5) and methanol (8:17 v/v). Validation of the method showed it to be precise, accurate, and linear over the concentration range of analysis, with a limit of detection of 2 ng for benazepril. The method was applied to the analysis of three different binary commercial formulations.

El-Gindy *et al.* described a liquid chromatographic method for the simultaneous determination of benazepril hydrochloride and hydrochlorothiazide [25]. The liquid chromatography was based on the separation of the two drugs on a reversed phase ODS column at ambient temperature, using a mobile phase consisting of acetonitrile and water (7:13) and adjusting to pH 3.3 with acetic acid. Quantitation was achieved with ultraviolet detection at 240 nm based on peak area, with linear calibration curves at concentration ranges of 10–60 and 12.5–75 μg/mL for benazepril hydrochloride and hydrochlorothiazide, respectively. The method was successfully applied to the determination of the two drugs in laboratory prepared mixtures and in commercial tablets. No chromatographic interference from tablet excipients was found.

Dappen *et al.* studied the performance of a commercial (Polarmonitor, IBZ, Hannover, Germany) and a laboratory designed polarimetric detector in conjunction with a HPLC method based on a Nucleosil 100 RP-18 column, an ultraviolet detector, and aqueous 58% methanol as the mobile phase [33]. The polarimetric signal was proportional to the excess concentration of one enantiomer in a mixture of two, and measurement of the total concentration of both enantiomers using the ultraviolet detector then permitted the enantiomer ratio to be calculated. The unexpectedly strong polarimetric signals produced by some compounds (such as 1-phenylethyl alcohol) were attributed to RI-related effects generated by the eluting peaks. One example of a pharmaceutical application of polarimetric detection was the detection of the (*R*,*R*)-enantiomer impurity of benazepril (which itself is a pure (*S*,*S*)-enantiomer), with a detection limit of better than 0.5%. Another application was the determination of the diastereomeric ratio (not separable on a conventional reversed-phase column) in the preparation of 10-EDAM WS. This chromatographic system used a 5 μm Neocleosil 100 RP-18 reversed phase, packed into a 12.5 cm × 4.6 mm column, and was eluted using Veible buffer–water–methanol (17:25:58). The Vieble buffer is made of 90 mM potassium chloride and 10 mM hydrochloric acid at pH 2.06. The flow rate was 0.25 mL/min and temperature was 35°C.

Bonazzi *et al.* used an HPLC method and a second derivative ultraviolet spectroscopy method for the analysis of benazepril and other angiotensen-converting enzyme inhibitors [17]. For HPLC, 20 μL sample solutions containing the drug and an internal standard dissolved in 1:1 acetonitrile/20 mM sodium heptanesulfonate (pH 2.5) were used. HPLC was performed on a 5 μm Hypersil ODS column (25 cm × 4.5 mm) with a mobile phase mixture consisting of (A) 20 mM sodium heptanesulfonate (pH 2.5) and (B) 19:1 acetonitrile–tetrahydrofuran, eluted at a flow rate of 1 mL/min, and with detection at 215 nm. The A/B mixture used was 52:48 for benazepril. A low pH of 2.5 was essential to avoid peak splitting and band broadening.

Cirilli and La-Torre used a reversed-phase high performance liquid chromatographic method, on a chiral alpha-acid glycoprotein (AGP) column, for the stereoselective analysis of benazepril and its stereoisomers [34]. A 10 μL sample of benazepril, its enantiomer, two diastereomers, and a chiral active metabolite (benazeprilat) in ethanol (24 μg/mL) was injected onto a stainless steel chiral AGP based column (10 cm × 4 mm) or a Kromasil C-8 (5 μm) column (25 cm × 4.6 mm). Analytes were eluted (0.9 mL/min) with 0.01–0.06M potassium dihydrogen phosphate buffer (pH 3–7) containing various proportions of an organic modifier, and detected at 240 nm. A baseline separation of the five compounds was achieved in less than 35 min on an AGP column with 0.06 M-phosphate buffer (pH 4) and ethanol (17:3), or propan-2-ol (12.33:1). Detection limits (calculated as three times the signal-to-noise), when using propan-2-ol modifier, were 1.5 ng for benazepril, 5 ng for its enantiomer, the two diastereomers, and 0.6 ng for benazeprilat.

Gumieniczek and Przyborowski determined benazepril and cilazapril in pharmaceuticals by a high performance liquid chromatographic method [35]. Ground tablets equivalent to 25 mg of benazepril or cilazapril were extracted with methanol. Portions (20 μL) of the extract were analyzed by HPLC on a 10 μm LiChrosorb RP-18 column (25 cm × 4 mm) with phosphate buffer (pH 2.4)/acetonitrile (7:3) as the mobile phase (eluted at 1 mL/min), and detection at 211 nm. Quantitation was performed using enalapril maleate (4 mg/mL in methanol) as the internal standard. Calibration graphs were linear from 10–50 and 40–200 μg/mL of benazepril and cilazapril, respectively. Recoveries of the two drugs ranged from 96.34–102.04% and 103.08–107.96%, respectively. Relative standard deviations ($n = 5$) were less than 4.45% and less than 3.5 for benazepril and cilazapril, respectively.

Hassib *et al.* developed two chromatographic procedures for the simultaneous determination of benazepril hydrochloride and hydrochlorothiazide in laboratory made mixtures, and in pharmaceutical dosage forms (Cibadrex tablets) using reversed phase HPLC and TLC methods [24]. For reversed phase HPLC, a very sensitive, rapid, and selective method was developed. The linearity ranges were 32–448 ng/20 μL and 40–560 ng/20 μL for benazepril hydrochloride and hydrochlorothiazide, respectively. The corresponding recoveries were 99.38 ± 1.526 and 99.2 ± 1.123. The minimum detection limits were 7 ng/20 μL for benazepril and 14 ng/20 μL for hydrochlorothiazide. The method could be successfully applied for the determination of laboratory made mixtures and for pharmaceutical dosage forms. The results obtained were compared with those obtained by a spectrophotometric method.

Banoglu *et al.* reported a comparison of the spectrophotometric methods (absorbance ratio method and Vierordt's method) with HPLC for quantitative determination in the dissolution testing of benazepril hydrochloride and hydrochlorothiazide commercial tablets [18]. In the HPLC method, simultaneous determination of the two drugs from the dissolution medium was achieved using a mobile phase containing phosphate buffer (0.01 M, pH 6.2) and acetonitrile (65:35) on a Supelcoil LC-18 (4.6 × 250, 5.6 mm) reversed phase column. Dissolution tests of commercial tablets were carried out according to USP XXII method (0.1 M hydrochloric acid at 37 ± 0.5°C, and paddle rotation of 50 rpm). Comparison of the dissolution data from the HPLC and the two spectroscopic methods indicated good correlation between the methods. It was concluded that both the spectroscopic methods and the HPLC method are suitable for routine analyses of benazepril hydrochloride and hydrochlorothiazide in dissolution testing of commercial tablets.

Sreenivas *et al.* described a validated HPLC method for the assay of quinapril and benazepril hydrochloride in pharmaceuticals, and the purity evaluation of quinapril bulk drug substance [36]. Solutions of quinapril hydrochloride (2.5 μg/mL) and benazepril hydrochloride (1.5 μg/mL) were prepared by dissolving the compounds in mobile phase A (5:3:2 mixture of 0.05 M potassium dihydrogen phosphate (pH 4)/acetonitrile/methanol). The internal standard for quinapril hydrochloride was 0.3 mg/mL benazepril hydrochloride, and the internal standard for benazepril hydrochloride was 0.3 mg/mL quinapril hydrochloride. Extracts were prepared by grinding up to twenty Acupril 40 mg tablets and Lotensin 40 mg tablets, and extracting with methanol. Extracts were diluted to 0.2 mg/mL and 0.3 mg/mL for each drug substance. Samples (10 μL) were analyzed on a 5 μm Hypersil C-18 BDS column (25 cm × 4.6 mm) at 50°C, eluted with mobile phase A, and employing detection at 215 nm. The calibration curve was linear from 0.1 to 0.6 mg/mL of quinapril, and from 0.15 to 0.75 mg/mL of benazepril hydrochloride. Intra-day precision and accuracy was evaluated ($n = 3$) at concentrations of quinapril hydrochloride and benazepril hydrochloride of 0.15 mg/mL, 0.3 mg/mL, and 0.45 mg/mL. Relative standard deviations ranged from 0.57 to 0.97% and 0.3 to 0.88% for quinapril hydrochloride and benazepril hydrochloride, respectively. Interday precision and accuracy were calculated ($n = 3$) for three days. Relative standard deviations ranged from 0.65–0.94% and 0.74–1.19% for quinapril hydrochloride and benazepril hydrochloride, respectively. Recoveries from tablets ranged from 96.3 to 98.4% (relative standard deviations 0.32–0.43%) for quinapril, and 100.4 to 103.5% (relative standard deviations = 0.98–1.17%) for benazepril hydrochloride. The detection limit was 2.5 μg/mL for a 10 μL injection of quinapril hydrochloride and benazepril hydrochloride. The quantitation limit was 10 μg/mL for both quinapril hydrochloride and benazepril hydrochloride.

Gana *et al.* developed and validated a reversed-phase high performance liquid chromatographic method for the kinetic investigation of the chemical and enzymatic hydrolysis of benazepril hydrochloride [37]. Kinetic studies on the acidic hydrolysis of benazepril hydrochloride were carried out in 0.1 M hydrochloric acid solution at 50, 53, 58 and 63°C. Benazepril hydrochloride appeared stable in pH 7.4 phosphate buffer at 37°C, and showed susceptibility to *in vitro* enzymatic hydrolysis with porcine liver esterase (PLE) in a pH 7.4 buffered solution at 37°C.

Benazeprilat appeared to be the major degradation product in both the chemical and enzymatic hydrolysis studies. Statistical evaluation of the HPLC methods revealed their good linearity and reproducibility. Relative standard deviation was less than 4.76, while detection limits for benazepril hydrochloride and benazeprilat were 13×10^{-7} and 9×10^{-7}, respectively. Treatment of kinetic data of the acidic hydrolysis was carried out by nonlinear regression analysis, and k values were determined. The kinetic parameters of the enzymatic hydrolysis were determined by nonlinear regression analysis of the data using the Michaelis–Menten equation. Chromatographic separation was performed on a reversed-phase BDS C_{18} column (25 cm × 3 mm i.d., 5 μm particle size) Shandon Scientific Ltd. The mobile phase consisted of 0.020 M disodium hydrogen phosphate adjusted to pH 3 with phosphoric acid and methanol (40:60, v/v), which was filtered through a 0.20 μm nylon membrane filter and degassed under vacuum prior to use. A flow rate of 0.35 mL/min with a column inlet pressure of 2000 psi was used in order to separate the drug and its degradation products.

Graf *et al.* used an enzymatic method for the determination of benazeprilat in plasma and urine [38]. For this determination, urine and plasma samples were diluted 1:20 with phosphate buffer (pH 8.3) and heated at 75°C for 5 min to inactivate endogenous angiotensin-converting enzyme (ACE). Plasma samples were then incubated in blank plasma in 0.25 M disodium hydrogen phosphate/0.75 M sodium chloride (adjusted to pH 8.3 with 1 M-phosphoric acid) for 12 min before the addition of hippuryl-L-histidyl-L-leucine as substrate and further incubation for 80 min. The reaction was stopped with methanolic phosphoric acid, whereupon acetonitrile and mandelic acid (internal standard) were added, and the mixture centrifuged. A 15 μL portion of the supernatant solution was analyzed on a column (25 cm × 4.6 mm) of LiChrosorb-NH_2 (10 μm) with a mobile phase (2.3 mL/min) consisting of 1:4 acetonitrile/10 mM potassium dihydrogen phosphate (adjusted to pH 3 with 1 M phosphoric acid). Detection was effected on the basis of the UV absorbance at 228 nm. Urine was treated as for plasma with rabbit lung ACE solution in place of blank plasma, [^{3}H]-hippuryl-glycyl-glycine as the substrate, and 2 M hydrochloric acid saturated with sodium sulfate to stop the reaction. The released [^{3}H]-hippuric acid was extracted into a water-immiscible scintillation cocktail and counted without separation of the phases. Mean recoveries of benazeprilat were 97.1–104% and 92.9–116% from plasma and urine, respectively, with a coefficient of variation ($n = 8$) of 8% for 20 nM benazeprilat. Detection limits were 10 and 20 nM benazeprilat in urine and plasma, respectively. The results agreed well with those obtained by GC–MS.

Cakir *et al.* reported the use of a high performance liquid chromatographic method and a spectrophotometric method for the analysis of benazepril hydrochloride and hydrochlorothiazide [39].

Barbato *et al.* reported the analysis of benazepril and six other angiotensin converting enzyme inhibitors by a HPLC method [40]. The influence of different organic modifiers and counter-ions in the mobile phase at different pH values has been investigated, allowing the identification of the best experimental conditions for the analysis of the drug and the other ACE inhibitors. The synergistic effect of triethylamine and propanol in the eluting phase appeared to be a critical factor in order to obtain satisfactory peak shapes and resolution. A reversed phase Spherisorb 5 ODS-2 stainless steel column (25 cm × 4.6 mm) (Phase Separation-Clwid-UK) was employed. The mobile phases were constituted by phosphate buffer at various pH values, and contained counter ions at different percentages in mixtures with different amounts of acetonitrile, methanol, tetrahydrofuran, or *n*-propanol organic modifiers. Cetyltrimethylammonium bromide (CTAB), sodium dodecyl sulfate, *n*-butylamine, and triethylamine were used as counter ions. The analyses were performed at room temperature. The HPLC was performed on a Waters liquid chromatographs equipped with a Model 7125 Rheodyne injection valve and a Model 486 UV detector set at 240 nm.

4.7.4 Capillary electrophoresis methods

Gotti *et al.* developed a capillary electrophoresis method for the determination of benazepril hydrochloride and other angiotensin-converting enzyme inhibitors in pharmaceutical tablets [41]. Since a free solution capillary electrophoresis system failed to achieve a complete separation of closely related compounds (benazepril and others), alkylsulfonic additives (sodium heptane sulfonate and 10 camphor-sulfonic acid) were added to the running buffer. This resulted in an improved separation, suggesting a favorable effect of ion-pairing interactions between analytes and additives. The separations were carried out in acidic medium and a systematic investigation of electrophoresis parameters was made to evaluate the performance of the selected additives. Under optimized conditions, benazepril and ramipril in their commercial dosage forms were determined, confirming the applicability of the developed capillary electrophoresis approach to the analysis of pharmaceutical samples. The results were compared with those obtained applying a previously described and validated HPLC method.

Hillaert and Van den Bossche used capillary electrophoresis method for the separation of benazepril among eight other angiotensin-converting enzyme inhibitors [42]. The method was investigated with respect to the pH of the running buffer, the organic modifiers, and plausible surfactants. The most critical parameter was found to be the pH of the running buffer. The addition of sodium dodecyl sulfate had a negative influence on the peak symmetry, and the selectivity was not improved. The separation of the eight compounds can be performed by means of two phosphate buffers (each 100 mM) at pH 7 and 6.25, respectively. The combination is necessary for the selective identification of structurally related substances because of their similar pKa values.

The same authors also applied capillary electrophoresis to the study of benazepril hydrochloride and several angiotensin-converting enzyme inhibitors [43]. Separation of the compounds was performed by means of two phosphate buffers (each 0.1 M) at pH 7 and 6.25, respectively [42]. Due to the highest selectivity of the first mentioned running buffer, the same system has been applied for the quantification of benazepril and other compounds in their corresponding pharmaceutical formulations. It was found that the possibility of simultaneous identification and quantification of the active ingredient in the finished products was especially attractive, and that excipients do not adversely affect the results. This article deals with the validation of some parameters of the quantitative analysis, namely linearity, precision, accuracy, and robustness [43].

Hillaert *et al.* used a statistical experimental design to optimize a capillary electrophoresis separation method for benazepril and other angiotensin-converting enzyme inhibitors [44]. Because a free solution capillary electrophoresis system did not achieve a complete separation of these compounds in one run, the usefulness of alkylsulfonates as ion-pairing agents was investigated. After preliminary investigation to determine the experimental domain and the most important factors, a three-level full-factorial design was applied to study the impact of pH and molarity of the ion-pairing agent on the separation. Improved separations were obtained, suggesting a favorable effect of ion-pairing interactions between analytes and the additive. However, it remained impossible to separate all of these in a single run, so a combination of two systems was still necessary to enable selective identification of all structurally-related substances.

5. STABILITY

Benazepril is an ester prodrug, whose ester linkage can be cleaved chemically or enzymatically by hydrolysis or proteins displaying an esterase-like activity. It should be a good substrate for the porcine liver esterase, but it can be resistant to acidic hydrolysis. The kinetics of the acidic and enzymatically catalyzed hydrolysis of benazepril was studied using a HPLC method [37]. The acid-hydrolysis was studied in 0.1 M HCl at 50, 53, 58, and 63°C. The kinetics of the degradation (benazeprilat being the major degradation product) followed first-order kinetics, with reaction rate constants around 10 h^{-1} and half-lives ranging from 420 to 1400 h. The activation energy was found to be 19.9 kcal/mol. The enzymatic hydrolysis was carried out at 37°C, in phosphate buffer of pH 7.4. It was found that benazepril remained stable for at least 30 min.

6. ABSORPTION AND PHARMACOKINETICS

6.1 Absorption, metabolism and fate

Following oral administration, at least 37% of a dose of benazepril is absorbed. Benazepril is almost completely metabolized in the liver to benazeprilat. Peak plasma concentrations of benazeprilat following an oral dose of benazepril have been achieved after 1–2 h in the fasting state, or after 2–4 h in the non-fasting state. Both benazepril and benazeprilat are found to be about 95% bound to plasma protein [45–48].

Activation of benazepril involves cleavage of the ethoxy carbonyl group by hepatic esterase [49, 50]. After oral administration, peak plasma concentrations of benazepril and benazeprilat occur within half an hour and one to one and a half hours, respectively [49, 51]. Benazepril is extensively metabolized with less than 1% of dose being excreted in the urine as unchanged benazepril [49, 52]. The primary urinary metabolites of benazepril were benazeprilat (accounting for 17% of the oral dose), and glucuronide conjugates of benazepril and benazeprilat (accounting for 4% and 8% of the dose).

6.2 Excretion

Benazeprilat, the active metabolite of benazepril, is excreted mainly in the urine, with only about 11–12% being excreted in the bile. The effective half-life for accumulation of benazeprilat is 10–11 h following administration of multiple doses of benazepril. The elimination of benazeprilat is slowed in renally impaired individuals. Small amounts of benazepril and benazeprilat are distributed into breast milk [2].

The main route of elimination of benazeprilat is by way of the kidneys, although biliary excretion may also occur [49,51]. No clinically significant metabolic drug interactions with benzepril have yet been described [49].

7. ACKNOWLEDGEMENT

The authors would like to thank Mr. Tanvir A. Butt for his secretarial assistance in preparing this manuscript.

8. REFERENCES

1. ***The Merck Index***, 12th edn., S. Budavari, ed., Merck and Co, N.J., p. 172 (1996).
2. ***Martindale, The Complete Drug Reference***, 32nd edn., K. Parfitt, ed., The Pharmaceutical Press, Massachusetts, p. 808, 827 (2002).
3. ***Remington's: The Science and Practice of Pharmacy***, 19th edn., Volume II, A.R. Gennaro, ed., Mack Publishing Co., Pennsylvania, p. 950 (1995).
4. ***Index Nominum 2000, International Drug Directory***, 17th edn., Swiss Pharmaceutical Society, ed., Medpharm GmbH Scientifc Publishers Stuttgart, p. 97 (2000).
5. J.W.H. Watthey, *Eur. Pat., Appl.*, ***72***, 352 (1983) (Ciba-Geigy).
6. J.W.H. Watthey, *U.S. Pat.*, ***4,410***, 520 (1983) (Ciba-Geigy).
7. J.W.H. Watthey, J.L. Stanton, M. Desai, J.E. Babiarz, and B.M. Finn, *J. Med. Chem.*, ***28***, 1511 (1985).
8. J.R. Wade, D.M. Hughes, A.W. Kelman, C.A. Howie, and P.A. Meredith, *J. Pharm. Sci.*, ***82***, 471 (1993).
9. D.W. Cushman and H.S. Cheung, *Biochem. Pharmacol.*, ***20***, 1637 (1971).
10. I. Tikkanen, F. Fyhrquist, and T. Forslund, *Clin. Sci.*, ***67***, 273 (1984).
11. H. Hichens, E.L. Hand, and W.S. Mulcahy, *Ligand Quarterly*, ***4***, 43 (1981).
12. A. Sioufi, F. Pommier, G. Kaiser, and J.P. Dubois, *J. Chromatogr.*, ***434***, 239 (1988).
13. S. Khalil and S. Abd El-Aliem, *J. Pharm. Biomed. Anal.*, ***27***, 25 (2002).
14. N. Erk and F. Onur, *Analusis*, ***25***, 161 (1997).

15. I.E. Panderi, *J. Pharm. Biomed. Anal.*, ***21***, 257 (1999).
16. F.A. El-Yazbi, H.H. Abdine, and R.A. Shaalan, *J. Pharm. Biomed. Anal.*, ***20***, 343 (1999).
17. D. Bonazzi, R. Gotti, V. Andrisano, and V. Cavrini, *J. Pharm. Biomed. Anal.*, ***16***, 431 (1997).
18. E. Banoglu, Y. Ozkan, and O. Atay, *Farmaco*, ***55***, 477 (2000).
19. N. Erk, *J. Pharm. Biomed. Anal.*, ***20***, 155 (1999).
20. A. El-Gindy, A. Ashour, L. Abdel-Fattah, and M.M. Shabana, *J. Pharm. Biomed. Anal.*, ***25***, 299 (2001).
21. E. Dinc, *Anal. Lett.*, ***35***, 1021 (2002).
22. F. Belal, I.A. Al-Zaagi, and M.A. Abounassif, *Farmaco*, ***55***, 425 (2000).
23. F. Belal, I.A. Al-Zaagi, and M.A. Abounassif, *J. AOAC Int.*, ***84***, 1 (2001).
24. S.T. Hassib, Z.A. El-Sherif, R.I. El-Bagary, and N.F. Youssef, *Anal. Lett.*, ***33***, 3225 (2000).
25. A. El-Gindy, A. Ashour, L. Abdle-Fattah, and M.M. Shabana, *J. Pharm. Biomed. Anal.*, ***25***, 171 (2001).
26. K.M. Sereda, T.C. Hardman, M.R. Dilloway, and A.F. Lant, *Anal. Proc.*, ***30***, 371 (1993).
27. F. Pommier, F. Boschet, and G. Gosset, *J. Chromatogr.*, ***783***, 199 (2003).
28. G. Kaiser, R. Ackermann, W. Dieterle, and J.P. Dubois, *J. Chromatogr.*, ***419***, 123 (1987).
29. H.H. Maurer, T. Kraemer, and J.W. Arlt, *Ther. Drug Monit.*, ***20***, 706 (1998).
30. I.E. Panderi and M. Parissi-Poulou, *J. Pharm. Biomed. Anal.*, ***21***, 1017 (1999).
31. T. Radhakrishna, D.S. Rao, K. Vyas, and G.O. Reddy, *J. Pharm. Biomed. Anal.*, ***22***, 641 (2000).
32. L. Manna, L. Valvo, and S. Alimonti, *Chromatographia*, ***53*** (Suppl), S271 (2001).
33. R. Dappen, P. Voigt, F. Maystre, and A.E. Bruno, *Anal. Chim. Acta*, ***282***, 47 (1993).
34. R. Cirilli and F. La-Torre, *J. Chromatogr.*, ***818***, 53 (1998).
35. A Gumieniczek and L. Przyborowski, *J. Liq. Chromatogr. Rel. Technol.*, ***20***, 2135 (1997).
36. R.D. Sreenivas, M.K. Srinivasu, C.L. Narayana, and G.O. Reddy, *Indian Drugs*, ***37***, 80 (2000).
37. M. Gana, I. Panderi, M. Parissi-Poulou, and A. Tsantili-Kokoulidou, *J. Pharm. Biomed Anal.*, ***27***, 107 (2002).

38. P. Graf, F. Frueh, and K. Schmid, *J. Chromatogr., Biomed. Appl.*, ***69***, 353 (1988).
39. B. Cakir, O. Atay, and U. Tamer, *J. Fac. Pharm. Gazi Univ.*, ***17***, 43 (2000) through reference 21.
40. F. Barbato, P. Morrica, and F. Quaglia, *Farmaco*, ***49***, 457 (1994).
41. R. Gotti, V. Andrisano, V. Cavrini, C. Bertucci, and S. Fulanetto, *J. Pharm. Biomed. Anal.*, ***22***, 423 (2000).
42. S. Hillaert and W. Van den Bossche, *J. Chromatogr.*, ***895***, 33 (2000).
43. S. Hillaert and W. Van den Bossche, *J. Pharm. Biomed. Anal.*, ***25***, 775 (2001).
44. S. Hillaert, Y. Vander Heyden, and W. Van den Bossche, *J. Chromatogr.*, ***978***, 231 (2002).
45. G. Kaiser, R. Ackermann, S. Brechbuhler, and W. Dieterle, *Biopharm. Drug Dispos.*, ***10***, 365 (1989).
46. G. Kaiser, R. Ackermann, and A. Sioufi, *Am. Heart J.*, ***117***, 746 (1989).
47. G. Kaiser, R. Ackermann, W. Dieterle, C.J. Durnin, J. McEwen, K. Ghos, A. Richens, and I.B. Holmes, *Eur. J. Clin. Pharmacol.*, ***38***, 379 (1990).
48. N.J. Macdonald, H.L. Elliott, D.M. Hughes, and J.L. Reid, *Br. J. Clin. Pharmacol.*, ***36***, 201 (1993).
49. ***Metabolic Drug Interactions***, Reń H. Levy, and others. eds., Lippincott Williams & Wilkins, Philadelphia, USA 2000, p. 369.
50. E.K. Jackson. "Diuretics", in: ***Goodman, and Gilman's the Pharmacological Basis of Therapeutics***, 9th edn., J.G. Hardman and L.E. Limbird, eds., McGraw Hill, New York, p. 685 (1996).
51. J.A. Balfour and K.L. Goa, *Drugs*, ***42***, 511 (1991).
52. F. Waldmeier, G. Kaiser, R. Ackermann, J.W. Faigle, J. Wagner, A. Barner, and K.C. Lasseter, *Xenobiotica*, ***21***, 251 (1991).

Ciprofloxacin: Physical Profile

Mohammed A. Al-Omar

Department of Pharmaceutical Chemistry
College of Pharmacy, King Saud University
P.O. Box 2457, Riyadh-11451
Kingdom of Saudi Arabia

PROFILES OF DRUG SUBSTANCES,
EXCIPIENTS, AND RELATED
METHODOLOGY – VOLUME 31
DOI: 10.1016/S0000-0000(00)00000-0

CONTENTS

1. **General Information** . . . 164
   - 1.1 Nomenclature . . . 164
     - 1.1.1 Systematic chemical names . . . 164
     - 1.1.2 Nonproprietary names . . . 165
     - 1.1.3 Proprietary names . . . 165
     - 1.1.4 Synonyms . . . 165
   - 1.2 Formulae . . . 165
     - 1.2.1 Empirical formula, molecular weight, CAS number . . . 165
     - 1.2.2 Structural formula . . . 165
   - 1.3 Elemental analysis . . . 165
   - 1.4 Appearance . . . 165
2. **Physical Characteristics** . . . 165
   - 2.1 Solution pH . . . 165
   - 2.2 Solubility characteristics . . . 166
   - 2.3 Optical activity . . . 166
   - 2.4 X-Ray powder diffraction pattern . . . 166
   - 2.5 Thermal methods of analysis . . . 170
     - 2.5.1 Melting behavior . . . 170
     - 2.5.2 Differential scanning calorimetry . . . 170
   - 2.6 Spectroscopy . . . 170
     - 2.6.1 Ultraviolet spectroscopy . . . 170
     - 2.6.2 Vibrational spectroscopy . . . 171
     - 2.6.3 Nuclear magnetic resonance spectrometry . . . 172
   - 2.7 Mass spectrometry . . . 174
3. **Stability and Storage** . . . 177
4. **References** . . . 178

1. GENERAL INFORMATION

1.1 Nomenclature

1.1.1 Systematic chemical names [1, 2]

1-Cyclopropyl-6-fluoro-1,4-dihydro-4-oxo-7-(1-piperazinyl)-3-quinolinecarboxylic acid.

1.1.2 Nonproprietary names
Ciprofloxacin

1.1.3 Proprietary names [2, 3, 4]
Aceto, Baycip, Ciflox, Cifluran, Ciloxan, Ciplox, Ciprinol, Cipro, Ciprobay, Ciproxan, Ciproxin, Flociprin, Septicide, Velmonit, Xorpic.

1.1.4 Synonyms [3]
Bay-q-3939.

1.2 Formulae

1.2.1 Empirical formula, molecular weight, CAS number [3, 4]
$C_{17}H_{18}FN_3O_3$ 331.35 [0085721-33-1].

1.2.2 Structural formula

HN N N F COOH O

1.3 Elemental analysis

C = 61.62% H = 5.48% F = 5.73% N = 12.68% O = 14.49%.

1.4 Appearance [1, 3]

Ciprofloxacin is obtained as a light yellow crystalline powder.

2. PHYSICAL CHARACTERISTICS

2.1 Solution pH

A 2.5% solution in water has a pH of 3.0–4.5 [3].

2.2 Solubility characteristics

Ciprofloxacin is practically insoluble in water, very slightly soluble in dehydrated alcohol and in dichloromethane, and soluble in dilute acetic acid [3].

2.3 Optical activity

Since ciprofloxacin has no centers of dissymmetry, it does not exhibit optical activity.

2.4 X-Ray powder diffraction

The X-ray powder diffraction pattern of ciprofloxacin was obtained using a Simons XRD-5000 diffractometer, which is shown in Figure 1. Table 1 contains a compilation of values for the observed scattering peaks (in units of degrees 2-θ), the interplanar d-spacings (units of Å), and the relative intensities associated with the powder pattern.

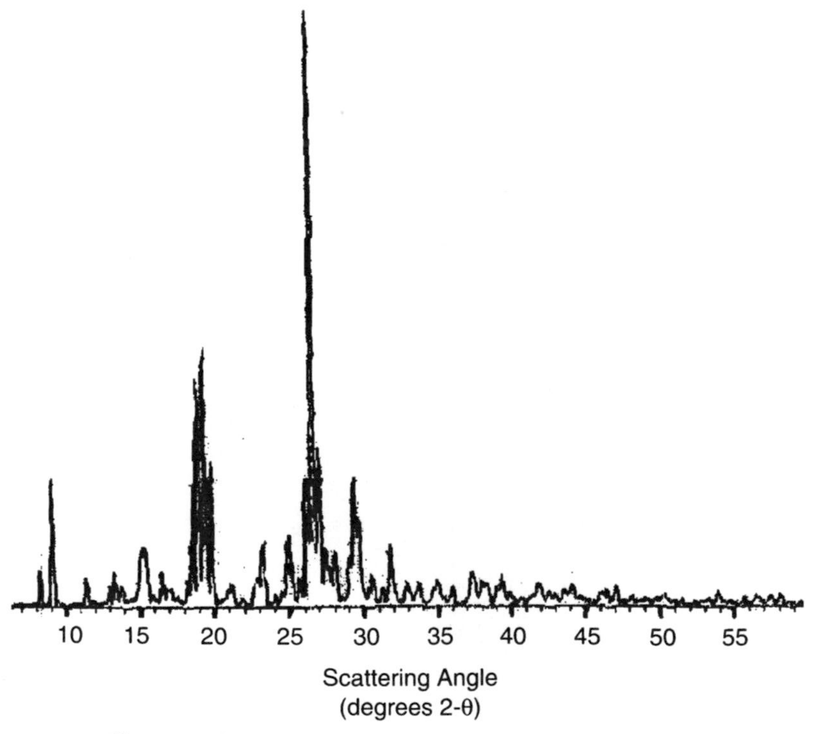

Figure 1. X-ray powder diffraction pattern of ciprofloxacin.

Table 1

Crystallographic Data from the X-Ray Powder Diffraction Pattern of Ciprofloxacin

Scattering angle (degrees 2θ)	***d*-Spacing (Å)**	**Relative intensity (%)**	**Scattering angle (degrees 2θ)**	***d*-Spacing (Å)**	**Relative intensity (%)**
8.185	10.7928	5.94	9.065	9.7475	21.43
11.348	7.7913	4.75	12.843	6.8870	3.41
13.247	6.6780	5.91	13.700	6.4583	3.39
15.139	5.8474	9.95	15.390	5.7527	9.95
16.451	5.3840	5.90	18.144	4.8853	4.40
18.478	4.7976	11.76	18.913	4.6882	37.85
19.37	4.5864	43.48	19.810	4.4779	23.85
21.104	4.2063	4.03	21.820	4.0697	1.34

(continued)

Table 1 (continued)

Scattering angle (degrees 2θ)	***d*-Spacing (Å)**	**Relative intensity (%)**	**Scattering angle (degrees 2θ)**	***d*-Spacing (Å)**	**Relative intensity (%)**
22.575	3.9354	5.10	23.192	3.8321	13.76
23.935	3.7148	1.94	24.372	3.6491	2.44
24.783	3.5895	11.12	25.094	3.5458	8.16
25.581	3.4793	4.50	26.112	3.4098	15.18
26.545	3.3551	100.00	26.978	3.3023	26.33
27.360	3.2570	9.73	28.015	3.1823	10.33
28.929	3.0838	7.98	29.304	3.0452	21.44
29.585	3.0169	14.82	30.458	2.9324	5.99
31.120	2.8715	2.89	31.747	2.8163	10.09

32.786	2.7293	4.01	33.614	2.6639	3.78
34.916	2.5675	4.14	35.893	2.4998	3.30
37.335	2.4065	5.42	38.237	2.3518	3.68
38.919	2.3122	2.96	39.304	2.2904	4.99
40.576	2.2215	1.27	41.682	2.1651	3.71
44.047	2.0542	3.26	45.987	1.9719	2.13
46.324	1.9584	2.75	46.939	1.9341	3.38
51.353	1.7777	1.17	53.777	1.7032	2.44
55.569	1.6524	1.47	57.358	1.6051	1.64
57.924	1.5907	1.70			

2.5 Thermal methods of analysis

2.5.1 Melting behavior

Ciprofloxacin is found to melt in the range of about 318–320°C [2].

2.5.2 Differential scanning calorimetry

The differential scanning calorimetry (DSC) thermogram of ciprofloxacin was obtained using a DuPont TA-9900 thermal analyzer interfaced with a DuPont Data Unit. The thermogram shown in Figure 2 was obtained at a heating rate of 10°C/min, and was run from 40 to 400°C. The sole observed thermal event was the melting endotherm which was observed at 322°C.

2.6 Spectroscopy

2.6.1 Ultraviolet spectroscopy

The UV spectrum of ciprofloxacin dissolved in methanol was recorded using a Shimadzu ultraviolet–visible Spectrophotometer 1601 PC, and is

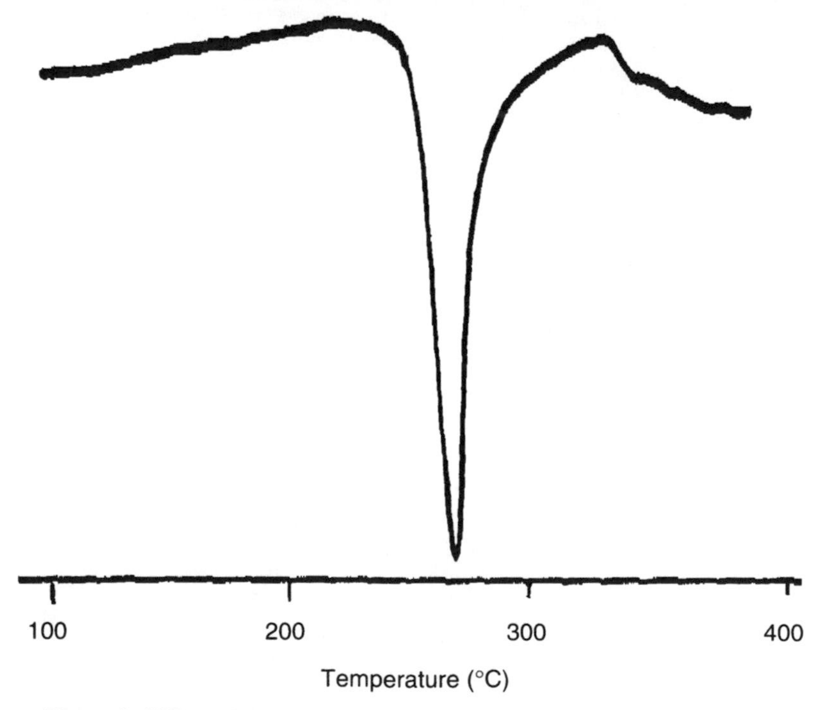

Figure 2. Differential scanning calorimetry thermogram of ciprofloxacin.

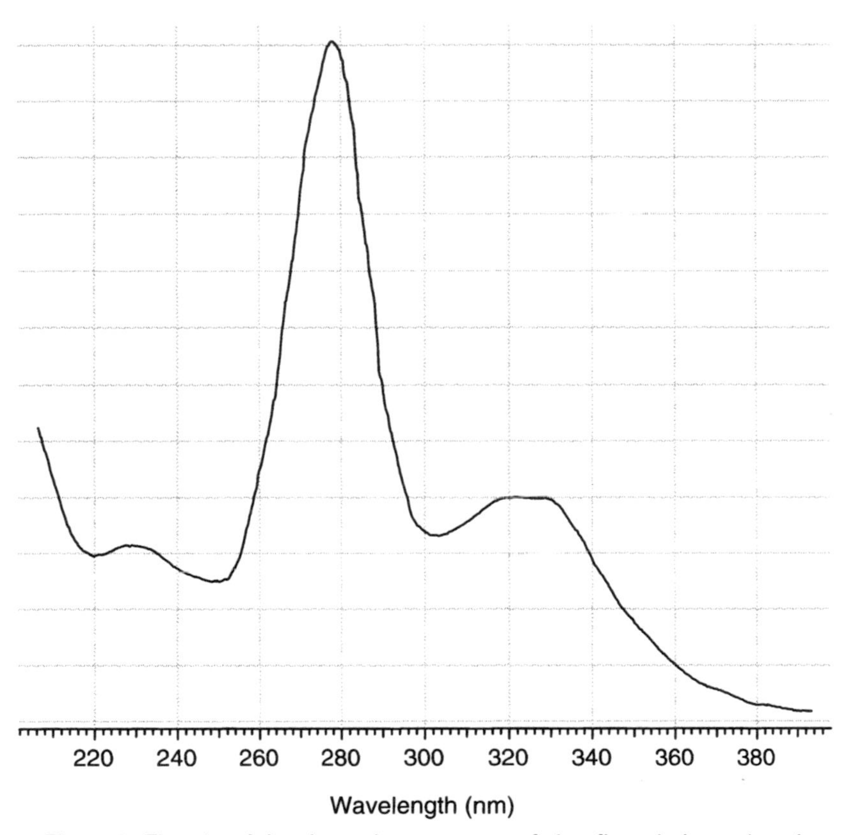

Figure 3. The ultraviolet absorption spectrum of ciprofloxacin in methanol.

shown in Figure 3. The spectrum of ciprofloxacin was found to exhibit the following characteristics at the three observed maxima:

Wavelength maximum (nm)	$A_{[1\%,\ 1\ cm]}$	Molar absorptivity (L mol^{-1} cm^{-1})
320.6	1.95	129
280	5.76	382
228	1.57	104

2.6.2 Vibrational spectroscopy

The infrared absorption spectrum of ciprofloxacin was obtained in a KBr pellet using a Perkin Elmer infrared spectrophotometer. The infrared spectrum is shown in Figure 4, and the principal peaks were noted at 3490, 3320, 2930, 2840, 1696, 1605, 1480, 1435 cm^{-1}. The assignments for the major infrared absorption bands are summarized in Table 2.

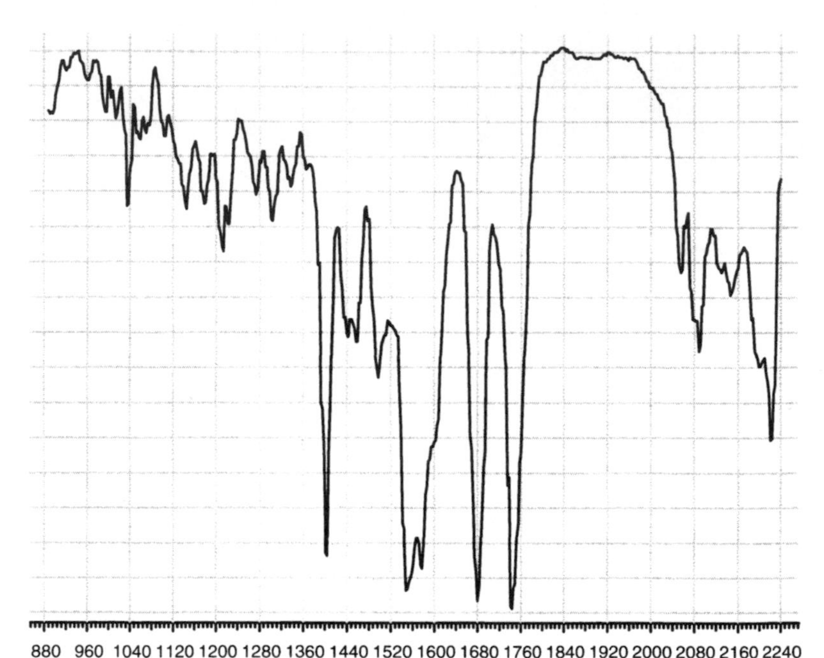

Figure 4. The infrared absorption spectrum of ciprofloxacin obtained in a KBr pellet.

2.6.3 Nuclear magnetic resonance spectrometry

2.6.3.1 ^{1}H-NMR spectrum

The proton NMR spectrum of ciprofloxacin was obtained using a Bruker Advance system, operating at 300, 400, and 500 MHz. The sample was dissolved in D_2O, and tetramethylsilane (TMS) was used as the internal standard. The proton NMR spectrum is shown in Figure 5, and assignments for the ^{1}H-NMR resonance bands of ciprofloxacin are found in Table 3.

2.6.3.2 ^{13}C-NMR spectrum

The carbon-13 NMR spectrum of ciprofloxacin was obtained using a Bruker Advance system operating at 75, 100, and 125 MHz. Standard Bruker Software was used to obtain DEPT spectra. The sample was dissolved in D_2O, and tetramethylsilane (TMS) was used as the internal standard. Assignments for the various carbons of ciprofloxacin are shown in Table 4.

Table 2

Assignments for the Infrared Absorption Bands of Ciprofloxacin

Frequency (cm^{-1})	Assignments
3490	O–H stretch
3320	N–H stretch of piperazinyl moiety
2930	Aliphatic C–H stretch
2840	N–C stretch
1696	C=O stretch of carboxyl group
1605	C=O stretch of quinoline
1480, 1435	C–N stretch

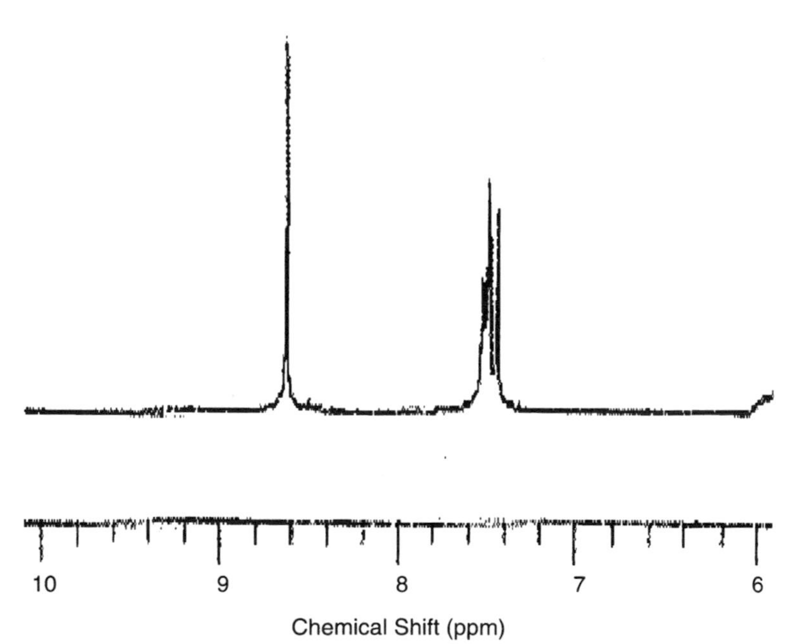

Figure 5. The ^{1}H-NMR spectrum of ciprofloxacin in D_2O.

Table 3

Assignments for the observed Resonance Bands in the ^{1}H-NMR Spectrum of Ciprofloxacin

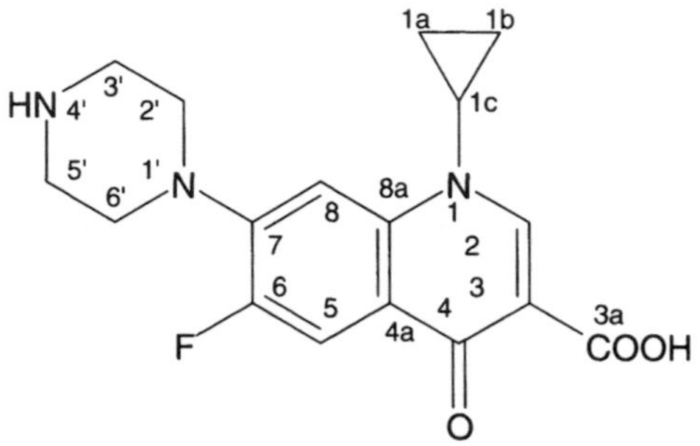

Proton atoms	Chemical shift (ppm relative to TMS)	Multiplicity (s: singlet, d: doublets)	Number of proton atoms
H-2	8.63	s	1
H-5	7.46	d	1
H-8	7.52	d	1
H-2′,6′ or 3′,5′	3.66	d	2
H-2′,6′ or 3′,5′	3.56	d	2
H-1a	1.22	s	1
H-1b, 1c	1.47	d	2

2.7 Mass spectrometry

The electron impact (EI) spectrum of ciprofloxacin is presented in Figure 6, and was recorded using a Shimadzu PQ-5000 GC-MS spectrometer. The spectrum shows a mass peak ($M^{\cdot}$) at *m*/*z* 332, and a base peak at *m*/*z* 288 resulting from the loss of the group. All these and

Table 4

Assignments for the Observed Resonance Bands in the ^{13}C-NMR Spectrum of Ciprofloxacin

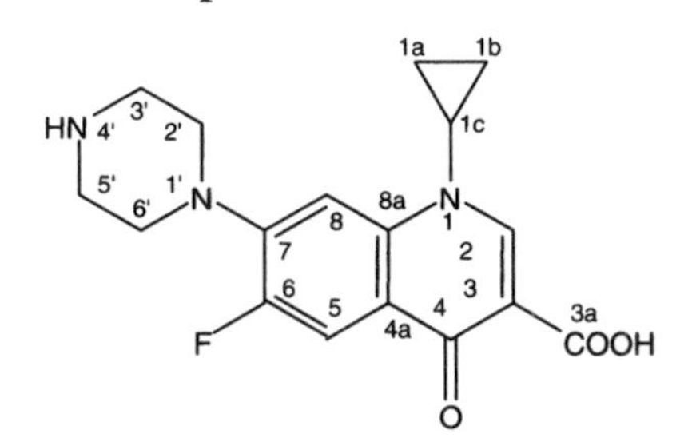

Carbon atoms	Chemical shift (ppm relative to TMS)	Assignment at carbon number	Carbon atoms	Chemical shift (ppm relative to TMS)	Assignment at carbon number
C-1a, C-1b	36.90	2	C-5	111.26	1
C-1c	43.96	1	C-6	107.14	1
C-2	139.48	1	C-7	145.40	1

(*continued*)

Table 4 (continued)

Carbon atoms	Chemical shift (ppm relative to TMS)	Assignment at carbon number	Carbon atoms	Chemical shift (ppm relative to TMS)	Assignment at carbon number
C-3	155.60	1	C-2′, C-6′	43.96	2
C-3a	169.31	1	C-3′, C-5′	47.05	2
C-4	176.12	1	C-8	119.04	1
C-4a	148.69	1	C-8a	152.26	1

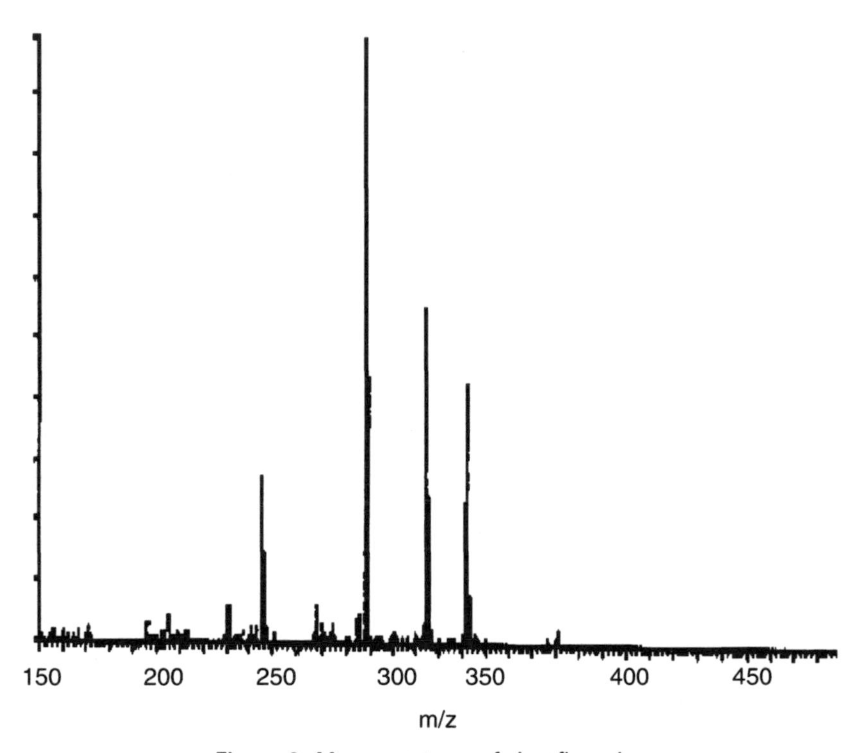

Figure 6. Mass spectrum of ciprofloxacin.

other proposed fragmentation patterns of the drug are presented in Table 5.

3. STABILITY AND STORAGE [3, 5]

Solutions of ciprofloxacin are light sensitive and should be protected from light and freezing [3]. When the concentrate formulated for intravenous injection, or the 1.2 g pharmacy bulk package, is diluted with 5% dextrose injection or 0.9% sodium chloride injection to a final concentration of 0.5–2 mg/mL, the resultant solution is stable for upto 14 days when stored at room temperature or when refrigerated at 2–8°C [5].

The commercially available injection for intravenous infusion that contains 2 mg/mL in 5% dextrose is provided in a plastic container fabricated from a specially formulated polyvinyl chloride (PVC). Tammilehto *et al.* used thin-layer chromatography to study the degradation of ciprofloxacin hydrochloride solutions after these were irradiated by a high pressure mercury lamp [6].

Table 5

Assignments for the Fragmentation Pattern Observed in the Mass Spectrum of Ciprofloxacin

Mass number (m/z)	**Relative intensities**	**Structural assignment**
332	45	$C_{17}H_{18}FN_3O_3]^{++}$·mass peak (M·)
314	55	$M–H_2O$
288	100	$M–CO_2$
245	30	$M–(CO_2 \cdot C_2H_5N)$

4. REFERENCES

1. ***Remington's: Pharmaceutical Sciences***, 20th edn., A.R. Gennaro, ed., Mack Publishing Co., Pennsylvania, p. 1539 (2000).
2. ***The Merck Index***, 12th edn., S. Budavari, ed., Merck and Co., NJ, p. 2374 (1996).
3. ***Martindale, The Complete Drug Reference***, 33rd edn., S.C. Sweetman, ed., The Pharmaceutical Press, Chicago, p. 182 (2002).
4. ***Index Nominum 2000: International Drug Directory***, 17th edn., Swiss Pharmaceutical Society, Medpharm Scientific Publishers, Ontario, p. 239 (2000).
5. ***Drug Information,*** 95 edn., G.K. McEvoy, ed., American Society of Health-System Pharmacists, p. 493 (1995).
6. S. Tammilehto, H. Salomies, and K. Torniainen, *J. Planar. Chromatogr. Mod. TLC*, *7*, 368 (1994).

Ciprofloxacin: Analytical Profile

Mohammed A. Al-Omar

Department of Pharmaceutical Chemistry
College of Pharmacy, King Saud University
P.O. Box 2457, Riyadh-11451
Kingdom of Saudi Arabia

PROFILES OF DRUG SUBSTANCES,
EXCIPIENTS, AND RELATED
METHODOLOGY – VOLUME 31
DOI: 10.1016/S0000-0000(00)00000-0

CONTENTS

1. **Compendial Methods of Analysis** 180
1.1 Identification 180
1.2 Compendial analyses 181
1.2.1 Appearance of solution 181
1.2.2 Fluoroquinolonic acid content 181
1.3 Related substances: 181
1.4 Heavy metals 183
1.5 Loss on drying 183
1.6 Sulfated ash 183
1.7 Chlorine 183
1.8 Sulfate 184
1.9 Titrimetric assay method for bulk drug substance 184
1.10 Assay method for tablets 184
1.11 Assay method for intravenous infusion formulations 185
1.12 Impurities 185
1.12.1 Chromatographic purity 187
2. **Reported Methods of Analysis** 188
2.1 Titration methods 188
2.2 Spectrophotometric methods 189
2.3 Chemiluminescence methods 191
2.4 Electrochemical methods 192
2.5 Thin-layer chromatography 192
2.6 High performance liquid chromatographic methods 193
2.7 Gas chromatography 202
2.8 Capillary electrophoresis 202
2.9 Immunoassay methods 203
3. **References** 204

1. COMPENDIAL METHODS OF ANALYSIS

1.1 Identification

The European Pharmacopoeia [1] recommends the use of infrared absorption spectrophotometry for the identification of the pure drug substance. The procedure entails an examination by infrared absorption spectrophotometry, and a comparison with the spectrum obtained with the ciprofloxacin reference standard.

1.2 Compendial analyses

1.2.1 Appearance of solution

Dissolve 0.25 g of the substance to be tested in 0.1 M hydrochloric acid, and dilute to 20 mL with the same solvent. The solution is clear, and not more intensely colored than the reference solution GY4.

1.2.2 Fluoroquinolonic acid content

Examine by thin-layer chromatography, using as the coating substance a suitable silica gel with a fluorescent indicator having an optimal intensity at 254 nm.

> **Test solution**: Dissolve 50 mg of the substance to be examined in dilute ammonia R1 and dilute to 5 mL with the same solvent.
>
> **Reference solution**: Dissolve 10 mg of fluoroquinolonic acid CRS in a mixture of 0.1 mL of dilute ammonia R1 and 90 mL of water R, and dilute to 100 mL with water R. Dilute 2 mL of this solution to 10 mL with water R.

Apply 5 μL of each solution to the plate. At the bottom of a chromatographic tank, place an evaporating dish containing 50 mL of concentrated ammonia R. Close the tank and expose the plate to the ammonia vapor for 15 min. Withdraw the plate, transfer to a chromatographic tank, and develop over a path of 15 cm using a mobile phase mixture consisting of 10 volumes of acetonitrile R, 20 volumes of concentrated ammonia R, 40 volumes of methanol R, and 40 volumes of methylene chloride R. Allow the plate to dry in air and examine under 254 nm ultraviolet light. Any spot corresponding to fluoroquinolonic acid in the chromatogram obtained with the test solution is not more intense than the spot in the chromatogram obtained with the reference solution (0.2%).

1.3 Related substances [1, 2]

Examine the substance in question by liquid chromatography, using the following procedure:

> **Test solution (a)**: Add 0.2 mL of dilute phosphoric acid R to 25 mg of the substance to be examined, and dilute to 50.0 mL with the

mobile phase. Treat in an ultrasonic bath until a clear solution is obtained.

Test solution (b): Dilute 0.1 mL of test solution (a) to 50 mL with the mobile phase.

Reference solution (a): Dissolve 2.5 mg of ciprofloxacin impurity B CRS in the mobile phase and dilute to 50.0 mL with the mobile phase. This solution is also used to prepare the reference solution (d). Dilute 1.0 mL of this solution to 50 mL with mobile phase.

Reference solution (b): Dissolve 2.5 mg of ciprofloxacin impurity C CRS in mobile phase and dilute to 50.0 mL with mobile phase. This solution is also used to prepare reference solution (d). Dilute 1 mL of this solution to 50 mL with mobile phase.

Reference solution (c): Dissolve 2.5 mg of ciprofloxacin impurity D CRS in mobile phase and dilute to 50 mL with mobile phase. This solution is also used to prepare reference solution (d). Dilute 1 mL of this solution to 50.0 mL with mobile phase.

Reference solution (d): Mix 0.1 mL of test solution (a), 1 mL of reference solution (a), 1.0 mL of reference solution (b), 1 mL of reference solution (c) and dilute to 50 mL with mobile phase.

The chromatographic procedure may be carried out using:

- Chromatographic column: a stainless steel column, 0.25 m long and 4.6 mm internal diameter, packed with base-deactivated octadecylsilyl silica for chromatography R (5 μm particles). The column is maintained at a temperature of 40°C.

- Mobile Phase: a mixture of 13 volumes of acetonitrile R and 87 volumes of 0.025 M phosphoric acid, previously adjusted to pH 3.0 with triethylamine R. The flow rate is to be about 1.5 mL/min.

- Detector: a spectrophotometer set at a wavelength of 278 nm.

Inject separately 50 μL of test solution (b) and 50 μL of reference solutions (a), (b), (c), and (d). When the chromatogram is recorded using the conditions described above, the retention time of ciprofloxacin is about 9 min. Adjust the sensitivity of the system so that the height of the ciprofloxacin impurity C peak in the chromatogram obtained with reference solution (d) is at least 40% of the full scale deflection of the recorder. The test is not valid unless in the chromatogram obtained with reference solution (d), the resolution between the peaks corresponding to ciprofloxacin impurity B and ciprofloxacin impurity C is at least 1.3, and the resolution between the peaks corresponding to ciprofloxacin and ciprofloxacin impurity D is at least 3.

Inject separately 50 μL of test solution (a) and 50 μL of reference solutions (b) and (c), and record the chromatograms for twice the retention time of ciprofloxacin. In the chromatogram obtained with test solution (a), the areas of the peaks corresponding to ciprofloxacin impurity C and ciprofloxacin impurity D are not greater than the corresponding peaks in the chromatogram obtained with reference solutions (b) and (c) (*i.e.*, not more than 0.2%). The areas of any other secondary peaks are not greater than the peak in the chromatogram obtained with reference solution (b) (*i.e.*, not more than 0.2%). The sum of the areas of all the peaks, apart from the principal peak, is not greater than 2.5 times the area of the peak in the chromatogram obtained with reference solution (b) (*i.e.*, not more than 0.5%). Disregard any peak whose area is less than 0.25 times the area of the peak in the chromatogram obtained with reference solution (b).

1.4 Heavy metals [1]

Dissolve 0.5 g of the substance to be tested in dilute acetic acid R, dilute to 30 mL with the same solvent, and then add 2 mL of water R. The filtrate complies with limit test E for heavy metals (not more than 20 ppm). The comparison stand is prepared using 5 mL of lead standard solution (2 ppm Pb) R.

1.5 Loss on drying [1]

The sample does not contain more than 1.0% total volatile content when 1.000 g is dried under vacuum at 120°C.

1.6 Sulfated ash [1]

The sample does not contain more than 0.1% sulfated ash, determined on 1.0 g.

1.7 Chlorine

The turbidity exhibited by the test solution does not exceed that of the standard solution (0.02%) [3].

1.8 Sulfate

The turbidity exhibited in the tube containing the test solution does not exceed that of the tube containing the standard solution (0.04%) [3].

1.9 Titrimetric assay method for bulk drug substance

Dissolve 0.300 g of the substance under test in 80 mL of anhydrous acetic acid R. Titrate with 0.1 M perchloric acid, using potentiometric means to determine the end-point. In the titration, each milliliter of 0.1 M perchloric acid is equivalent to 33.14 mg of $C_{17}H_{18}FN_3O_3$.

1.10 Assay method for tablets [2]

The tablet assay is performed using a liquid chromatographic method, which requires the following solutions. Weigh and powder 20 tablets. For solution (1), add a quantity of powdered tablets containing the equivalent of 2 g of ciprofloxacin to 750 mL of water, mix for 20 min, add sufficient water to produce 1000 mL, and mix again. Centrifuge a portion of the resulting suspension, and dilute the clear supernatant liquid with sufficient water to produce a solution containing the equivalent of 0.05% w/v of ciprofloxacin. Solution (2) contains 0.058% w/v of ciprofloxacin hydrochloride EPCRS in the mobile phase. Solution (3) contains 0.025% w/v of ciprofloxacin impurity C EPCRS (the ethylenediamine compound) in solution (2). For solution (4), dilute 1 volume of solution (3) to 100 volumes with the mobile phase.

The chromatographic procedure may be carried out using a stainless steel column (25 cm × 4.6 mm) packed with a stationary phase C (7 mm) (Nucleosil C18 is suitable), whose temperature is to be maintained at 40°C. The mobile phase is eluted at a flow rate of 1.5 mL/min, and consists of a mixture of 13 volumes of acetonitrile and 87 volumes of a 0.245% w/v solution of orthophosphoric acid (the pH of which has been adjusted to 3.0 with triethylamine). Analytes are detected on the basis of their UV absorption at 278 nm. The Assay is not valid unless in the chromatogram obtained with solution (3), the resolution factor between the peaks due to ciprofloxacin and ciprofloxacin impurity C is at least 3.

Calculate the content of $C_{17}H_{18}FN_3O_3$ in the tested tablets using the declared content of $C_{17}H_{19}ClFN_3O_3$ in ciprofloxacin hydrochloride

EPCRS. Each mg of $C_{17}H_{19}ClFN_3O_3$ is equivalent to 0.9010 mg of $C_{17}H_{18}FN_3O_3$. The content of ciprofloxacin should be 95.0–105.0% of the label claim.

1.11 Assay method for intravenous infusion formulations [2]

Ciprofloxacin intravenous infusion is a sterile solution of ciprofloxacin lactate (prepared by the interaction of ciprofloxacin and lactic acid) in sodium chloride intravenous infusion solution.

The assay of an IV formulation is performed using a liquid chromatographic method, which requires the following solutions. For solution (1), dilute a quantity of the intravenous infusion with sufficient mobile phase to produce a solution containing the equivalent of 0.05% w/v of ciprofloxacin. Solution (2) contains 0.058% w/v of ciprofloxacin hydrochloride EPCRS in mobile phase. Solution (3) contains 0.025% w/v of ciprofloxacin impurity C EPCRS (ethylenediamine compound) in solution (2). For solution (4), dilute 1 volume of solution (3) to 100 volumes with the mobile phase. The chromatographic procedure may be carried out using a stainless steel column (12.5 cm × 4 mm) packed with stationary phase C (5 mm) (Nucleosil C18 is suitable). The mobile phase is eluted at a flow rate of 1.5 mL/min, and consists of a mixture of 13 volumes of acetonitrile and 87 volumes of a 0.245% w/v solution of orthophosphoric acid (the pH of which has been adjusted to 3.0 with triethylamine). Analytes are detected on the basis of their UV absorption at 278 nm. The assay is not valid unless in the chromatogram obtained with solution (3), the resolution factor between the peaks due to ciprofloxacin and the ciprofloxacin impurity C is at least 1.5.

Calculate the content of $C_{17}H_{18}FN_3O_3$ in the intravenous infusion formulation using the declared content of $C_{17}H_{19}ClFN_3O_3$ in ciprofloxacin hydrochloride EPCRS. Each milligram of $C_{17}H_{19}ClFN_3O_3$ is equivalent to 0.9010 mg of $C_{17}H_{18}FN_3O_3$. The ciprofloxacin content should be 95.0–105.0% of the label claim.

1.12 Impurities [1, 3]

The European Pharmacopoeia [1] (Section 1.3) and the US Pharmacopoeia [3] describe liquid chromatographic methods for the determination of five related substances, denoted as substances A, B, C,

D, and E. The following structure is used to define substances A and C:

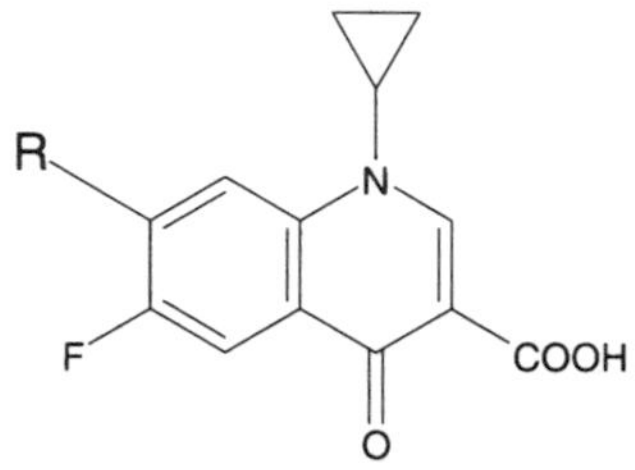

Substance A: $R = Cl$; systematic name is 7-chloro-1-cyclopropyl-6-fluoro-4-oxo-1,4-dihydroquinoline-3-carboxylic acid (fluoroquinolic acid).

Substance C: $R = NH–[CH_2]_2–NH_2$; systematic name is 7-[(2-aminoethyl)amino]-1-cyclopropyl-6-fluoro-4-oxo-1,4-dihydroquinoline-3-carboxylic acid (also known as the ethylenediamine compound).

The following structure is used to define substances B and E:

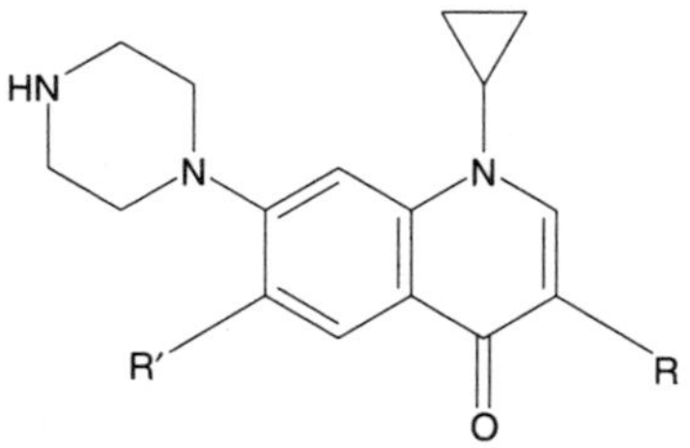

Substance B: $R = COOH$ and $R' = H$; systematic name is 1-cyclopropyl-4-oxo-7-(piperazin-1-yl)-1,4-dihydroquinoline-3-carboxylic acid (also known as the desfluoro compound).

Substance E: $R = H$ and $R' = F$; systematic name is 1-cyclopropyl-6-fluoro-7-(piperazin-1-yl)quinolin-4(1*H*)-one (also known as the decarboxylated compound).

Substance D: 7-Chloro-1-cyclopropyl-4-oxo-6-(piperazin-1-yl)-1,4-dihydroquinoline-3-carboxylic acid.

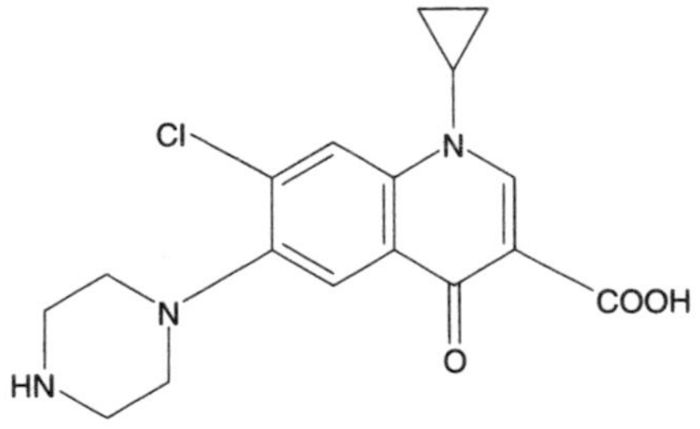

1.12.1 Chromatographic purity [3]

The chromatographic purity of ciprofloxacin is determined using the following general procedure:

Standard Preparation Solution: Transfer about 25 mg of USP ciprofloxacin RS, accurately weighed, to a 50-mL volumetric flask. Add 0.2 mL of 7% phosphoric acid to dissolve, dilute to volume with mobile phase, and mix.

Resolution Solution: Dissolve a quantity of USP ciprofloxacin ethylenediamine analog RS in the Standard Preparation Solution, to obtain a solution containing about 0.5 mg/mL.

Assay Preparation Solution: Transfer about 25 mg of ciprofloxacin, accurately weighed, to a 50-mL volumetric flask. Add 0.2 mL of 7% phosphoric acid to dissolve, dilute to volume with mobile phase, and mix.

Mobile Phase: Prepare a filtered and degassed 87:13 v/v mixture of 0.025 M phosphoric acid (previously adjusted to a pH of 3.0 ± 0.1 with triethylamine) and acetonitrile.

Chromatographic System: The liquid chromatograph is equipped with a 278-nm detector, and with a 4-mm $\times$ 25-cm column that contains packing L1. The column is maintained at a temperature of 40 ± 1°C, and the flow rate is about 1.5 mL/min.

System Suitability: (A) Chromatograph the Resolution Solution, and record the responses as directed under Procedure. The retention time for ciprofloxacin is between 6.4 and 10.8 min. Relative retention times are about 0.7 for ciprofloxacin ethylenediamine analog, and 1.0 for ciprofloxacin. The resolution between the ciprofloxacin ethylenediamine analog peak and the ciprofloxacin peak is not less than 6. (B) Chromatograph the Standard Preparation Solution. The column efficiency, determined from the ciprofloxacin peak, is not less than 2500 theoretical plates, the tailing factor for the ciprofloxacin peak is not more than 4.0, and the relative standard deviation for replicate injections is not more than 1.5%.

Procedure: Separately inject equal volumes (about 10 μL) of the Standard Preparation and the Assay Preparation Solutions into the chromatograph, record the chromatograms, and measure the responses for the major peaks. Calculate the quantity (in mg) of $C_{17}H_{18}FN_3O_3$ in the sample of ciprofloxacin taken using the formula 50 C (r_U/r_S), in which C is the concentration (units of mg/mL) of USP ciprofloxacin RS in the Standard Preparation Solution, and r_U and r_S are the ciprofloxacin peak responses obtained from the chromatograms of the Assay Preparation and Standard Preparation Solutions, respectively.

Calculations: Calculate the percentage of each impurity peak in the chromatogram obtained from the assay preparation taken using the formula, $100r_i/r_t$, in which r_i is the response of each impurity peak, and r_t is the sum of the responses of all the peaks. Not more than 0.4% of ciprofloxacin ethylenediamine analog and not more than 0.2% of any other individual impurity peak is found, and the sum of all impurity peaks is not more than 0.7%.

2. REPORTED METHODS OF ANALYSIS

Several methods were reported in the literature for the determination of ciprofloxacin in pharmaceutical formulations and in biological fluids.

2.1 Titration methods

Four DNA topisomerase (gyrase) inhibitors, including ciprofloxacin (alone or in mixture with other drugs), were determined by a titration method [4]. Tablets containing ciprofloxacin were extracted three times with 30 mL of methanol. A portion of the resulting solution was mixed with 2 mL water, and that solution was titrated with aqueous

5 mM NaOH, 5 mM tetrabutyl-ammonium hydroxide, or 5 mM $AgNO_3$ by adding 0.1 mL increments at 2 min intervals. The end-point was detected using the solution conductance.

Zhang *et al.* determined ciprofloxacin lactate in injectable solutions by acid dye biphasic titration [5]. A 200 mg dried sample was dissolved in and diluted with water to a volume of 100 mL, and then a 2 mL aliquot of the solution was mixed with 10 mL Na_2HPO_4/citric acid buffer (pH 7) and 15 mL $CHCl_3$. This solution was titrated with 0.5 mM-bromothymol blue, with agitation, until a light-blue end-point appeared in the aqueous phase. Recoveries were 99.9–102%, with a relative standard deviation ($n=4$) of 0.22–0.24.

Kilic *et al.* selected five different antibiotics, including ciprofloxacin, and reported a titration method based on the use of tetrabutylammonium hydroxide as the titrant in a non-aqueous assay method [6]. Ciprofloxacin was dissolved in water and diluted to 25 mL with pyridine. A 20 mL portion of the resulting solution was titrated with 0.04 M tetrabutyl-ammonium hydroxide in methanol/propan-2-ol at 25°C under nitrogen in a jacketed glass reaction cell. End-point detection was performed with an Orion 720A digital pH ion-meter, equipped with a combination pH electrode that contained a saturated solution of anhydrous methanolic KCl in the reference. The method was applied to pharmaceutical preparations, and enabled recoveries in the range of 99.95–101.53%, with a RSD of 0.5–1.09%.

2.2 Spectrophotometric methods

Rizk *et al.* developed a sensitive and selective derivative UV-spectrophotometric method for the determination of three fluoroquinolone compounds, including ciprofloxacin, in formulations and spiked biological fluids [7]. The method depends on the complexation of Cu(II) with the studied compounds in an aqueous medium. A linear correlation was established between the amplitude of the peak and the drug concentration over the range of 35–120 ng/mL. The detection limit was reported as 1.3 ng/mL. The method was used for the determination of the ciprofloxacin bulk drug substance and its tablet formulation, with an overall percentage recovery of 99.22 ± 0.55 to 100.33 ± 1.60.

Ciprofloxacin in both the tablets and blood serum was determined by a fluorimetric method that was based on the intrinsic luminescence of complexed terbium(III) [8]. A standard solution of ciprofloxacin (0.1 mM) was mixed with 2 mL of a 3 mM Tb(III) solution and 0.1 M sodium acetate/0.1 M acetic acid buffer (pH 6). The mixture was diluted to 10 mL with water, and after 30 s the fluorescence intensity was measured at a wavelength of 545 nm (using an excitation wavelength of 325 nm). The calibration curve was linear from 13 ng/mL to 1 μg/mL of ciprofloxacin, with a detection limit of 10 ng/mL. The recoveries were 91.3–106.5% in the tablets with an RSD ($n=5$) of 0.8%, and 93.9–105.2% in the blood serum.

Navalon *et al.* developed a method for the determination of trace amounts of ciprofloxacin, based on solid-phase spectrofluorimetry [9]. The relative fluorescence intensity of ciprofloxacin fixed on Sephadex SP C-25 gel was measured directly after packing the gel beads in a 1 mm silica cell, using a solid-phase attachment. The wavelengths of excitation and emission were 272 and 448 nm, respectively. The linear concentration range of the compound was 0.3–10 ng/mL, with a RSD of 1.2% (for a level of 4 ng/mL) and a detection limit of 0.1 ng/mL. This method was used for the determination of ciprofloxacin in human urine and serum samples with a recovery of 100% in all cases.

Amin described a simple and sensitive spectrophotometric method for the determination of three gyrase inhibitors, including ciprofloxacin, in pharmaceutical formulations [10]. This method is based on the formation of an ion pair with Sudan III in 40% v/v aqueous acetone. The color of the dye changes at 566 nm in the presence of ciprofloxacin. Beer's law is obeyed over the range of 0.4–10.4 μg/mL, and the results showed good recovery ($100 \pm 1.7\%$) with a RSD of 1.08%.

An analytical procedure for the determination of ciprofloxacin in serum without previous extraction has been developed by Djurdjevic *et al.* [11]. The determination was carried out at pH 3, using iron(III) nitrate as the chromogenic agent with the addition of sodium dodecyl sulfate. Absorbance values were measured at 430 nm. The linearity range was between 0.5 and 20 μg/mL, with a detection limit of 0.2 μg/mL.

Five drugs used as antibacterial agents, including ciprofloxacin, were measured spectrophotometrically by Avadhanulu *et al.* [12]. Ciprofloxacin was determined in the tablets after extraction with aqueous 50% acetone on the basis of its absorbance at 524 nm. The recovery and RSD for ciprofloxacin was 99.3–100.2% and 0.3–0.55%, respectively.

Abdel-Gawad *et al.* described a spectrophotometric method for the determination of ciprofloxacin in its pure form and in its tablet formulations that made use of charge-transfer complexation reactions [13]. In this method, either 5 mM 2,3-dichloro-5,6-dicyano-*p*-benzoquinone (**I**), 5 mM 7,7,8,8-tetracyanoquinodimethane (**II**), or 5 mM *p*-chlorail (**III**) was added to a methanolic solution of ciprofloxacin. After 15 min at 25°C for (**I**), or at 60°C for (**II**) and (**III**), the solution was diluted to 10 mL with acetonitrile, and the absorbance was measured at 460, 843, or 550 nm for (**I**), (**II**), and (**III**), respectively. Beer's law was obeyed over the concentration range of 5–50 μg/mL of ciprofloxacin, with a RSD of 1.4% ($n=7$).

The effect of pH on the fluorescence of ciprofloxacin and norfloxacin has been studied, and the ground-state microscopic dissociation constants were determined [14]. After their acidic solutions (pH 2–4) were allowed to stand for 5 min at 25°C, the optimum wavelength for fluorescence determinations was found to be 445 nm (excitation at 280 or 331 nm) of these compounds.

Ciprofloxacin was determined in tablets and intravenous infusion solutions by an oxidative spectrophotometric method [15]. A yellow–orange complex was obtained upon reaction of ciprofloxacin with 5% ammonium sulfamate, 5 N H_2SO_4, and 0.1 M cerium(IV) ammonium sulfate. The complex was extracted into $CHCl_3$, dried with anhydrous Na_2SO_4, and measured at 345 nm within up to 4 h. Beer's law was obeyed over the range of 12–120 μg/mL, and the molar absorptivity was found to be 5090.

2.3 Chemiluminescence methods

Aly *et al.* [16] reported a rapid and sensitive chemiluminescence (CL) method for the determination of three fluoroquinolone derivatives, including ciprofloxacin, in both pharmaceutical dosage forms and in biological fluids. The method is based on the CL reaction of the drugs with tris(2,2′-bipyridyl)ruthenium(II) and cerium(IV) in sulfuric acid medium. The CL intensity was proportional to the concentration of ciprofloxacin in solution over the range 0.05–6 μg/mL, and the limit of detection was reported to be 26 nM.

A flow-injection CL method was developed by Rao *et al.* for the determination of fluoroquinolones, including ciprofloxacin, based on the

CL reaction of sulfite with cerium(IV) [17]. The linear range for ciprofloxacin-CL is 0.4–30 μg/mL, with a RSD of 2.1–2.6% ($n = 10$).

Ciprofloxacin hydrochloride was determined in capsules and tablets using 5 mM Na_2SO_3 in a flow-injection chemiluminescence system [18]. The system was eluted at a flow rate of 6 mL/min, with a carrier/reagent stream consisting of 0.4 mM $CeSO_4 \cdot 2(NH_4)_2SO_4$ containing 50 mM H_2SO_4. The CL signal was measured in an on-line flow cell. The calibration graph was linear over the range of 1–20 mg/L of ciprofloxacin, with a detection limit of 0.27 mg/L and a method recovery of 96–108% [18].

2.4 Electrochemical methods

The voltammetric behavior of the Ni(II) complexes of three 4-quinolone antibacterial compounds in tablets, including ciprofloxacin, was studied using direct current (DCt), differential pulse (DPP) and alternating current polarography (ACt) [19]. A well-defined, cathodic wave was detected at pH 5.5 for ciprofloxacin. The current–concentration relationship was found to be linear over the range $2\text{–}6.4 \times 10^{-5}$ M and $0.8\text{–}5.6 \times 10^{-5}$ M for ciprofloxacin using DCt and DPP modes, respectively. Limits of detection (S/N = 2) for these species were found to be about 2×10^{-7} M. The average percent recovery for ciprofloxacin was 99.58 ± 0.72 to 100.50 ± 0.79, and 99.50 ± 0.71 to 100.17 ± 0.29, when using DCt and DPP, respectively.

Oscillopolarography was used to determine the ciprofloxacin content in tablets [20]. The method was performed with measurement of the second-derivative peak height at −1.16 V. The calibration graph for the drug was found to be linear from 0.985 to 19.7 μM, with recoveries of 95.8–101.4%.

Ciprofloxacin hydrochloride, in pharmaceutical raw material and capsules, has been analyzed by oscillopolarography [21]. A sample that was mixed with 2 mL of 0.5 M KH_2PO_4/KOH buffer (pH 6.9), and diluted with water, yielded a reductive peak potential at −1.51 V vs. SCE. The calibration graph was linear from 0.1 to 20 μM, with an average recovery of 101.1% and a RSD of 1.14%.

2.5 Thin-layer chromatography

Mixtures of antibacterial agents (including ciprofloxacin hydrochloride) together with *p*-aminophenol (internal standard) in 0.1 M HCl were analyzed by thin-layer chromatography (TLC) [22]. The system used a silica gel GF254 plate, and a mobile phase consisting of chloroform/methanol/concentrated ammonium hydroxide (15:10:3). The chromatogram was examined under a UV lamp at 254 nm, and it was found that all spots were cleanly separated.

Quinolones (including ciprofloxacin) have been shown to be separable by TLC using thin layers of Diol-silica adsorbent, and solutions of di-(2-ethylhexyl)-orthophosphoric acid (HDEHP, an ion-pairing reagent) in polar solvents as the mobile phase [23]. Retention and selectivity in the adsorption-ion-association system can be controlled by adjusting the concentration of HDEHP (typically 5–10%), or by changing the polar diluents.

Wang *et al.* reported the simultaneous determination of trace amounts of ciprofloxacin, pefloxacin, and norfloxacin in serum and urine by TLC-fluorescence spectrodensitometry [24]. The drug substances were applied to the silica gel plates, and developed to 9 cm with $CHCl_3$/methanol/toluene/CH_2Cl_2/aqueous NH_3 (27:46:17:5:5). Fluorescence spots were observed upon excitation at 254 nm, while the fluorescence spectrodensitometric measurements were performed at 400 nm (excitation at 278 nm). The calibration graph was linear upto 75 ng/spot for ciprofloxacin, with an RSD that was less than 8.6%. The recovery was 96–108%.

The degradation of ciprofloxacin hydrochloride solution after irradiation by a high pressure mercury lamp was studied using TLC [25]. Samples of solution were taken and analyzed on 10×10 cm silica gel 60F254 HPTLC plates, after applying samples 5 mm apart at opposite edges of the plate and eluting in a horizontal HPTLC developing chamber. The plate was developed from opposite edges with acetonitrile/10% ammonia containing 0.3 M ammonium chloride (13:5). Plates were scanned in absorbance/reflectance mode at 283 nm. The working range was 10–100 ng/spot, and the RSD values ($n = 11$) were 2.4 and 1.5% for the 20 and the 80 ng/spot, respectively.

2.6 High performance liquid chromatographic methods

Numerous high performance liquid chromatography (HPLC) methods have been reported, and the salient features of these are summarized in Table 1.

Table 1

Reported HPLC Methods for Ciprofloxacin

No.	Material	Column	Mobile phase	Internal standard	Flow rate	Detection wavelength	Ref.
1	Raw material	Waters C18 (5 μm) 15 cm × 4.6 mm i.d.	Phosphate buffer –0.98 mg/mL sodium heptane sulfonate pH 2.4/ methanol (13:7)	*p*-Aminobenzoic acid	1 mL/min	277 nm	22
2	Human plasma	LiChrospher 60 RP (5 μm)	5% Acetic acid/ methanol/acetoni-trile (18:1:1)	Lomefloxacin	NA	280 nm	26
3	Influenza vaccine	Purospher RP-18e (5 μm) 12 cm × 3.5 mm i.d.	Acetonitrile/water/ phosphoric acid (85%) (60:340:1)	NA	0.6 mL/min	280 nm	27
4	Ointment	Water Nova-pak C18 (10 μm) 15 cm × 3.9 mm i.d.	50 mM citric acid/ acetonitrile (82:16) pH 3.5	NA	1 mL/min	277 nm	28

5	Urine	Micro Bondapak-NH_2 (30 cm × 3.9 mm i.d.)	Methanol/ethyl acetate/water/acetic acid (50:10:40:1)	Lomefloxacin	0.7 mL/min	285 nm	29
6	Plasma and Mueller-Hinton broth	Kromasil C8 (5 µm) 15 cm × 4.6 mm i.d.	Acetonitrile/20 mM- KH_2PO_4 (pH 3.5) (3:97)	NA	1 mL/min	278/418 nm	30
7	Dosage forms	Purosher-RP-18e (12.5 cm × 4 mm i.d.)	Methanol/20 mM-H_3PO_4 (1:3)	NA	0.8 mL/min	220 nm	31
8	Dosage forms	LiChrospher 100 RP-18 (5 µm) 25 cm × 4.6 mm i.d.	100 mM-tertrabutylammonium hydroxide in acetonitrile/water (93:7)/25 mM-H_3PO_4 in acetonitrile/water (93/7)	NA	1 mL/min	444/491 nm	32

(*continued*)

Table 1 (continued)

No.	Material	Column	Mobile phase	Internal standard	Flow rate	Detection wave-length	Ref.
9	Plasma and urine	YMC pack A-3-2 (5 μm) 15 cm × 6 mm)	5% Acetic acid/ acetonitrile/ methanol (9:5:5)	Lomefloxacin	1 mL/min	280 nm	33
10	Human serum	Lichrosorb RP-18 (5 μm) 25 cm × 4 mm i.d.	Acetonitrile/metha-nol/0.4 M citric acid (1:3:6)	Theophylline	NA	275 nm	34
11	Tablets and capsules	Nacalai Cosmosil 5C18 MS (25 cm × 4.6 mm i.d.)	Methanolic 3 mM SDS/40 mM-phosphate buffer/ acetonitrile (5:11:4) pH 3.5	Piroxicam	0.8 mL/ min	257 nm	35
12	Human aqueous humor	Novapak C18 car-tridge (4 μm) 10 cm × 8 mm i.d.	Methanol/acetoni-trile/0.4 M citric acid (3:1:10)	Pipemidic acid	1 mL/min	278/450 nm	36

13	Raw material	LiChrospher 100 C18 (5 μm) 25 cm × 4 mm i.d.	0.1 M-tetrabutylammonium bromide in 25 mM H_3PO_4 of pH 3.89/acetonitrile (93:7)	NA	NA	278 nm	37
14	Impurities	RP-C-18 (10 μm) 30 cm × 3.5 mm i.d.	Methanol/water/acetic acid (840:159:1)	NA	1 mL/min	254 nm	38
15	Dosage forms	Shimpack CLC-ODS (15 cm × 6 mm i.d.)	Tetrabutylammonium hydroxide buffer/acetonitrile (9:1)	NA	1 mL/min	280 nm	39
16	Clinical specimens	Spheri-5-OD-5A (25 cm × 4.6 mm)	0.1 M-KH_2PO_4/acetonitrile (1:1)	NA	1 mL/min	280 nm	40
17	Plasma, whole blood and erythrocytes	Micro Bondapak C18 (10 μm) 30 cm × 3.9 mm i.d.)	0.1 M H_3PO_4-45% KOH (pH 2.5) mixed 3:1 with acetonitrile	Difloxacin	1 mL/min	278/470 nm	41

(continued)

Table 1 (continued)

No.	Material	Column	Mobile phase	Internal standard	Flow rate	Detection wave-length	Ref.
18	Human plasma	MB C18 radial pak (10 µm) 10 cm × 8 mm i.d.	0.1 M-$(NH_4)_2HPO_4$ (pH 2.5)/acetonitrile/methanol (80:13:7)	Quinine bisulfate	4 mL/min	277/453 nm	42
19	Tablets	Micro Bondapak C18 (10 µm) 30 cm × 3.9 mm i.d.	Acetonitrile/methanol/acetic acid/10 mM KH_2PO_4 (15:12:0.3:73)	Cephalothin	2 mL/min	NA	43
20	Dosage forms	Micro Bondapak C18 (10 µm) 30 cm × 3.9 mm i.d.	Methanol/water/triethylamine (250:250:1) pH 2.4	Nalidixic acid	1.8 mL/min	254 nm	44
21	Serum and pro-static	Phenyl Spherisorb (3 µm) 10 cm × 4.6 mm i.d.	Water/acetonitrile/methanol (83:15:2)	NA	1 mL/min	178.6 nm	45

22	Clinical specimens	Ultropac LiChrosorb RP-18 (10 µm) 25 cm × 4 mm)	Acetonitrile/0.4 M-citric acid (1:5)	Pipemidic acid	1 mL/min	340 or 275 nm	46
23	Serum, urine and sputum	Micro-Pak MCH 10 (10 µm) 30 cm × 4 mm i.d.	Acetonitrile/0.1 M-KH_2PO_4 (pH 2.5; 19:81)	NA	0.8 mL/min	NA	47
24	Serum and urine	Micro Bondapak C18 (10 µm) 30 cm × 3.9 mm i.d.	14% Methanol/5%acetonitrile/0.3%tetrabutylammonium hydroxide (pH 3)	1-(4-fluorophenyl)-7-(4-methyl piperazin-1-yl)	1.7 mL/min	278/470 nm	48
25	Body fluid	Poly(styrene-divinylbenzene) (5 µm) 15 cm × 4.6 mm i.d.	Acetonitrile/methanol/0.02 M trichloroacetic acid	NA	0.5 mL/min	277 nm	49
26	Biological fluids	Spherisorb ODS II (5 µm) 25 cm × 4 mm	0.025 M H_3PO_4 (pH 3)/acetonitrile (19:1)	NA	2 mL/min	277/445 nm	50

(continued)

Table 1 (continued)

No.	Material	Column	Mobile phase	Internal standard	Flow rate	Detection wave-length	Ref.
27	Dosage forms and bulk drug	LiChrospher C18 (5 μm) 25 cm × 4 mm i.d.	Water/acetonitrile/triethylamine (80:20:0.6) pH 3 with orthophosphoric acid	NA	1.5 mL/min	280 nm	51
28	Stability in tablets	LiChrospher C18 (5 μm) 25 cm × 4 mm i.d.	Methanol/0.245% H_3PO_4 (3:22) pH 3. or aqueous acetonitrile pH 2.5	NA	1.5 mL/min	278 nm	52

29	Biological material	Inertsil OSD-2 (5 µm) 25 cm × 4.6 mm i.d.	50% Aqueous methanol of pH2.5	Nalidixic acid	0.8 mL/min	320/545 nm	53
30	Human serum	Nucleosil C18 (3 µm) 3 cm × 4.6 mm i.d.	50 mM KH_2PO_4 (pH3)/water/0.1 M tetrabutylammonium bromide (pH 3)/acetonitrile (12:6:1:1)	Eprofloxacin	2 mL/min	277 nm	54

NA: Not available.

2.7 Gas chromatography

Gas chromatography was used to analyze the residual piperazine of ciprofloxacin and norfloxacin in pharmaceutical preparations [55]. The analyte was dissolved in water, and the solution analyzed by GC on a fused-silica column (30 m × 0.3 mm i.d.) coated with 5% cross-linked ph-Me silicone (3 μm). The temperature program consisted of an increase to 50°C, holding for 5 min, ramping to 180°C (ramp = 15°C/min), and finally holding for 10 min at 180 5°C/min. Nitrogen was used as the carrier gas (flow rate of 1 mL/min), detection was effected using flame ionization. Piperazine was determined in the two mixtures by mixing for 10 min with cyclohexane, filtering, partitioning the filtrate with water, heating the aqueous phase at 45°C for 10 min, and then analyzing as above. The calibration graph was linear for 0.4–10 ppm piperazine, with a RSD ($n = 10$) of 2.1% for 2 μg piperazine. The recovery from ciprofloxacin was 98.1–98.4%.

2.8 Capillary electrophoresis

Six quinolone antibiotics (including ciprofloxacin) were separated and determined by CE on fused-silica capillaries (57 cm × 75 μm i.d.; 50 cm to detector) at 25°C, with injection on the anode side, an applied voltage of 10 kV, and detection at 280 nm [56]. The buffer was 100 mM HEPES/acetonitrile (9:1). The calibration graphs were linear from 0.25 to 40 μg/mL and detection limits were approximately 0.25 ng/mL.

Sun and Wu described a capillary electrophoresis method for the determination of seven fluoroquinolones, including ciprofloxacin, in pharmaceutical formulations [57]. Pipemidic acid (10 mg) was added as an internal standard to 25 mg active ingredient dissolved in 100 mL of methanol. The clear solutions (4.1 nL) were hydrodynamically injected at 60 mbar for 3 s on fused silica capillaries (67 cm × 50 μm i.d.; effective length = 52 cm). The compounds were well separated in less than 8.5 min when using 65 mM sodium borate/35 mM sodium dihydrogen phosphate/60 mM sodium cholate of pH 7.3 in acetonitrile (72:28) as the running buffer, an applied voltage of +27 kV, and detection at 275 nm. The calibration graphs were linear from 25 to 300 μg/mL of the fluoroquinolones, with recoveries within 95–105% of their label claims.

A mixture of four fluoroquinolones (including ciprofloxacin hydrochloride) was determined by high performance capillary electrophoresis using caffeine as an internal standard [58]. A portion of the solution was introduced by pressure-differential sampling at 10 cm for 10 s, whereupon the drugs were separated and determined by high-performance CE. The method used a fused-silica capillary column (60 cm × 0.75 μm i.d.; effective length 55 cm), operated at an applied voltage of 18 kV, with 50 mM $Na_2B_4O_7/NaH_2PO_4$ buffer of pH 8.5 as the running buffer, and detection at 214 nm. The calibration graph for ciprofloxacin was linear from 20 to 100 mg/L, and the detection limits and RSD were 0.40–0.48 mg/L and less than 2%, respectively.

Ciprofloxacin and its metabolite desethyleneciprofloxacin were determined in human plasma using capillary electrophoresis in the presence of *N*-(1-naphthyl)ethylenediamine dihdrochloride as the internal standard [59]. The sample was injected hydrodynamically into the capillary (37 cm × 50 μm i.d.), and the running buffer was 0.1 M H_3PO_4/0.1 M H_3BO_3 adjusted to pH 2.3 with triethylamine. The applied voltage was 28 kV, and laser-induced fluorescence detection at 450 nm (excitation at 325 nm) was used. The limits of quantitation were 20 and 10 μg/L for ciprofloxacin and its metabolite (desethyleneciprofloxacin), respectively, with a RSD ($n = 6$) being less than 9%.

Fourteen quinolone antibacterial compounds (including ciprofloxacin) were separated by fused-silica column (59 cm × 50 μm i.d.; 43 cm to detector) at 30 kV in less than 8 min and monitored at 260 nm [60]. In this method, a background electrolyte of pH 7.3 containing 32 mM borate, 39 mM cholate, 8 mM heptanesulfonate, and 18 mM phosphate (as the sodium salts), modified with 28% acetonitrile, was used as running buffer.

Ciprofloxacin was separated from its impurities and determined by free-solution capillary electrophoresis in a fused-silica tube (72 cm × 50 μm i.d.) operated at 20 kV and 30°C with a 50 mM borate buffer acidified with H_3PO_4 to pH 1.5 and detection at 272 nm [61]. The linear calibration range was 5 μg/mL–0.75 mg/mL of ciprofloxacin, and the elution time was 14.8 min. The limit of determination was 1.6 μMm and the RSD was 4.9% ($n = 10$). Six impurities were detected.

2.9 Immunoassay methods

A high performance immunoaffinity chromatography (HPIAC) column, containing covalently bound anti-sarafloxacin antibodies, was developed to determined fluoroquinolones in bovine serum [62]. In this method, the HPIAC column was used to capture ciprofloxacin with other fluoroquinolones, while allowing the remainder of the serum components to elute to waste. After binding to HPIAC column, the fluoroquinolones were eluted directly onto a reversed-phase column for final separation of the compounds prior to their fluorescence detection (observation at 444 nm after excitation at 280 nm). The recovery of the method exceeded 95%, and its RSD was less than 7%.

Snitkoff *et al.* reported a specific immunoassay for monitoring the levels of ciprofloxacin in human samples [63]. Serum was incubated at 37°C with shaking for 2 h, with rabbit anti-ciprofloxacin keyhole limpet haemocyanin primary antibody in microtiter wells coated with ciprofloxacin-BSA conjugate. Further incubation with horseradish peroxidase-labeled goat anti-rabbit IgG was carried out. The samples were shaken in 0.1 M citric buffer (pH 4.5), plates were incubated for 10 min, and then read at 450 nm. Calibration graphs were linear for 10 pg/mL to 10 ng/mL of ciprofloxacin, and the within-day RSD was less than 10%.

Holtzapple *et al.* developed an immunoassay method for determination of four fluoroquinolone compounds (including ciprofloxacin) in liver extracts [64]. In this method, an immunoaffinity capture SPE column was used, that contained anti-sarafloxacin antibodies covalently cross-linked to protein G. After interfering liver matrix compounds had been washed away, the bound ciprofloxacin was eluted directly onto the HPLC column. The HPLC system used a 5 μm Inertsil phenyl column (15 cm × 4.6 mm i.d.), with 0.1 M-glycine hydrochloride/acetonitrile (17:3) as the mobile phase (eluted at a rate of 0.7 mL/min). Fluorimetric detection at 444 nm was used after excitation at 280 nm. The recovery of ciprofloxacin ranged from 85.7 to 93.5%, and the detection limit was 0.47 ng/mL.

3. REFERENCES

1. ***European Pharmacopoeia***, 4th edn., Council of Europe, Strasbourg, p. 926 (2002).

2. ***British Pharmacopoeia***, 16th edn., Volume I, on line, HMSO Publication, Ltd., London, CD (2000).

3. ***United States Pharmacopoeia***, 23rd edn., United States Pharmacopoeial Convention, Inc., Rockville, MD, USA, p. 375 (1995).

4. F. Belal, M. Rizk, F.A. Aly, and N.M. El-Enany, *Chem. Anal.*, ***44***, 763 (1999).

5. S.E. Zhang, Z.X. Sun, and Z.L. Sun, *Yaown-Fenxi-Zazhi.*, ***16***, 402 (1996).
6. E. Kilic, F. Koseoglu, and M.A. Akay, *J. Pharm. Biomed. Anal.*, ***12***, 347 (1994).
7. M. Rizk, F. Belal, F. Ibrahim, S. Ahmed, and Z.A. Sheribah, *J. AOAC. Int.*, ***84***, 368 (2001).
8. S.Q. Wu, Q.E. Cao, Y.K. Zhao, W.J. Zhang, X.G. Chen, and Z.D. Hu, *Fenxi. Huaxue*, ***28***, 1462 (2000).
9. A. Navalon, O. Ballesteros, R. Blanc, and J.L. Vilchez, *Talanta*, ***52***, 845 (2000).
10. A.S. Amin, *Mikrochim. Acta*, ***134***, 89 (2000).
11. P. Djurdjevic, M. Todorovic, M.J. Stankov, and J. Odovic, *Anal. Lett.*, ***33***, 657 (2000).
12. A.B. Avadhanulu, Y.R.R. Mohan, J.S. Srinivas, and Y. Anjaneyulu, *Indian Drugs*, ***36***, 296 (1999).
13. F.M. Abdel-Gawad, Y.M. Issa, H.M. Fahmy, and H.M. Hussein, *Mikrochim. Acta*, ***130***, 35 (1998).
14. Z.Y. Huang, H.P. Huang, R.X. Cai, Z.X. Lin, T. Korenaga, and Y.E. Zeng, *Anal. Sci.*, ***13*** (suppplement), 77 (1997).
15. P.V. Bharat, G. Rajani, and S. Vanita, *Indian Drugs*, ***34***, 497 (1997).
16. F.A. Aly, S.A. Al-Tamimi, and A.A. Alwarthan, *Talanta*, ***53***, 885 (2001).
17. Y. Rao, Y. Tong, X.R. Zhang, G.A. Luo, and W.R.G. Baeyens, *Anal. Lett.*, ***33***, 1117 (2000).
18. Y.D. Liang, J.Z. Li, and Z.J. Zhang, *Fenxi. Huaxue.*, ***25***, 1307 (1997).
19. M.S. Rizk, F. Belal, F.A. Ibrahim, S.M. Ahked, and Z.A. Sheribah, *Electroanalysis (NY)*, ***12***, 531 (2000).
20. J.C. Zhao, R.X. Shi, X.F. Kang, J.F. Song, and Z.A. Guo, *Fenix. Shiyanshi.*, ***18***, 1 (1999).
21. J.W. Di and M. Jin, *Fenix. Shiyanshi.*, ***14***, 33 (1995).
22. Y.J. Niu and S.X. Zhang, *Yaown-Fenxi-Zazhi.*, ***21***, 204 (2001).
23. E. Soczewinski and M. Wojciak-Kosior, *J. Planar. Chromatogr. Mod. TLC.*, ***14***, 28 (2001).
24. P.L. Wang, Y.L. Feng, and L.A. Chen, *Microchem. J.*, ***56***, 229 (1997).
25. S. Tammilehto, H. Salomies, and K. Torniainen, *J. Planar. Chromatogr. Mod. TLC*, *7*, 368 (1994).
26. M.T. Maya, N.J. Goncalves, N.B. Silva, and J.A. Morais, *J. Chromatogr. B: Biomed.*, ***755***, 305 (2001).

27. P. Forlay-Frick, Z.B. Nagy, and J. Fekete, *J. Liq. Chromatogr. Relat. Technol.*, ***24***, 827 (2001).
28. K.J. Wang, S.Y. Cai, and H.S. Liu, *Yaown-Fenxi-Zazhi.*, ***21***, 43 (2001).
29. D.N. Tipre and A.V. Kasture, *Indian Drugs*, ***37***, 148 (2000).
30. B.B. Ba, D. Ducint, M. Fourtillan, and M.C. Saux, *J. Chromatogr. B: Biomed.*, ***714***, 317 (1998).
31. G. Battermann, K. Cabrera, S. Heizenroeder, and D. Lubda, *LaborPraxis*, ***22***, 30 (1998).
32. J.A. Hernandez-Arteseros, J. Barbosa, R. Compano, and M.D. Prat, *Chromatographia*, ***48***, 251 (1998).
33. M. Kamberi, K. Tsutsmi, T. Kotegawa, K. Nakamura, and S. Nakano, *Clin. Chem.*, ***44***, 1251 (1998).
34. I.N. Papadoyannis, V.F. Samanidou, and K.A. Georga, *Anal. Lett.*, ***31***, 1717 (1998).
35. Y.P. Chen, C.Y. Shaw, and B.L. Chang, *Yaowu. Shipin. Fenxi.*, ***4***, 155 (1996).
36. N.E. Bssci, A. Bozkurt, D. Kalayci, and S.O. Kayaalp, *J. Pharm. Biomed. Anal.*, ***14***, 353 (1996).
37. J. Barbosa, R. Berges, and V. Sanz-Nebot, *J. Chromatogr. A.*, ***719***, 27 (1996).
38. S. Husain, S. Khalid, V. Nagaraju, and R. Nageswara-Rao, *J. Chromatogr. A.*, ***705***, 380 (1995).
39. R. Jain, and C.L. Jain, *Liquid Chromatography and Gas Chromatography*, ***10***, 707 (1992).
40. G. Mack, *J. Chromatogr. Biomed. Appl.*, ***120***, 263 (1992).
41. P. Teja. Isavadharm, D. Keeratithakul, G. Watt, H.K. Webster, and M.D. Edsterin, *Ther. Drug. Monit.*, ***13***, 263 (1991).
42. A. El-Yazigi and S. Al-Rawithy, *Ther. Drug. Monit.*, ***12***, 378 (1990).
43. J. Parasrampuria and V. Das-Gupta, *Drug. Dev. Ind. Pharm.*, ***16***, 1597 (1990).
44. R.T. Sane, D.V. Patel, S.N. Dhumal, V.R. Nerukar, P.S. Mainkar, and D.P. Gangal, *Indian Drugs.*, ***27***, 248 (1990).
45. K. Tyczkowska, K.M. Hedeen, D.P. Aucoin, and A.L. Aronson, *J. Chromatogr. Biomed. Appl.*, ***85***, 337 (1989).
46. C.Y. Chan, A.W. Lam, and G.L. French, *J. Antimicrob. Chemother.*, ***23***, 597 (1989).
47. C.M. Myers and J.L. Blumer, *J. Chromatogr. Biomed. Appl.*, ***66***, 153 (1987).
48. W.M. Awni, J. Clarkson, and D.R.P. Guay, *J. Chromatogr. Biomed. Appl.*, ***63***, 414 (1987).

49. G.J. Krol, A.J. Noe, and D. Beermann, *J. Liq. Chromatogr.*, ***9***, 2897 (1986).
50. W. Gau, H.J. Ploschke, K. Schmidt, and B. Weber, *J. Liq. Chromatogr.*, ***8***, 485 (1985).
51. S.O. Thoppil and P.D. Amin, *J. Pharm. Biomed. Anal.*, ***22***, 699 (2000).
52. F. Hudrea, C. Grosset, J. Alary, and M. Bojita, *Biomed. Chromatogr.*, ***14***, 17 (2000).
53. A. Rieutord, L. Vazquez, M. Soursac, P. Prognon, J. Blais, P. Bourget, and G. Mahuzier, *Anal. Chim. Acta.*, ***290***, 215 (1994).
54. L. Pou-Clave, F. Campos-Barreda, and C. Pascual-Mostaza, *J. Chromatogr. Biomed. Appl.*, ***101***, 211 (1991).
55. K.N. Ramachandran and G.S. Kumar, *Talanta*, ***43***, 1269 (1996).
56. T. Perez-Ruiz, C. Martinez-Lozano, A. Sanz, and E. Bravo, *Chromatographia.*, ***49***, 419 (1999).
57. S.W. Sun and A.C. Wu, *J. Liq. Chromatogr. Relat. Technol.*, ***22***, 281 (1999).
58. Yin and Y.T. Wu, *Yaown. Fenxi. Zazhi.*, ***17***, 371 (1997).
59. K.H. Bannefeld, H. Stass, and G. Blaschke, *J. Chromatogr. Biomed. Appl.*, ***692***, 453 (1997).
60. S.W. Sun and L.Y. Chen, *J. Chromatogr. A.*, ***766***, 215 (1997).
61. K.D. Altria and Y.L. Chanter, *J. Chromatogr. A.*, ***652***, 459 (1993).
62. C.K. Holtzapple, S.A. Buckley, and L.H. Stanker, *J. Chromatogr. B: Biomed. Appl.*, ***754***, 1 (2001).
63. G.G. Snitkoff, D.W. Grabe, R. Holt, and G.R. Bailie, *J. Immunoassay.*, ***19***, 227 (1998).
64. C.K. Holtzapple, S.A. Buckley, and L.H. Stanker, *J. Agric. Food. Chem.*, ***47***, 2963 (1999).

Ciprofloxacin: Drug Metabolism and Pharmacokinetic Profile

Mohammed A. Al-Omar

Department of Pharmaceutical Chemistry
College of Pharmacy, King Saud University
P.O. Box 2457, Riyadh-11451
Kingdom of Saudi Arabia

PROFILES OF DRUG SUBSTANCES,
EXCIPIENTS, AND RELATED
METHODOLOGY – VOLUME 31
DOI: 10.1016/S0000-0000(00)00000-0

CONTENTS

1. **Uses, Applications, and Associated History** 210
2. **Absorption and Bioavailability** 211
3. **Metabolism** 211
4. **Excretion** 212
5. **References** 213

1. USES, APPLICATIONS, AND ASSOCIATED HISTORY

Since 1980, many effective antimicrobial drugs of the synthetic fluoroquinolone series have been developed for the treatment of bacterial infections in humans. Ciprofloxacin is one of the most widely used drugs of this group, and exhibits bactericidal effects by inhibition of DNA gyrase [1]. The drug is structurally related to other quinolones, including cinoxacin, enoxacin, lomefloxacin, nalidixic acid, norfloxacin, and pefloxacin. This family of compounds contains a 4-quinolone nucleus with a nitrogen at position-1, a carboxyl group at position-3, and a ketone moiety at position-4. Ciprofloxacin is a fluoroquinolone carboxylic acid, since it contains a fluorine atom at position-6, and a carboxylic acid moiety at position-3 of the 4-quinolone nucleus. Fluorinated quinolones have a wider spectrum of activity and more potency as compared to non-fluorinated quinolone [2]. The piperazine group at position-7 of the 4-quinolone nucleus is responsible for the antipseudomonal activity of ciprofloxacin. The drug also contains a cyclopropyl group at position-1, which enhances its antimicrobial activity.

Ciprofloxacin is commercially available as the monohydrate phase of its hydrochloride salt, and has been formulated for oral and ophthalmic administration. The compound has also been developed as the lactate salt for use in intravenous administration [3].

Ciprofloxacin is approved for use in the treatment of bone and joint infections, infectious diarrhea caused by *Shigella* or *Campylobacter*, lower respiratory tract infections, skin infections, and urinary tract infections. It is the drug of choice for the treatment of infections caused by *Campylobacter jejuni*. In addition, it has found off-label use as an alternative drug for the treatment of gonorrhea, salmonella, and yersinia infections [3, 4]. In general, ciprofloxacin is active against susceptible gram-negative and gram-positive aerobic bacteria, so therefore it should not be used alone for mixed aerobic–anaerobic bacterial infections [4, 5].

2. ABSORPTION AND BIOAVAILABILITY

All formulations of ciprofloxacin are rapidly and well absorbed from the GI tract following oral administration, and the compound undergoes minimal first-pass metabolism [3]. The presence of food in the GI tract decreases the rate, but not the extent, of drug absorption. Antacids containing magnesium, aluminum, and/or calcium, decrease the oral bioavailability of ciprofloxacin [6]. The oral bioavailability of ciprofloxacin is 50–85%, and peak serum concentrations of the drug are generally attained within 0.5–2.3 h. Peak serum concentrations, and areas under serum concentration-time curves (AUC), are slightly higher in geriatric patients than in younger adults. In adults who receive a single 200 mg dose of ciprofloxacin by intravenous (IV) injection over 10 min, serum concentrations of the drug immediately following the injection average 6.3–6.5 μg/mL, and serum concentrations 1 and 2 h later average 0.87 and 0.1 μg/mL, respectively [4].

Ciprofloxacin is widely distributed into body tissues and fluids following oral and IV administration. Highest concentrations of the drug generally are found in bile, lung, kidney, liver, gall bladder, uterus, seminal fluid, prostatic tissue and fluid, tonsils, endometrium, fallopian tubes, and ovaries [4, 7, 8]. It also distributed into bone, aqueous humor, sputum, saliva, nasal secretions, skin, muscle, adipose tissue, and cartilage [4, 7, 9].

The apparent volume of distribution of ciprofloxacin is 2–3.5 L/kg, and the apparent volume of distribution at steady state is 1.7–2.7 L/kg. Only low concentrations of ciprofloxacin are distributed into cerebrospinal fluid (CSF). Peak CSF concentrations may be 6–10% of peak serum concentrations. The drug is moderately bound to serum protein (16–43%), and crosses the placenta and is produced with the mammary milk [6, 10].

3. METABOLISM

At least four active metabolites have been identified, namely desethylene-ciprofloxacin, sulfo-ciprofloxacin, oxo-ciprofloxacin, and *N*-acetyl-ciprofloxacin. Oxo-ciprofloxacin appears to be the major urinary metabolite, and sulfo-ciprofloxacin the primary fecal metabolite [4, 11].

Carolyn *et al.* have reported two additional ultraviolet-absorbing metabolites in urine specimens [12]. The drug is partially metabolized in the liver by modification of the piperazinyl group. Oxo-ciprofloxacin and *N*-acetyl-ciprofloxacin microbial activities are comparable to norfloxacin, and desethylene-ciprofloxacin is comparable to nalidixic acid for certain organisms. About 40–50% of an oral dose is excreted unchanged in the urine, and 15% as metabolites. Upto 70% of a parental dose may be excreted unchanged, and 10% as metabolites within 24 h [13, 14].

Ciprofloxacin and its metabolite desethylene-ciprofloxacin were determined in human plasma using capillary electrophoresis in the presence of N-(1-naphthyl) ethylenediamine dihydrochloride as an internal standard. Krol *et al.* [15] have thoroughly investigated the biotransformation of ciprofloxacin in body fluids using high performance liquid chromatography (HPLC) and proposed the sequence of metabolic pathways illustrated in Figure 1.

4. EXCRETION

Ciprofloxacin is eliminated by renal and non-renal mechanisms. The drug is partially metabolized in the liver by modification of the piperazinyl group to atleast four metabolites. These metabolites, which have been identified as desethylene-ciprofloxacin, sulfo-ciprofloxacin, oxo-ciprofloxacin, and *N*-acetyl-ciprofloxacin, have microbiological activities that are less than that of the parent drug, but may be similar to or greater than that of some other quinolones [4, 8, 9].

Ciprofloxacin and its metabolites are excreted in urine by both glomerular filtration and by tubular secretion. Following oral administration of a single 250-, 500-, or 750-mg dose in adults with normal renal function, 15–50% of the dose is excreted in urine as unchanged drug, and 10–15% as metabolites within 24 h. 20–40% of the dose is excreted in feces as the unchanged drug and metabolites within 5 days [4, 5, 16, 17]. Most, but not all, of the unchanged ciprofloxacin in feces appears to result from biliary excretion [4, 5].

Renal clearance of ciprofloxacin averages 300–479 mL/min in adults with normal renal function, and the drug is 16–43% bound to serum protein *in vitro*. It crosses the placenta and is distributed into the amniotic fluid in humans. The usual human dosage has not revealed evidence of harm to the fetus.

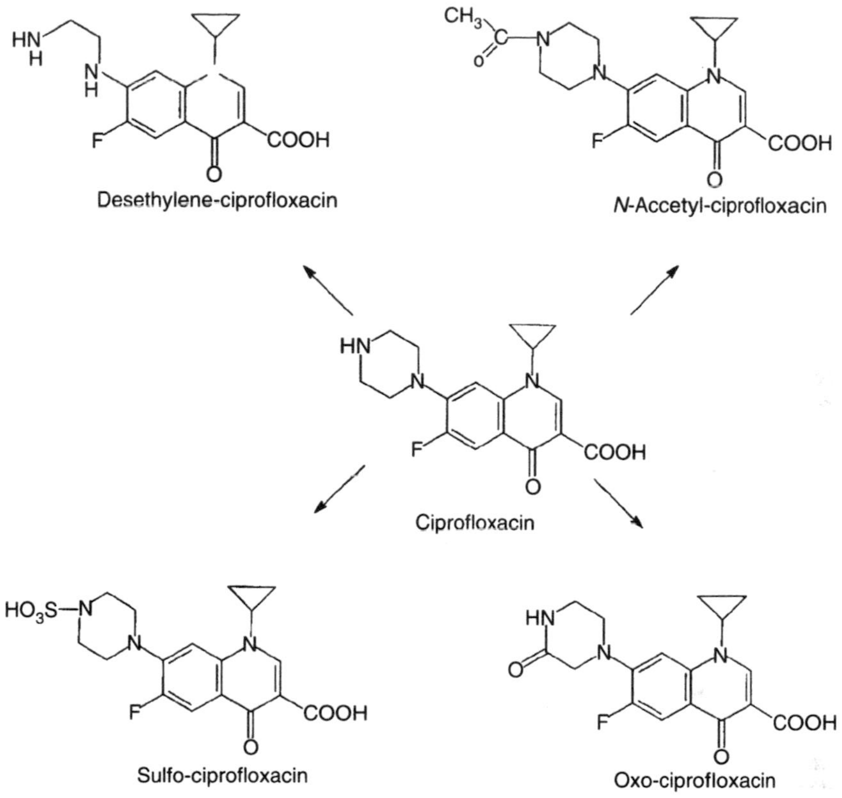

Desethylene-ciprofloxacin

N-Accetyl-ciprofloxacin

Ciprofloxacin

Sulfo-ciprofloxacin

Oxo-ciprofloxacin

Figure 1. Biotransformation of ciprofloxacin in body fluids, as proposed by Krol *et al.* [15].

The drug is also distributed into human-milk [4, 5, 10]. In lactating women who received 3 doses every 12 h of 750 mg of ciprofloxacin, the concentration of the drug in milk that was obtained 2–4 h after a dose averaged 2.26–3.79 μg/mL. Concentrations in milk were higher than concomitant serum concentrations for up to 12 h after a dose [10]. The plasma half-life is about 3.5–4.5 h, and there is an evidence of modest accumulation [4].

5. REFERENCES

1. C.J. Gilles, R.A. Magonigle, W.T.R. Grinshaw, A.C. Tanner, J.E. Risk, M.J. Lynch and J.R. Rice, *J. Vet. Pharmacol. Therapy*, ***14***, 400 (1991).

2. J.W. Spoo and J.E. Riviere, ***Veterinary Pharmacology and Therapeutics***, 7th edn., in H.R. Adams, ed., Iowa State University Press, Ames, IA, p. 832 (1995).
3. ***Martindale, The Complete Drug Reference***, 33rd edn., S.C. Sweetman, ed., The Pharmaceutical Press, Chicago, p. 182 (2002).
4. ***Drug Information,*** 95th edn., G.K. McEvoy, ed., American Society of Health-System Pharmacists, p. 493 (1995).
5. ***Thomson MICROMEDEX(R), Healthcare Series***, Vol. 116, DE0858 (2002).
6. R.L. Davis, J.R. Koup, J. Williams-Warren *et al.*, *Antimicrob. Agents Chemotherapy*, ***28***, 74 (1985).
7. W. Wingender, K.H. Graefe, W. Gau *et al.*, *Eur. J. Clin. Microbiol.*, ***3***, 355 (1984).
8. M. Dan, N. Verbin, A. Gorea *et al.*, *Eur. J. Clin. Pharmacol.*, ***32***, 217 (1987).
9. D.C. Brittain, B.E. Scully, M.J. McElrath *et al.*, *J. Clin. Pharmacol.*, ***25***, 82 (1985).
10. ***Drugs in Pregnancy and Lactation***, 5th edn., G.G. Briggs, R.K. Freeman, and S.J. Yaffe, Williams & Wilkins, Baltimore, MD, p. 213 (1998).
11. H. Scholl, K. Schmidt and B. Weber, *J. Chromatogr. Biomed. Appl.*, ***416***, 321 (1987).
12. C.M. Myers and J.L. Blumer, *J. Chromatogr. Biomed. Appl.*, ***422***, 153 (1987).
13. M.A. Gonzalez, A.H. Moranchel, S. Duran *et al.*, *Clin. Pharmacol. Therapy*, ***37***, 633 (1985).
14. M.A. Gonzalez, A.H. Moranchel, S. Duran *et al.*, *Antimicrob. Agents Chemotherapy*, ***28***, 235 (1985).
15. G.J. Krol, G.W. Beck and T. Benham, *J. Pharm. Biomed. Anal.*, ***14***, 181 (1995).
16. K.H. Bannefeld, H. Stass and G. Blaschke, *J. Chromatogr. Biomed. Appl.*, ***692***, 453 (1997).
17. D. Paradis, F. Vallee, S. Allard *et al.*, *Antimicrob. Agents Chemotherapy*, ***36***, 2085 (1992).

Dipyridamole: Comprehensive Profile

A. Khalil, F. Belal and Abdullah A. Al-Badr

Department of Pharmaceutical Chemistry,
College of Pharmacy, King Saud University
P.O. Box 2457, Riyadh-11451, Saudi Arabia

PROFILES OF DRUG SUBSTANCES,
EXCIPIENTS, AND RELATED
METHODOLOGY – VOLUME 31
DOI: 10.1016/S0000-0000(00)00000-0

CONTENTS

1. **Description** 217
1.1 Nomenclature 217
1.1.1 Systematic chemical names 217
1.1.2 Nonproprietary names 217
1.1.3 Proprietary names 218
1.2 Formulae 218
1.2.1 Empirical formula, molecular weight, CAS number 218
1.2.2 Structural formula 218
1.3 Elemental analysis 218
1.4 Appearance 218
1.5 Uses and applications 219
2. **Methods of Preparation** 219
3. **Physical Characteristics** 220
3.1 Ionization constant 220
3.2 Solubility characteristics 220
3.3 X-ray powder diffraction pattern 221
3.4 Thermal methods of analysis 223
3.4.1 Melting behavior 223
3.4.2 Differential scanning calorimetry 223
3.5 Ultraviolet spectroscopy 223
3.6 Vibrational spectroscopy 224
3.7 Fluorescence spectrum 225
3.8 Nuclear magnetic resonance spectrometry 226
3.8.1 ^{1}H-NMR spectra 226
3.8.2 ^{13}C-NMR spectra 226
3.9 Mass spectrometry 227
4. **Methods of Analysis** 230
4.1 Compendial tests 230
4.1.1 USP 24 compendial tests 230
4.1.2 British pharmacopoeia compendial tests 245
4.2 Methods of analysis reported in the literature 250
4.2.1 Identification 250
4.2.2 Titrimetric methods 251
4.2.3 Potentiometric methods 251
4.2.4 Extraction-gravimetric method 252
4.2.5 Conductimetric method 252

4.2.6 Spectrophotometric methods 253
4.2.7 Electrochemical methods 264
4.2.8 Mass spectrometry 266
4.2.9 Chromatographic methods 267
5. **Stability** 276
6. **Pharmacokinetics and Metabolism** 276
7. **Acknowledgment** 276
8. **References** 276

1. DESCRIPTION

1.1 Nomenclature

1.1.1 Systematic chemical names [1–5]

2,2′,2″,2‴-[(4,8-Di-1-piperidinyl-pyrimido-[5,4-d]-pyrimidine-2,6-diyl)-dinitrilo]-tetrakisethanol.

2,6-Bis-(diethanolamino)-4,8-dipiperidinopyrimido-[5,4-d]-pyrimidine.

2,2′,2″,2‴-[(4,8-Dipiperidinopyrimidino-[5,4-d]-pyrimidine-2,6-diyl)-dinitrilo]-tetraethanol.

2,2′,2″,2‴-[(4,8-Di-piperidin-1-yl)-pyrimido-[5,4-d]-pyrimidine-2,6-diyl]-dinitrilo]-tetraethanol.

Ethanol, 2,2′,2″,2‴-[(4,8-di-1-piperidinylpyrimido-[5,4-d]-pyrimidine-2,6-diyl)-dinitrilo]-tetrakis.

2,6-Bis-[di-(2-hydroxyethyl)-amino]-4,8-dipieridinopyrimido-[5,4-d]-pyrimidine.

2,6-Bis-[di-(2-hydroxyethyl)-amino]-4,8-dipieridinopyrimidino-[5,4-d]-pyrimidine.

1.1.2 Nonproprietary names [1–3]

Dipyridamolum, Dipyridamole, Dipyridamol, Dipiridamol.

1.1.3 Proprietary names [1–5]

Agredamol, Angipec, Apo-Dipyridamole, Atrombin, Cardiwell, Cleridium, Cordantin, Coribon, Coronair, Coronamole, Coronarine, Corosan, Coroxin, Corantyl, Didamol, Diphar, Dipiridamol, Dipramol, Diprimol, Dipyridamole, Dipyrida, Dipyridan, Dipyrin, Drisentin, Functiocardon, Kardisentin, Miosen, Natyl, Novo-Dipiradol, Novodil, Penselin, Perazodin, Peridamol, Perkod, Persantin, Persantine, Persentic, Piroan, Prandiol, Protangix, Proxicor, Rombosit, Santhimon, Stinocor, Tinol, Trancocard, Trodamol, Tromboliz, Trombosentin, Trombostaz, Vazodi.

1.2 Formulae

1.2.1 Empirical formula, molecular weight, CAS number [3]

$C_{24}H_{40}N_8O_4$, 504.63 [58–32–2]

1.2.2 Structural formula

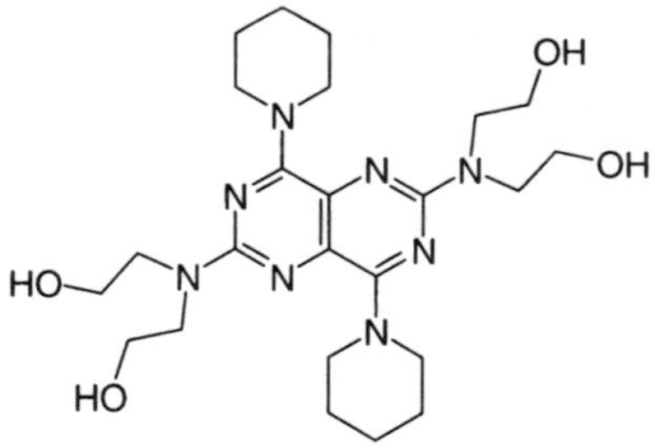

1.3 Elemental analysis [3]

Carbon	57.12%
Hydrogen	7.99%
Nitrogen	22.21%
Oxygen	12.68%

1.4 Appearance [2, 3]

Dipyridamole is obtained as deep yellow needles from ethyl acetate, or as an intensely yellow crystalline powder. Its solutions are yellow, and show strong blue-green fluorescence.

1.5 Uses and applications

Dipyridamole is an adenosine reuptake inhibitor and a phosphodiesterase inhibitor with antiplatelet and vasodilating activity. As a result, the compound is therefore used in thrombo-imbolic disorders. Orally administered dipyridamole is used in association with orally administered anticoagulants for the prophylactic treatment of thromboimbolism following cardiac valve replacement. It is also used as a coronary vasodilator [5–8].

Dipyridamole exerts its antiplatelet action by several mechanisms [7]. One of these is through the inhibition of phosphodiesterase enzyme in platelets, resulting in an increase in intraplatelet cyclic AMP and the consequent potentiation of the platelet inhibiting actions of prostacyclin. Another is the direct stimulation of the release of this eicosanoid by vascular endothelium, and the third is the inhibition of cellular uptake and metabolism of adenosine (thereby increasing its concentration at the platelet vascular interface).

The antiplatelet/antithrombotic activity of dipyridamole has been demonstrated in laboratory and in animal models, and has been shown to inhibit platelet aggregation and vessel-wall thrombogenesis [9–11]. Dipyridamole has been given either alone or with aspirin in the management of myocardial infarction and stroke. For the secondary prevention of stroke or transient ischemic attack, the drug may be given as a modified-release preparation in a dose of 200 mg twice daily. Dipyridamole administered intravenously results in a marked coronary vasodilation and is used in stress testing in patients with ischemic heart disease [5].

2. METHODS OF PREPARATION

A general outline of the procedure for synthesizing dipyridamole is shown in Scheme 1. Reaction of the pyrimidino pyrimidine-2,4,6,8-tetraol (**1**) with a mixture of phosphorous oxychloride and phosphorous pentachloride gives the tetrachloro derivative (**2**). The halogens at the peri positions 4 and 8 are more reactive to substitution than are the remaining halogen pairs 2 and 6, which are in effect the two positions of the pyrimidines. Thus, reaction with piperidine at ambient temperature gives the 4, 8 diamine (**3**). Subsequent reaction with bis-2-hydroxy ethylamine under more strenuous conditions gives dipyridamole (**4**) [12, 13].

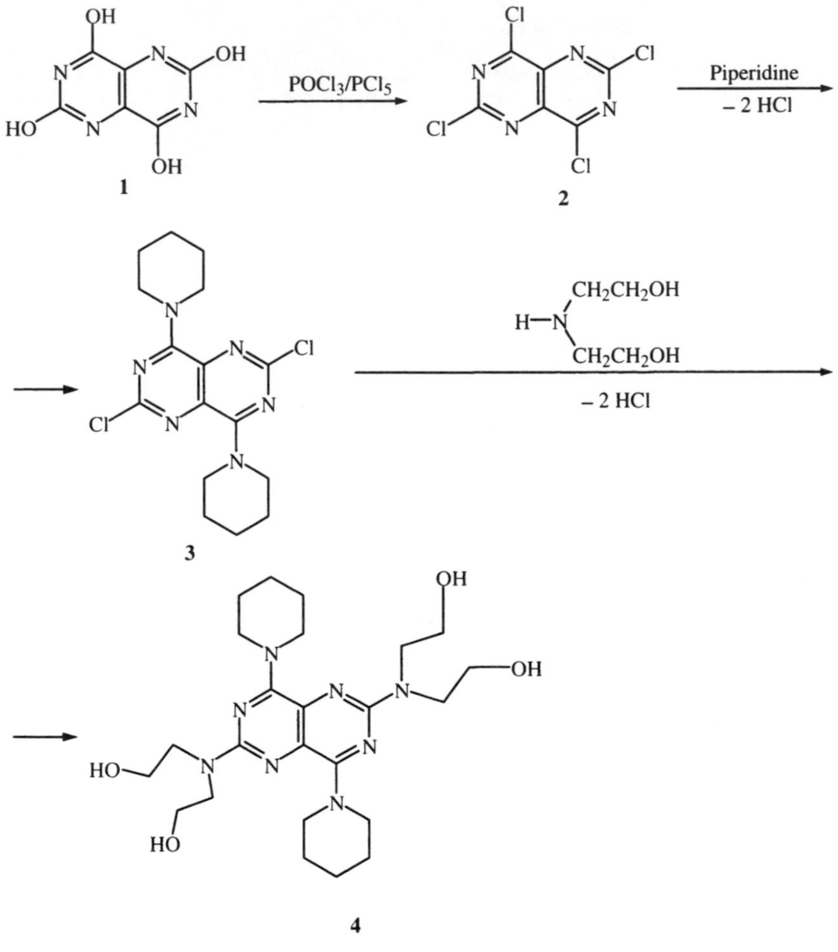

1 $\xrightarrow{POCl_3/PCl_5}$ 2 $\xrightarrow[-\,2\,HCl]{Piperidine}$ 3 $\xrightarrow[-\,2\,HCl]{H-N(CH_2CH_2OH)_2}$ 4

Scheme 1. General outline of the procedure for synthesizing dipyridamole.

3. PHYSICAL CHARACTERISTICS

3.1 Ionization constant [2, 3]

Dipyridamole exhibits a pKa value of 6.4.

3.2 Solubility characteristics [2, 3]

Dipyridamole is slightly soluble in water, and soluble in dilute acid having a pH of 3.3 or below. The bulk drug substance therefore dissolves in dilute solutions of mineral acids. It is also very soluble in methanol,

ethanol, chloroform, but not too soluble in acetone, benzene and ethyl acetate. Dipyridamole is practically insoluble in ether.

3.3 X-ray powder diffraction pattern

The X-ray powder diffraction pattern of dipyridamole was obtained using a Simons XRD–5000 diffractometer. The XRPD powder pattern of the drug is shown in Figure 1, and a summary of the crystallographic data deduced from the powder pattern is shown in Table 1.

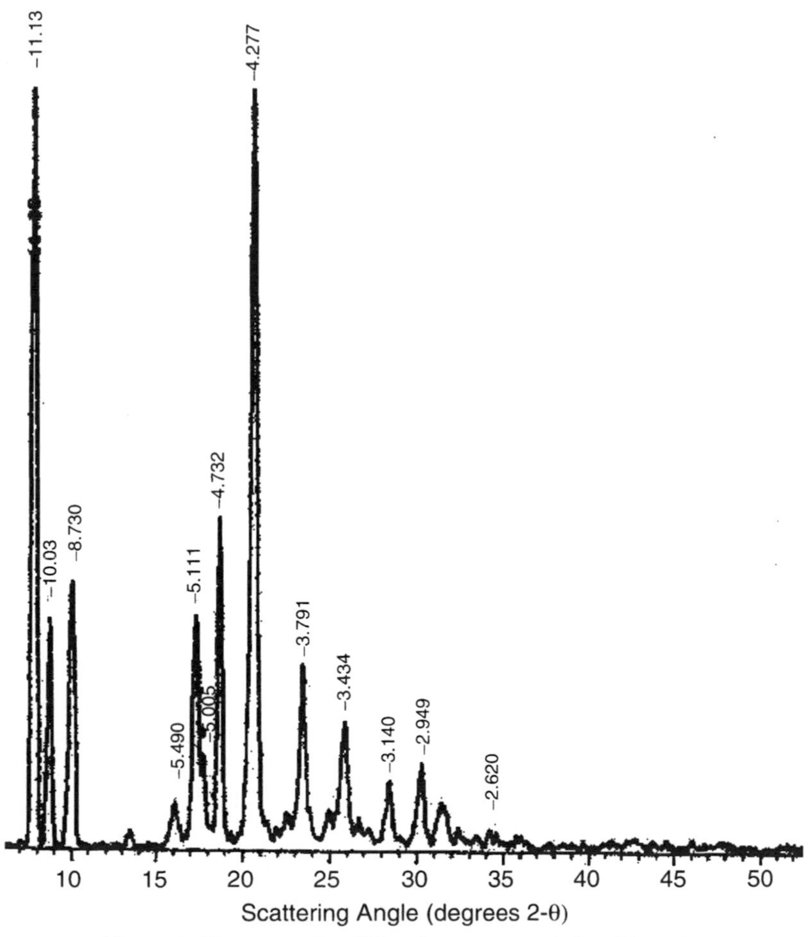

Figure 1. X-ray powder diffraction pattern of dipyridamole.

Table 1

The X-Ray Powder Diffraction Pattern of Dipyridamole

Scattering angle (degrees 2-θ)	***d*-spacing (Å)**	**Relative intensity (100*I/I_O)**
7.171	12.3173	1.07
7.987	11.0606	100.00
8.851	9.9825	36.13
10.187	8.6765	41.17
13.452	6.5767	1.66
16.107	5.4980	5.48
17.386	5.0964	36.73
17.764	4.9888	14.61
18.774	4.7228	25.89
20.798	4.2674	69.26
22.626	3.9266	3.51
23.500	3.7825	18.95
25.949	3.4308	15.02
27.306	3.2633	3.13
28.437	3.1361	6.40
30.333	2.9442	8.34
31.324	2.8533	5.07
31.649	2.8247	5.18
32.375	2.7631	2.16
36.076	2.4876	1.58
44.472	2.0355	0.81

3.4 Thermal methods of analysis

3.4.1 Melting behavior [2, 3]

Dipyridamole melts in the range of 164–167°C.

3.4.2 Differential scanning calorimetry

The differential scanning calorimetry characterization of dipyridamole was performed using a Dupont DSC model TA 9900 compiler thermal analyzer system. The DSC thermogram shown in Figure 2 was recorded at a heating rate of 10°C/min (under a nitrogen flow) over a temperature range of 50–450°C. The compound was found to melt at 169.32°C, with an enthalpy of fusion equal to 4.400 J/g.

3.5 Ultraviolet spectroscopy

The ultraviolet (UV/VIS) absorption spectrum of dipyridamole was recorded on a Shimadzu model 1604 PC UV/VIS spectrophotometer. The spectra were recorded at a concentration of 10 μg/mL, with the drug

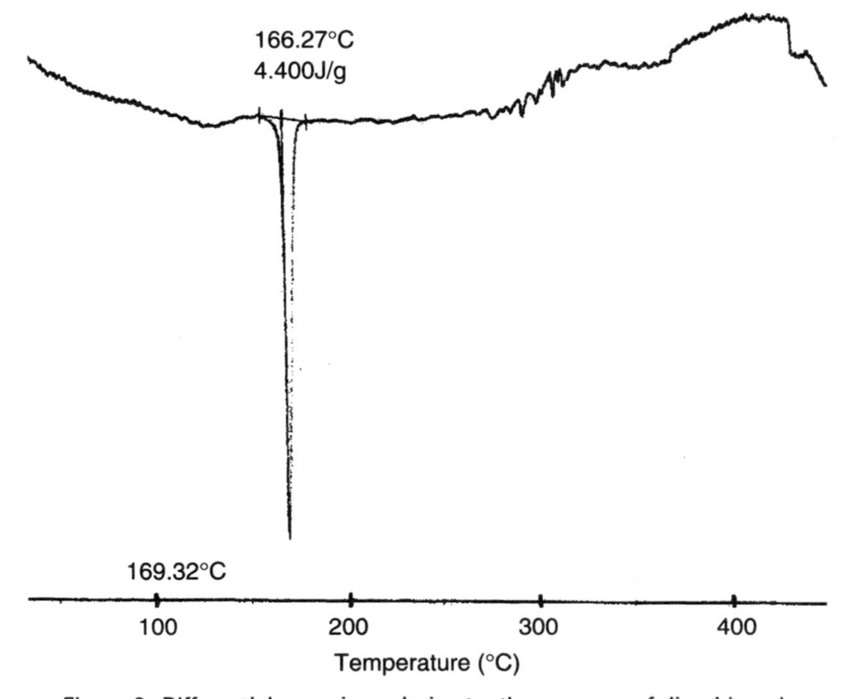

Figure 2. Differential scanning calorimetry thermogram of dipyridamole.

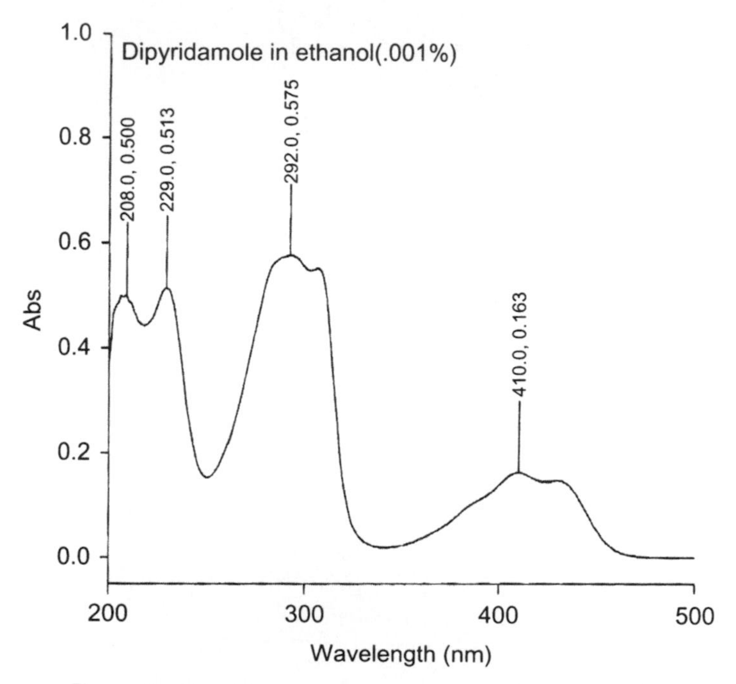

Figure 3. Ultraviolet absorption spectrum of dipyridamole.

being dissolved in ethanol, 0.1 M hydrochloric acid, and 0.1 M sodium hydroxide. The spectrum of dipyridamole in ethanol is shown in Figure 3. The values calculated for molar absorptivity and $A^{1\%}_{1cm}$ at different wavelengths in the three media are shown in Table 2.

Literature values reported for the drug dissolved in acidic methanol were at a wavelength maxima of 230 nm and 285 nm $A^{1\%}_{1cm} = 650$ a). In aqueous alkali, the wavelength maximum is observed at 295 nm [2].

3.6 Vibrational spectroscopy

The infrared absorption spectrum of dipyridamole was obtained using a Perkin-Elmer infrared spectrophotometer, with the drug being pressed in a KBr pellet. The IR spectrum is shown in Figure 4, and the principal lines are at 3460 (OH stretch), 2360, 2340, 1597 (C=C stretch), 1479, 1397, 1337, 1107, 965, 772, and 594 cm^{-1}.

Literature values reported for the principal peaks (KBr pellet) are at wavenumbers 1526, 1214, 1010, 1076, 1041, and 1251 cm^{-1} [2].

Table 2

Molar Absorptivity and $A^{1\%}_{1cm}$ Values from the Ultraviolet Absorption Spectrum of Dipyridamole in Different Media

Solvent	λ_{max} (nm)	$A^{1\%}_{1cm}$	Molar absorptivity (L/Mol/cm)
Ethanol	429	147	7.43×10^3
	409.5	163	8.23×10^3
	306.5	553	2.79×10^4
	292	573	2.89×10^4
	228.5	512	2.58×10^4
0.1 M NaOH	415.2	162	8.18×10^3
	294	580	2.93×10^4
	216.8	619	3.12×10^4
	212.8	756	3.82×10^4
	206.2	2429	1.22×10^5
0.1 M HCl	399.6	133	6.71×10^3
	283.8	568	7.87×10^4
	235.8	530	2.83×10^4

3.7 Fluorescence spectrum

The native fluorescence spectrum of dipyridamole was recorded in ethanol (5 μg/mL) using a Kontron spectrofluorimeter, Model SFM 25 A, equipped with a 150 W xenon-high pressure lamp and driven by a PC Pentium-II computer. As shown in Figure 5, the excitation maximum was found at 297 nm, and the emission maximum was located at 467 nm.

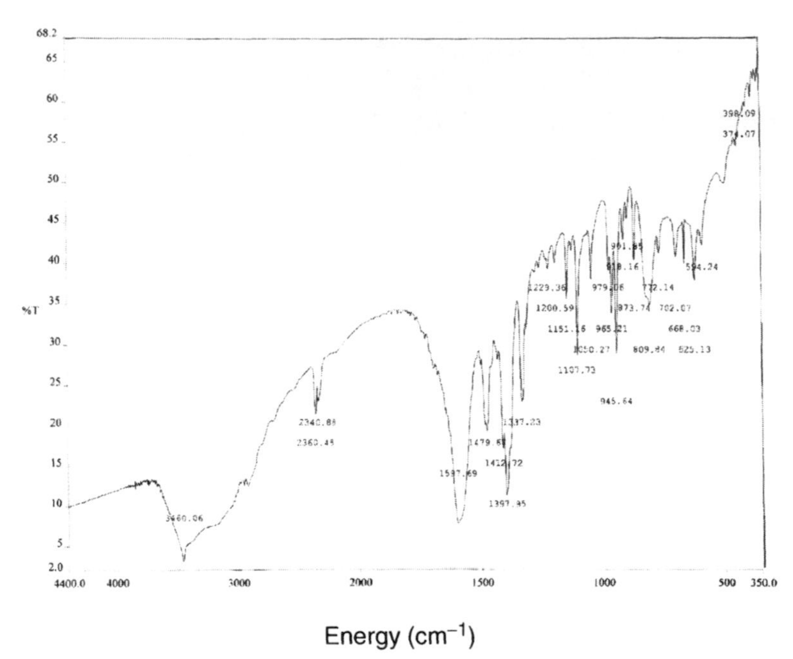

Figure 4. Infrared absorption spectrum of dipyridamole.

3.8 Nuclear magnetic resonance spectrometry

3.8.1 ^{1}H-NMR spectra

The ^{1}H-NMR spectrum of dipyridamole was obtained using a Bruker system operating at 300, 400, or 500 MHz. Standard Bruker software was used to execute recording of DEPT, COSY, and HETCOR spectra. The sample was dissolved in $CDCl_3$, and all resonance bands were referenced to tetramethylsilane (TMS) internal standard. The ^{1}H-NMR spectra are shown in Figures 6 and 7, while the assignments for the observed resonance bands are given in Table 3 along with an appropriate atom numbering system. The ^{1}H-NMR spectrum was also obtained using a JEOL instrument, using methanol-D_3 as the solvent and TMS as the reference standard. The spectra are shown in Figures 8 and 9, and assignments for the observed bands are also shown in Table 3. The COSY ^{1}H-NMR spectrum is shown in Figure 10.

3.8.2 ^{13}C-NMR spectra

The ^{13}C-NMR spectra of dipyridamole dissolved in $CDCl_3$ was obtained using a Bruker system operated at 75, 100, or 125 MHz. All resonance

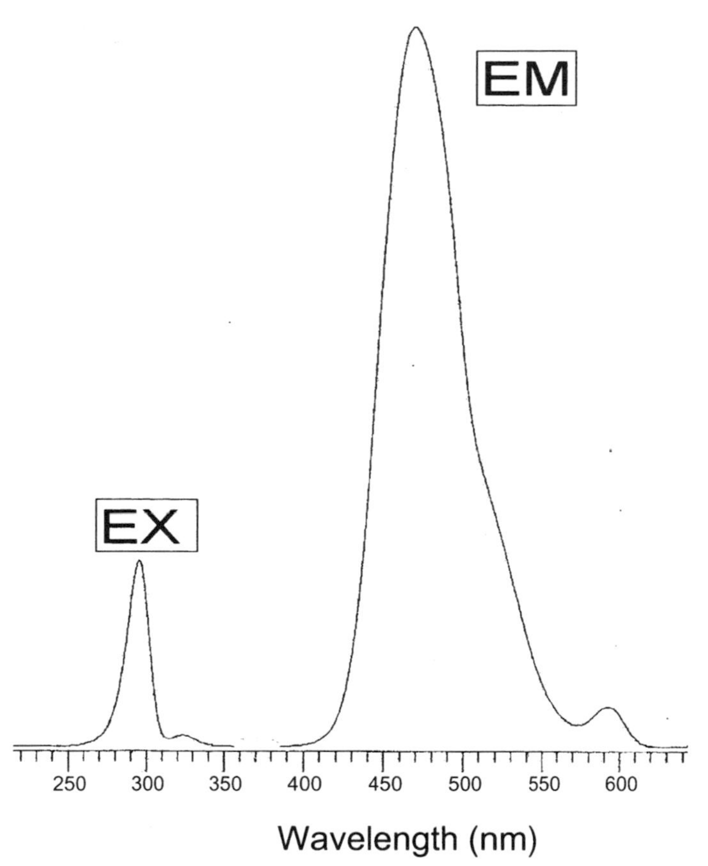

Figure 5. The fluorescence absorption spectrum of dipyridamole (EX = excitation spectrum, and EM = emission spectrum).

bands were referenced to tetramethylsilane (TMS) internal standard. The ^{13}C-NMR spectrum is shown in Figure 11, while the DEPT 135 ^{13}C-NMR spectrum is shown in Figure 12. The HMQC NMR spectra are shown in Figures 13 and 14, and the HMBC NMR spectrum is shown in Figure 15. Assignments for the observed bands are given in Table 4 together with an appropriate atom numbering system.

3.9 Mass spectrometry

The mass spectrum of dipyridamole was obtained using a Shimadzu PQ-5000 mass spectrometer. The parent ion was collided with helium as the

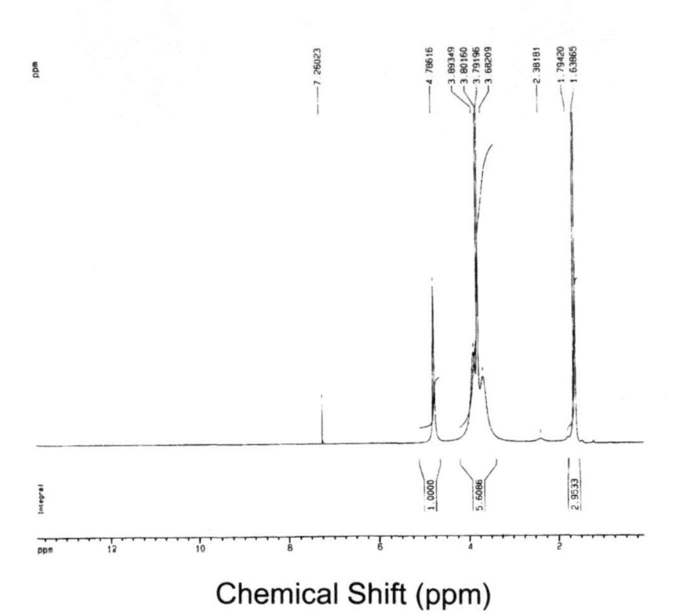

Figure 6. The ^{1}H-NMR spectrum of dipyridamole in $CDCl_3$.

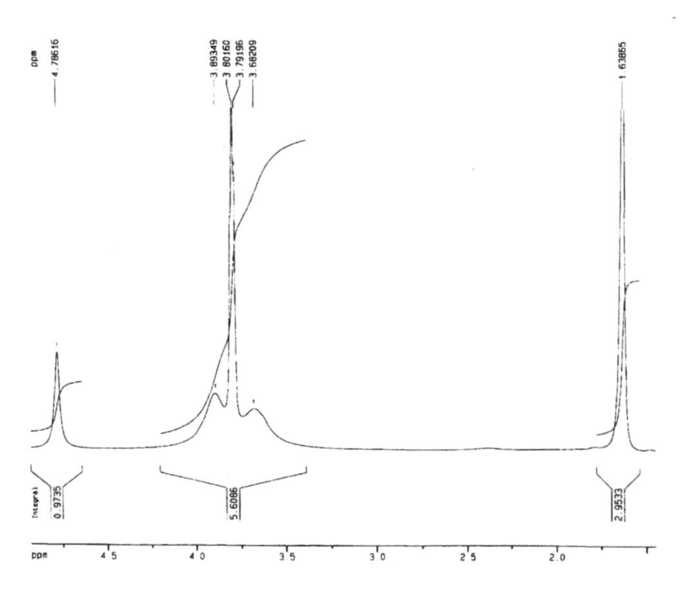

Figure 7. Expanded ^{1}H-NMR spectrum of dipyridamole in $CDCl_3$.

Table 3

Assignments for the Resonance Bands in the ^{1}H-NMR Spectrum of Dipyridamole

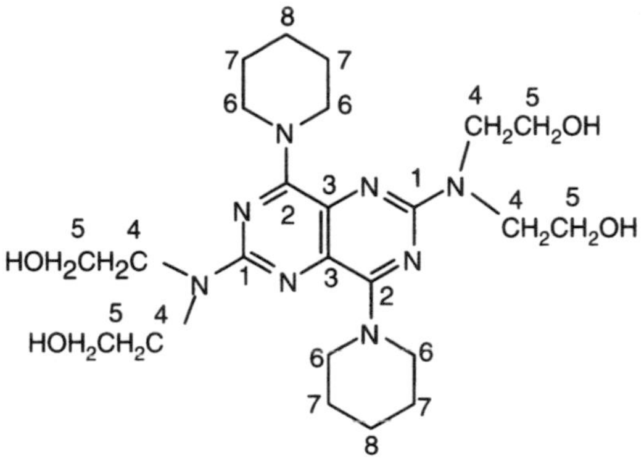

Chemical shift (ppm, relative to TMS)		**Number of protons**	**Multiplicity (m = multiplet)**	**Assignment (proton at carbon #)**
Bruker*	**Jeol***			
1.64	1.68	12	M	7 and 8
3.68	3.30	4	M	5 (OH)
3.79, 3.8	3.76	8	M	6
3.89	4.08	8	M	4
4.78	4.81	8	M	5

*NMR Instruments

carrier gas. Figure 16 shows the detailed mass fragmentation pattern, and Table 5 shows the proposed mass fragmentation pattern. The reported values of the principal peaks are at *m/z* values of 504, 473, 429, 505, 221, 474, 84, and 430 [2].

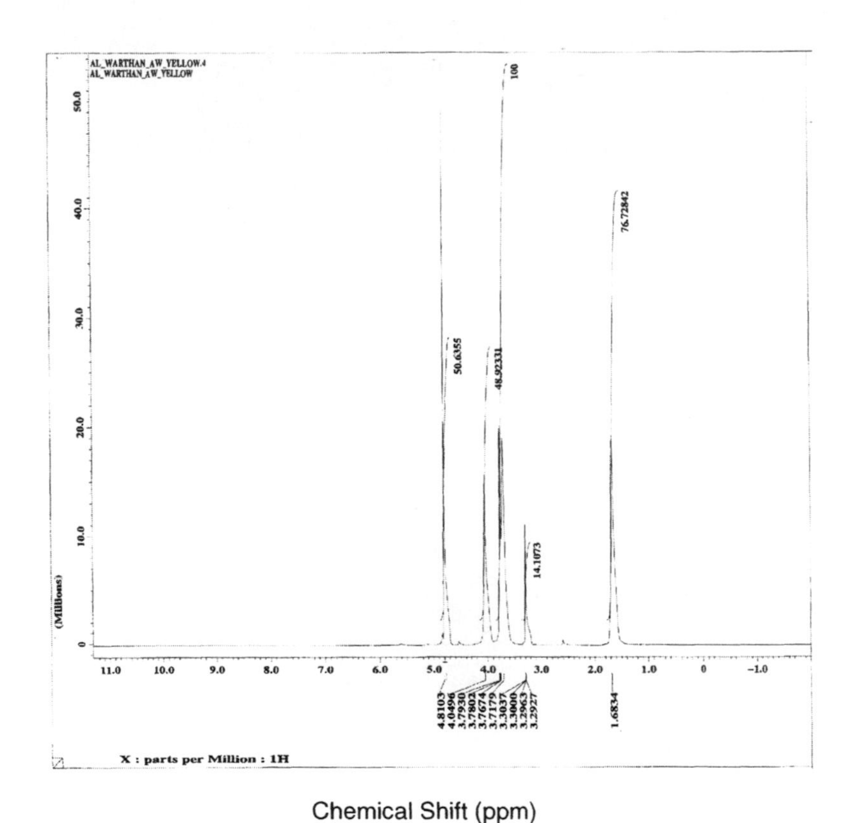

Figure 8. The ^{1}H-NMR spectrum of dipyridamole in deuterated methanol.

4. METHODS OF ANALYSIS

4.1 Compendial tests

4.1.1 USP 24 compendial tests [14]

4.1.1.1 Tests for dipyridamole bulk drug substance

Identification. When dipyridamole is tested according to General Method ⟨197 K⟩, the infrared absorption spectrum of the tested dipyridamole must be equivalent to that of the dipyridamole reference standard.

Melting range. When dipyridamole is tested according to General Method ⟨741⟩, the melting range of the tested dipyridamole is between

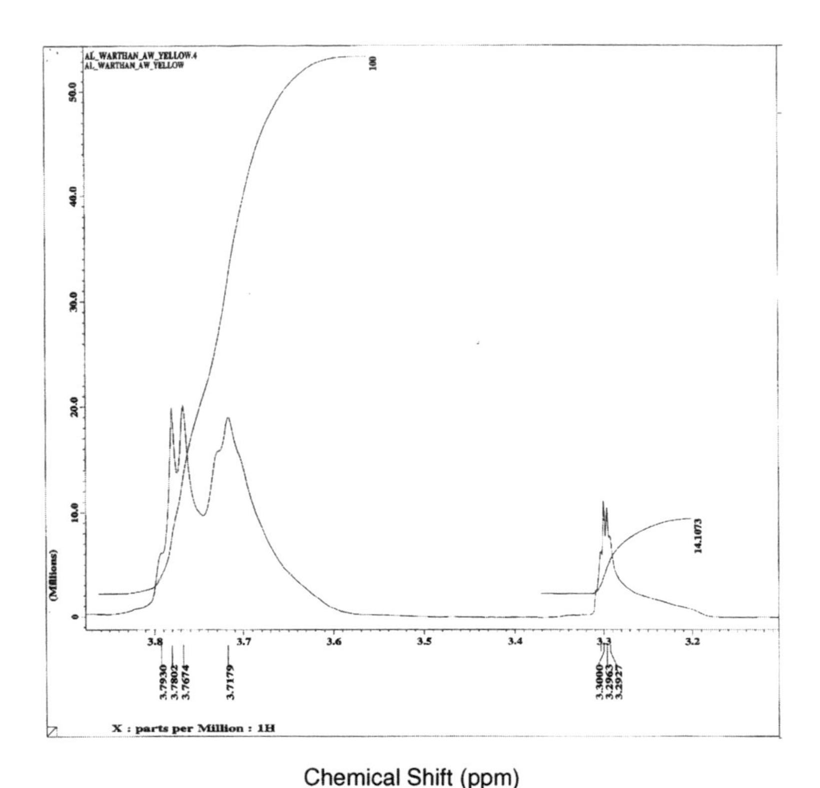

Chemical Shift (ppm)

Figure 9. Expanded ^{1}H-NMR spectrum of dipyridamole in deuterated methanol.

162°C and 168°C, but the range between the beginning and end of the melting does not exceed 2°C.

Loss on drying. When dipyridamole is dried according to General Method ⟨731⟩ (i.e., 105°C for 3 h, it loses not more than 0.2% of its weight.

Chloride content. Dissolve 500 mg of dipyridamole in 5 mL of alcohol and 2 mL of 2 N nitric acid, and then add 1 mL of silver nitrate TS. No turbidity or precipitate is produced.

Residue on ignition. When dipyridamole is tested according to General Method ⟨281⟩, it contains not more than 0.1% of non-ignitable residue.

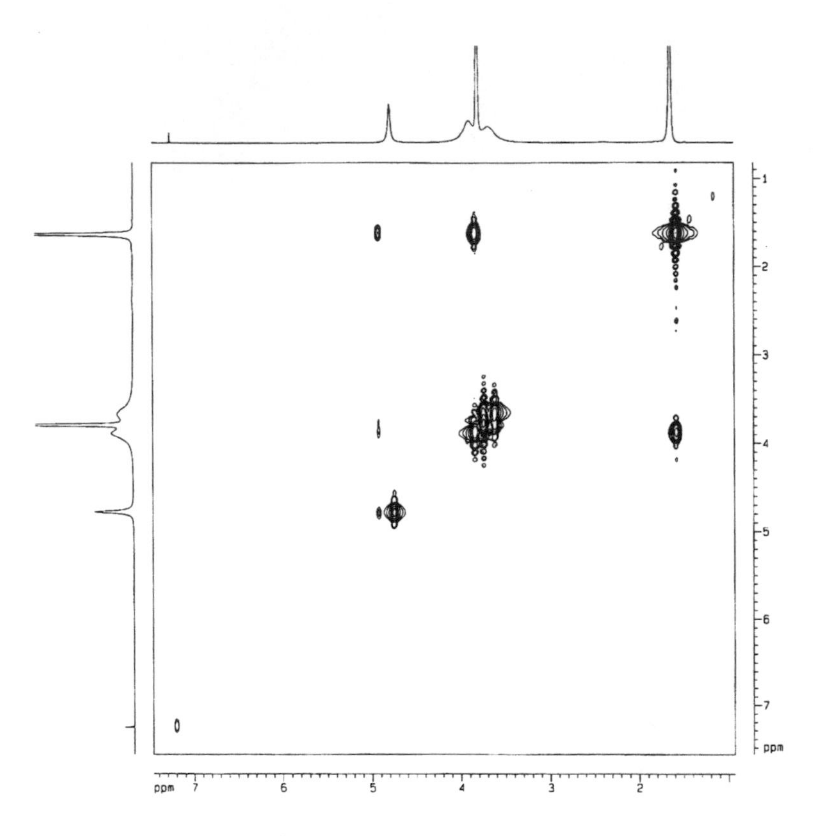

Chemical Shift (ppm)

Figure 10. COSY ^{1}H-NMR spectrum of dipyridamole in $CDCl_3$.

Heavy metals. When dipyridamole is tested according to General Method ⟨231⟩ (Method II), it contains not more than 0.001% heavy metal content.

Organic volatile impurities. When dipyridamole is tested according to General Method ⟨467⟩ (Method IV), it meets the requirements.

Chromatographic purity. The purity of dipyridamole is established using a HPLC method, which is defined by the following characteristics:

> *Mobile Phase*: Dissolve 250 mg of dibasic sodium phosphate in 250 mL of water and adjust with dilute phosphoric acid (1 in 3) to a

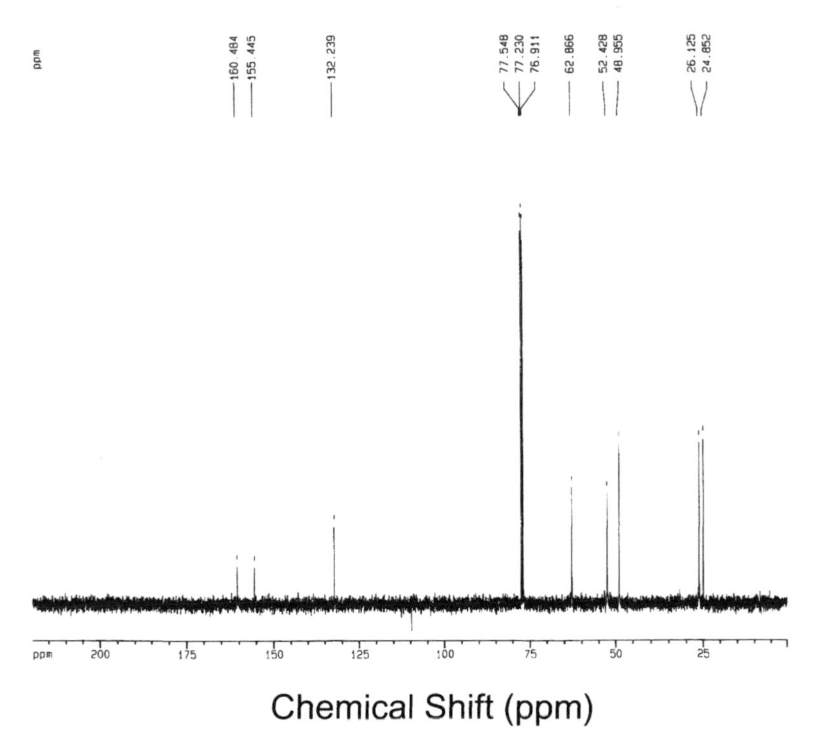

Figure 11. ^{13}C-NMR spectrum of dipyridamole in $CDCl_3$.

pH of 4.6. Add 750 mL of methanol, mix, filter through a 0.5 μm membrane filter, and degas. Adjustments can be made to the composition to enable System Suitability to be achieved.

Chromatographic System: The liquid chromatograph is equipped with a 288 nm detector, and a 3.9 mm × 30 cm column that contains packing L1. The flow rate is about 1.5 mL/min.

Test Preparation A: Prepare a solution of dipyridamole in methanol having a known concentration of 1 mg per mL.

Test Preparation B: Dilute 1 mL of Test preparation A with methanol to 100 mL, and mix.

Procedure: Inject 10 μL of Test Preparation B into the chromatograph by means of a sampling valve, adjusting the operating parameters so that the response of the main peak (retention time

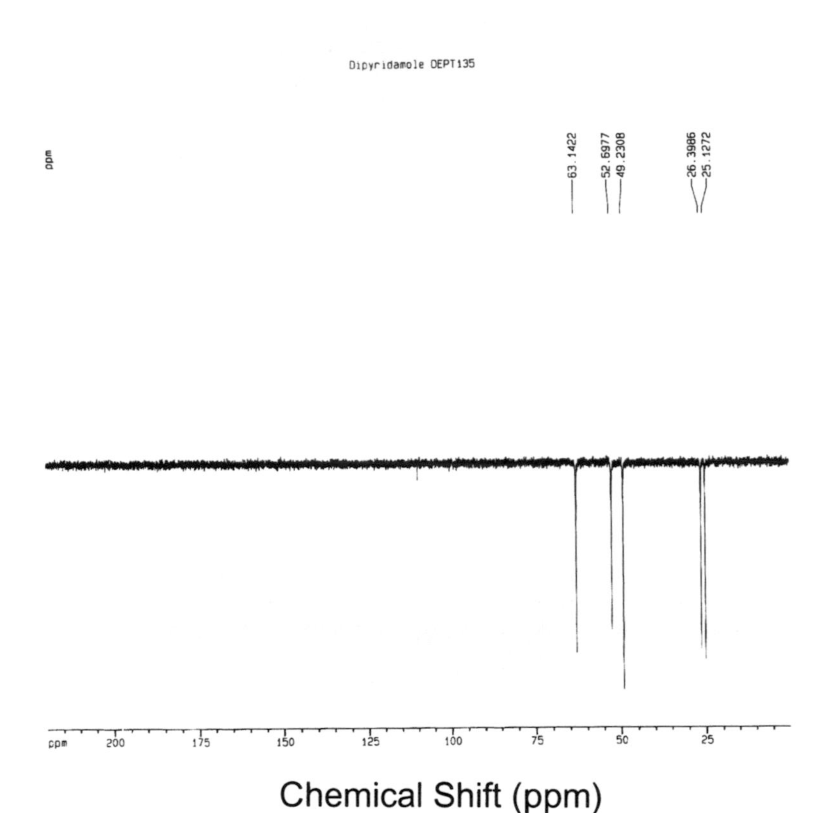

Figure 12. DEPT-135 ^{13}C-NMR spectrum of dipyridamole in $CDCl_3$.

about 6.5 min) is about 5% full scale. Inject 10 μL of Test Preparation A, and run the chromatograph for 10 min. The sum of responses of all the secondary peaks obtained from Test Preparation A is not greater than the response of the main peak obtained from Test Preparation B (i.e., 1.0%).

Assay. Transfer about 450 mg of dipyridamole, accurately weighed, to a 250 mL beaker, and dissolve in 50 mL of glacial acetic acid. Stir for 30 min, add 75 mL of acetone, and then stir for an additional 15 min. Titrate with 0.1 N perchloric acid VS to a potentiometric end point using a glass electrode and a silver–silver chloride reference electrode system. Perform a blank titration, and make any necessary correction. Each mL of 0.1 N perchloric acid is equivalent to 50.46 mg of $C_{24}H_{40}N_8O_4$.

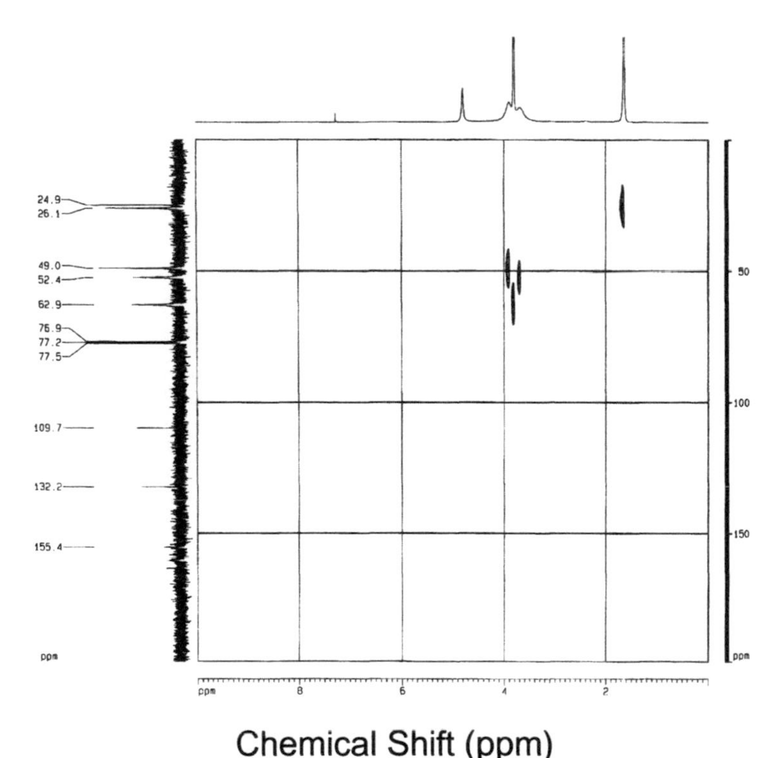

Figure 13. The HMQC NMR spectrum of dipyridamole in $CDCl_3$.

4.1.1.2 Tests for dipyridamole in formulations

Identification. Triturate a quantity of finely powdered tablets, containing the equivalent of about 100 mg of dipyridamole, with 10 mL of 0.1 N hydrochloric acid, and filter. Add 0.1 N sodium hydroxide to the filtrate until the solution is basic and a precipitate forms. Heat the mixture on a steam bath for 1 min, cool, filter, and dry the residue at 105°C for 1 h. The residue so obtained responds to the identification tests for dipyridamole.

Dissolution. Dissolution testing of dipyridamole tablets is to be performed according to the directives of General Method ⟨711⟩, with the inclusion of the following method characteristics:

Medium: 900 mL of 0.01 N hydrochloric acid.

Apparatus: Method 2 (paddles), rotation speed of 50 rpm.

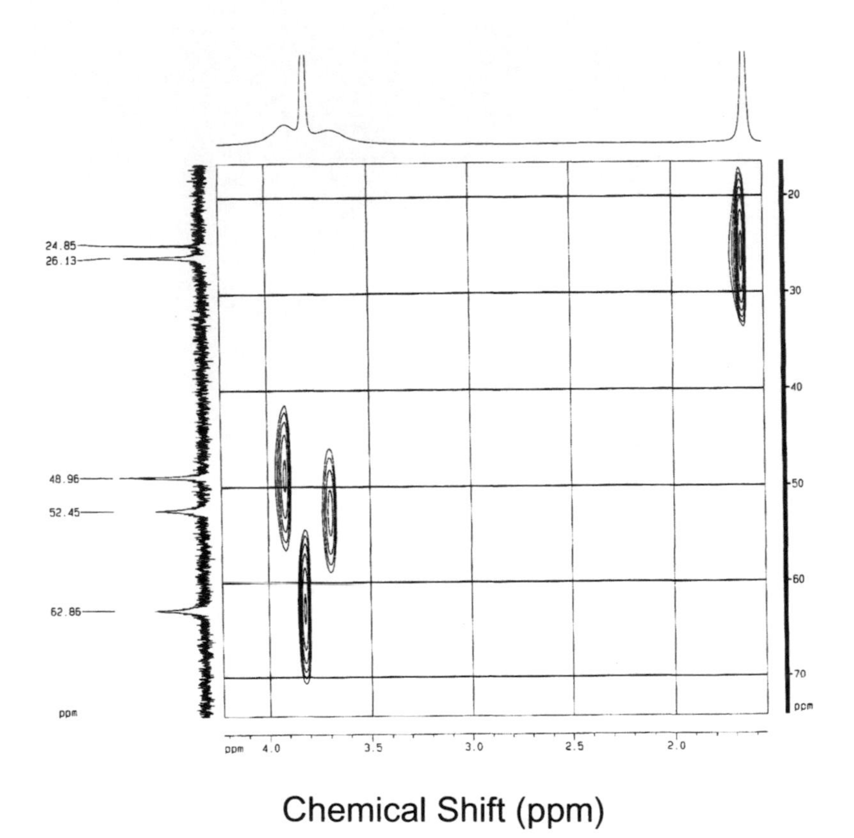

Figure 14. The HMQC-NMR spectrum of dipyridamole in $CDCl_3$.

Time: 30 min.

Procedure: Determine the amount of $C_{24}H_{40}N_8O_4$ dissolved by employing UV absorption at the wavelength of maximum absorbance at about 282 nm on filtered portions of the solution under test, suitably diluted with Dissolution Medium, if necessary. The comparison is to be made with a Standard solution having a known concentration of USP Dipyridamole RS in the same medium.

Tolerances: Not less than 70% (Q) of the labeled amount of $C_{24}H_{40}N_8O_4$ is dissolved in 30 min.

Content uniformity. The determination of Content Uniformity is made following General Method ⟨9057⟩. The test is conducted by transferring

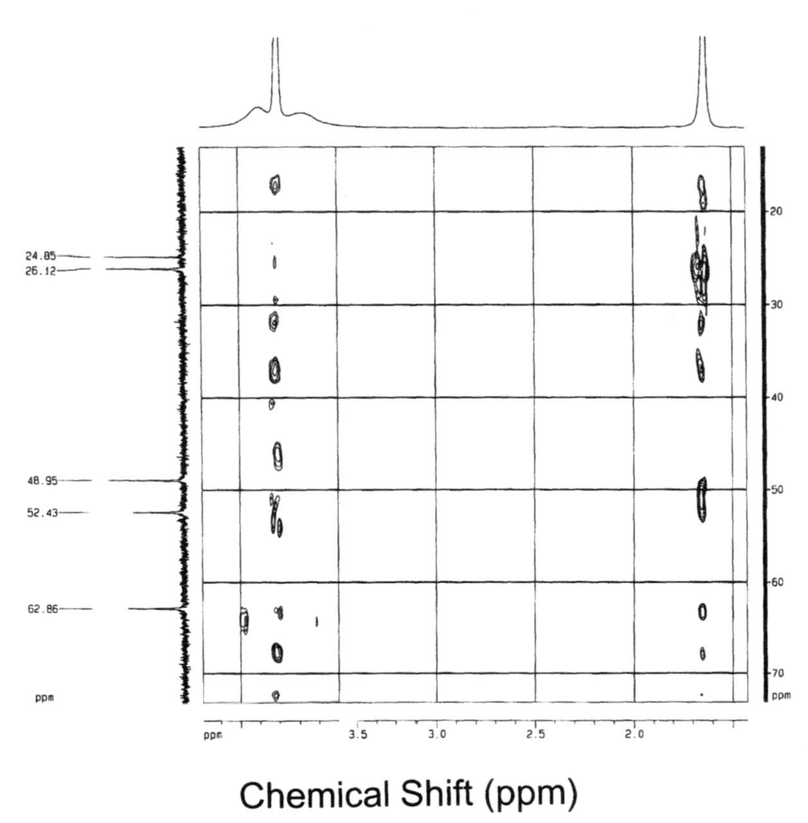

Figure 15. The HMBC-NMR spectrum of dipyridamole in $CDCl_3$.

1 tablet to a 100 mL volumetric flask, adding 50 mL of 1 N hydrochloride acid, heating in a steam bath for 5 min, and shaking by mechanical means for 30 min. The mixture is cooled to room temperature, diluted with 1 N hydrochloric acid to volume, and mixed. The resulting mixture is filtered (discarding the first 25 mL of the filtrate), and then an accurately measured portion of the subsequent filtrate is diluted with 1 N hydrochloric acid to provide a solution containing about 10 μg/mL of dipyridamole. Concomitantly, the absorbances of this solution and of a solution of USP Dipyridamole RS in the same medium having a known concentration of about 10 μg/mL in 1 cm cells at the wavelength of maximum absorbance (about 282 nm) using 1 N hydrochloric acid as the blank, are determined. The quantity, in units of mg, of

Table 4

Assignments for the Resonance Bands in the ^{13}C-NMR Spectrum of Dipyridamole

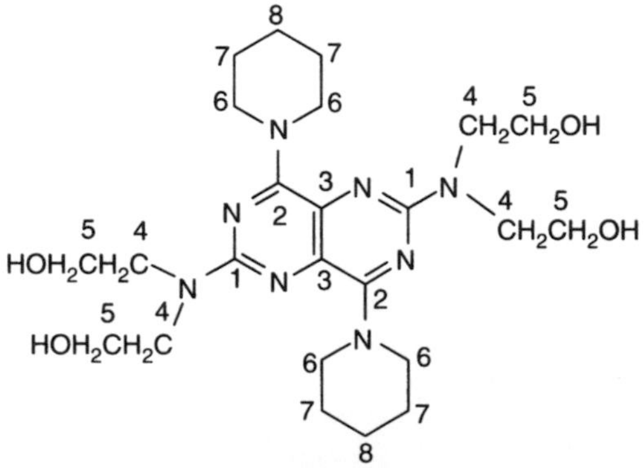

Chemical shift (ppm relative to TMS)	Assignment of carbon number	Number of carbons
160.48	1	2
155.44	2	2
132.30	3	2
62.86	5	4
52.6	6	4
48.9	4	4
26.12	7	4
24.85	8	2

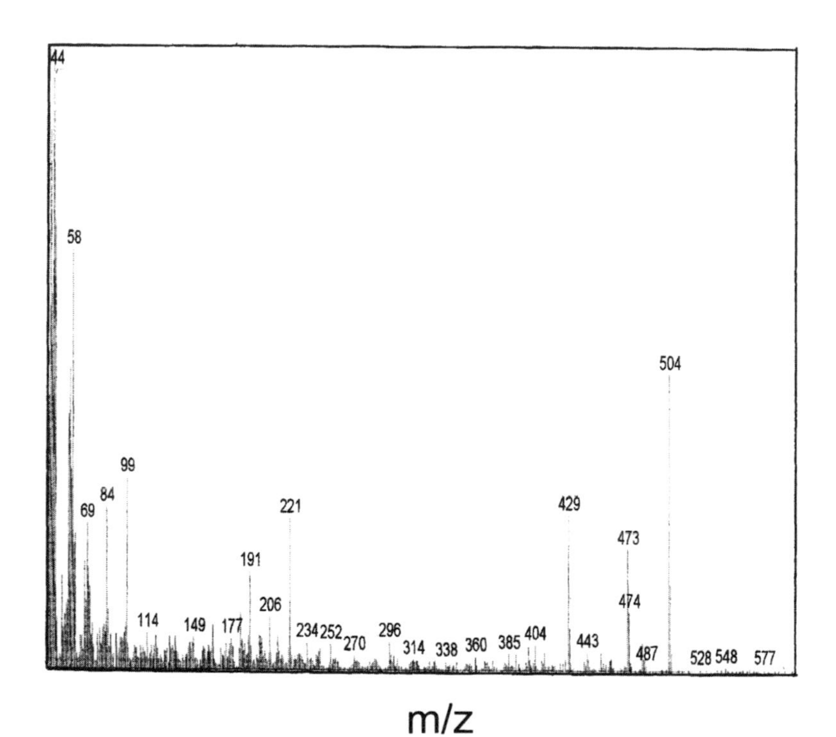

Figure 16. Mass spectrum of dipyridamole.

$C_{24}H_{40}N_8O_4$ in the tablet taken, is calculated by the formula:

$$\left(\frac{TC}{D}\right)\left(\frac{A_U}{A_S}\right)$$

where T is the labeled quantity (in mg) of dipyridamole in the tablet, C is the concentration (in µg/mL) of USP Dipyridamole RS in the Standard solution, D is the concentration (in µg/mL) of dipyridamole in the solution from the tablet based upon the labeled quantity per tablet and the extent of dilution, and A_U and A_S are the absorbances of the solution of the tablet and the Standard Solution, respectively.

Assay. The assay value of dipyridamole tablets is established using a HPLC method, which is defined by the following characteristics:

Mobile phase: Dissolve 250 mg of dibasic sodium phosphate in 250 mL of water, and adjust with dilute phosphoric acid (1 in 3) to

Table 5

Assignments for the Observed Mass Spectrum Fragments of Dipyridamole

m/z	Relative intensity	Fragment	
		Formula	**Structure**
504	48%	$C_{24}H_{40}N_8O_4$	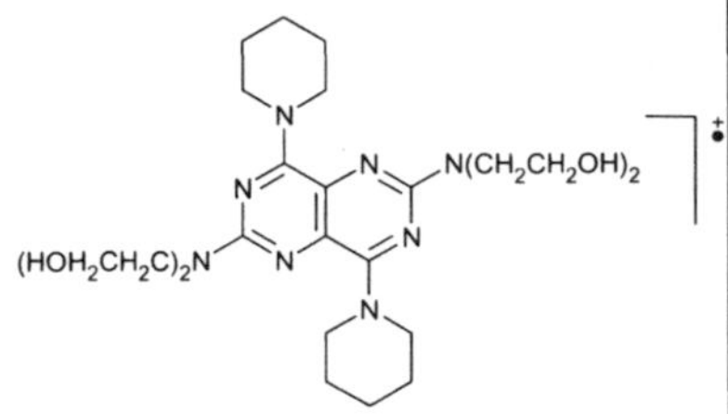
487	4%	$C_{24}H_{38}N_8O_3$	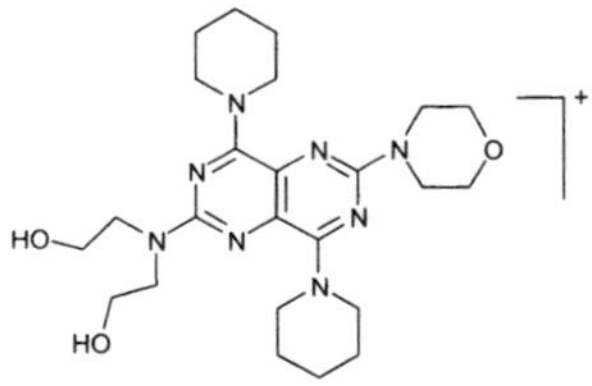
474	10%	$C_{23}H_{38}N_8O_3$	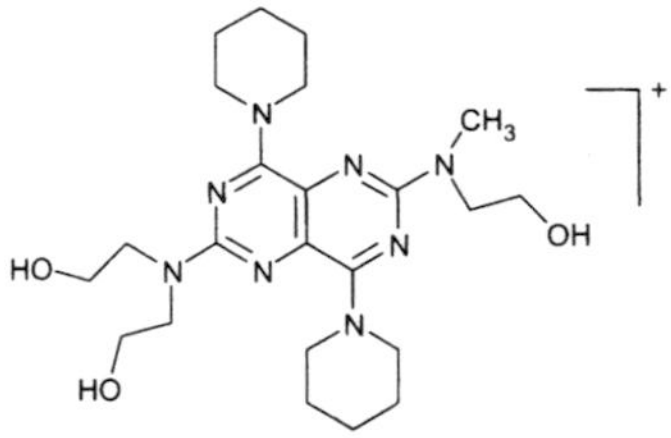
473	20%	$C_{23}H_{37}N_8O_3$	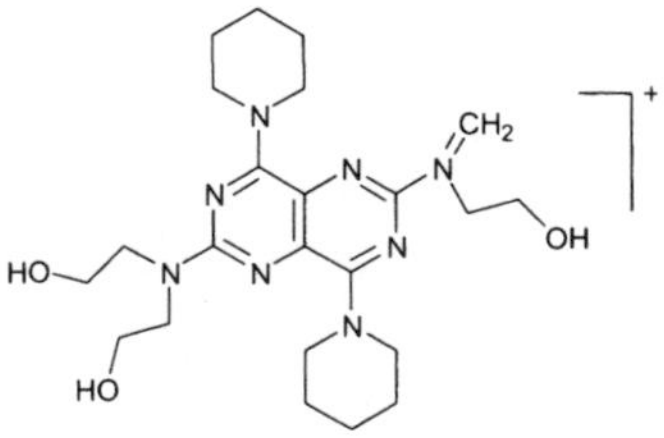

(*continued*)

Table 5 (continued)

m/z	Relative intensity	Fragment	
		Formula	Structure
443	3%	$C_{22}H_{35}N_8O_2$	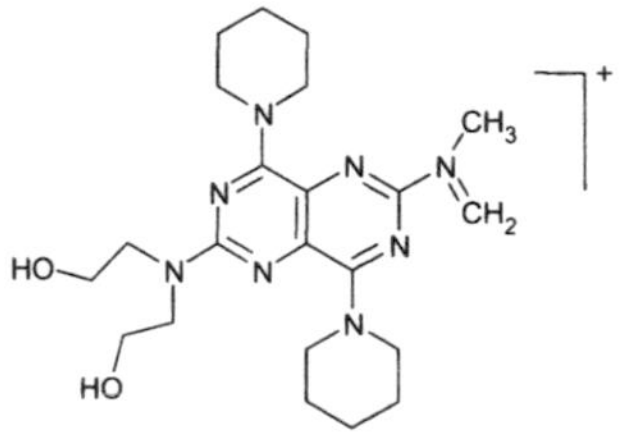
429	26%	$C_{21}H_{32}N_8O_2$	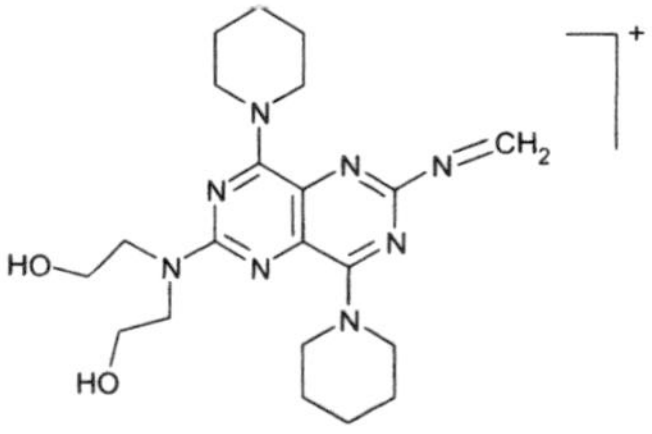
404	4%	$C_{19}H_{30}N_7O_3$	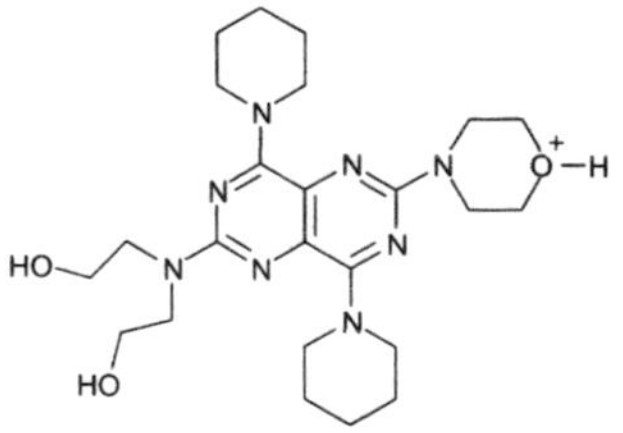
385	3%	$C_{17}H_{28}N_8O$	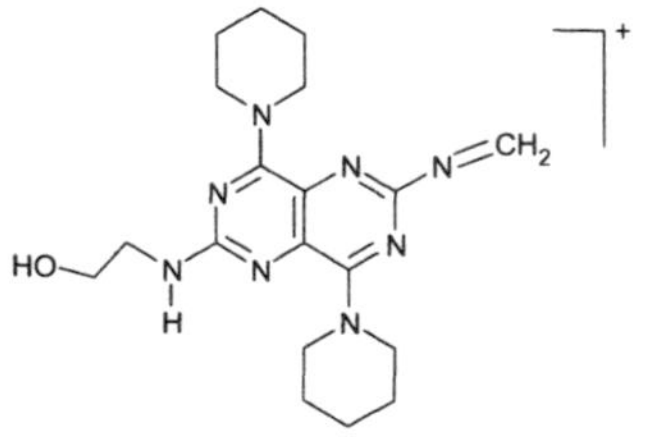

(continued)

Table 5 (continued)

m/z	Relative intensity	Fragment	
		Formula	**Structure**
360	3%	$C_{17}H_{26}N_7O_2$	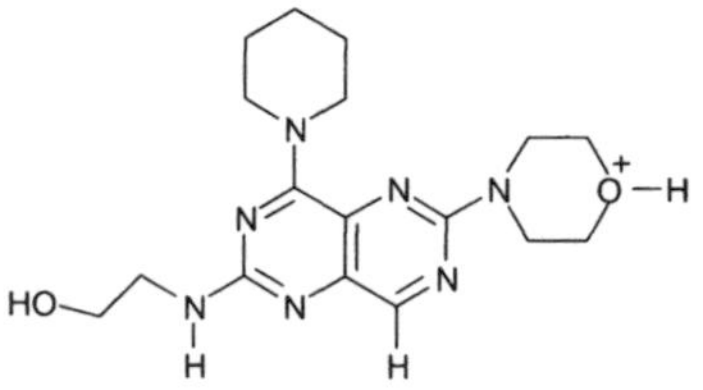 N N N N N N O + H HO N H H
314	2%	$C_{15}H_{18}N_6O_2$	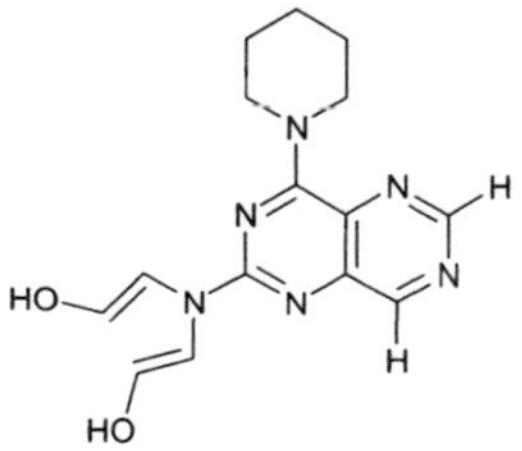 N N N H N N HO N H HO
296	8%	$C_{16}H_{20}N_6$	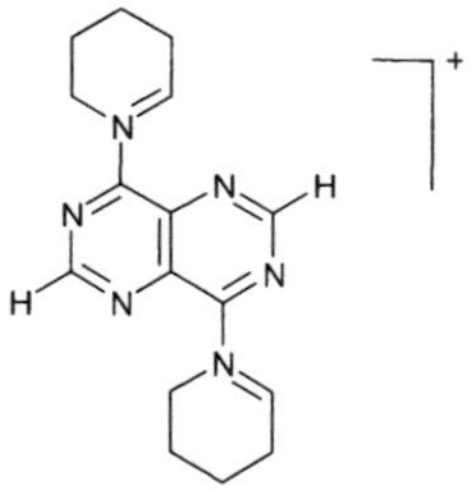 N N N H H N N N +
270	3%	$C_{13}H_{14}N_6O$	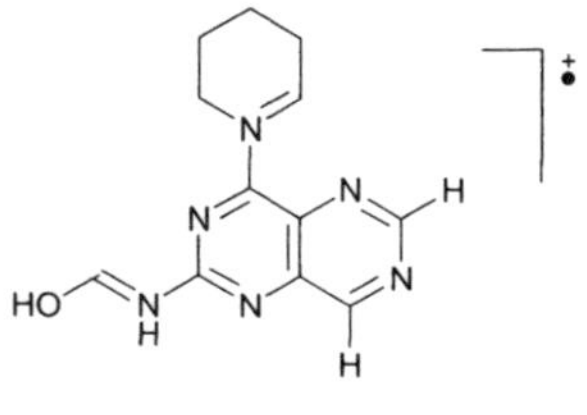 N N N H HO N H N N H +

(*continued*)

Table 5 (continued)

m/z	Relative intensity	Fragment	
		Formula	Structure
221	25%	$C_{11}H_{17}N_4O$	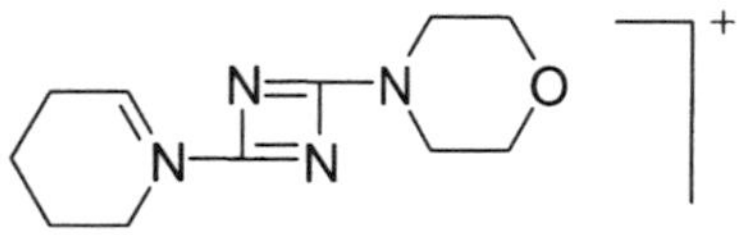
206	8%	$C_{13}H_{22}N_2$	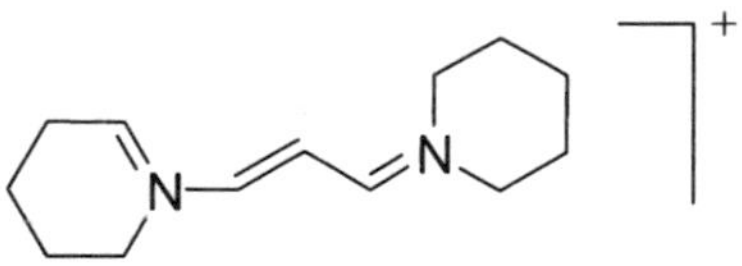
191	16%	$C_{10}H_{15}N_4$	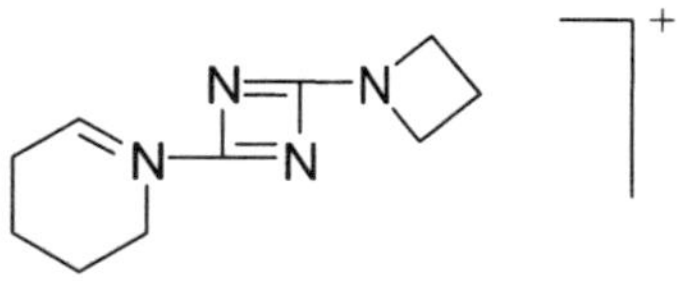
177	8%	$C_9H_{13}N_4$	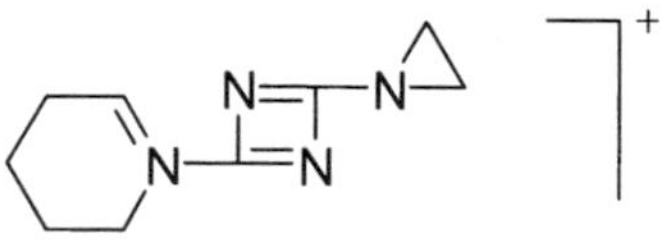
99	32%	$C_4H_6N_2O$	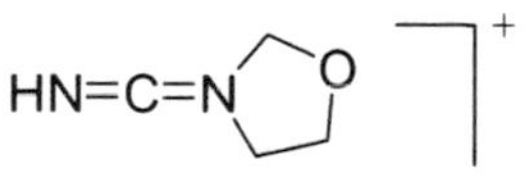 HN=C=N (ring with O)
84	25%	$C_5H_{10}N$	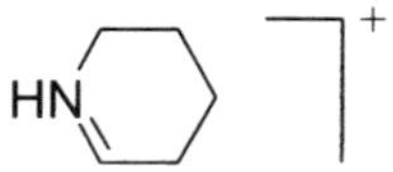 HN (ring)
69	23%	$C_3H_5N_2$	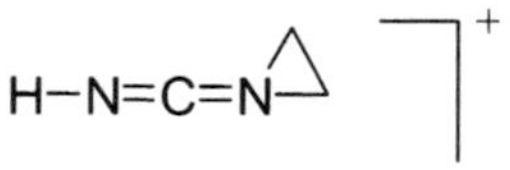 H–N=C=N (aziridine ring)

(*continued*)

Table 5 (continued)

m/z	Relative intensity	Fragment	
		Formula	Structure
58	70%	C_3H_8N	$[H(H_3C)N\text{-aziridine}]^+$
44	100%	C_2H_4O	$[H_2C{=}CH{-}OH]^+$
43	60%	C_2H_5N	$[H_2C{=}CH{-}CH_3]^+$
42	80%	C_2H_4N	$[H_2C{=}CH{-}CH_2]^+$

a pH of 4.6. Add 750 mL of methanol, mix, filter through a 0.5 μm membrane filter, and degas. Make adjustment, if necessary.

Standard preparation: Using an accurately weighed quantity of USP Dipyridamole RS, prepare a solution in Mobile phase having a known concentration of about 15 μg/mL.

Assay preparation: Transfer not less than 20 tablets to a 1000 mL volumetric flask, add 100 mL of water, and sonicate for 15 min. Add about 750 mL of methanol, and shake by mechanical means for 30 min. Dilute with methanol to volume, mix, and centrifuge. Dilute an accurately measured volume (V_S, in units of mL) of the clear supernatant solution quantitatively with Mobile phase to obtain a solution (V_A, in units of mL) containing about 15 μg/mL of dipyridamole.

Chromatographic system and system suitability: The liquid chromatograph system is equipped with a 288 nm detector and a 3.9 mm × 30 cm column that contains packing L1. The flow rate is about 1.5 mL/min. Chromatograph the standard preparation, and record the peak responses as directed under Procedure. The column efficiency determined from the analyte peak is not less than 1000 theoretical plates, the tailing factor for the analyte peak is not more than 2.0, and the relative standard deviation for replicate injections is not more than 2.0%.

Procedure: Separately inject equal volumes (about 50 μL) of the Standard preparation and the Assay preparation into the chromatograph, record the chromatograms, and measure the responses for the major peaks. Calculate the quantity (in mg) of $C_{24}H_{40}N_8O_4$ in the tablet taken by the formula:

$$C\left(\frac{V_A}{V_S}\right)\left(\frac{r_U}{r_S}\right)$$

in which C is the concentration (in μg/mL) of USP Dipyridamole RS in the Standard preparation, V_A is the volume (in mL) of the assay preparation, V_S is the volume (in mL) of supernatant solution taken for the assay preparation, and r_U and r_S are the peak responses obtained from the assay preparation and the Standard preparation, respectively.

4.1.2 *British pharmacopoeia compendial tests [4]*

4.1.2.1 Tests for dipyridamole bulk drug substance

Identification.

Test 1: The melting point range of the tested substance is in the range of 162°C–168°C.

Test 2: Dissolve 10 mg in a mixture of 1 volume of 0.1 M hydrochloric acid and 9 volumes of methanol R, and dilute to 50 mL with the same mixture of solvents. Dilute 5 mL of this solution to 100 mL with a mixture of 1 volume of 0.1 M hydrochloric acid and 9 volumes of methanol R. Examine between 220 nm and 350 nm, where the solution will exhibit absorption maxima at 232 nm and 284 nm. The ratio of absorbance measured at the 284 nm maximum to that measured at the 232 nm maximum is in the range of 1.25–1.45.

Test 3: Examine by infrared absorption spectrophotometry, comparing with the spectrum obtained with dipyridamole CRS. Examine the substances as pellets prepared using potassium bromide R.

Test 4: Dissolve about 5 mg in a mixture of 0.1 mL of nitric acid R and 2 mL of sulfuric acid R. An intense violet color is produced.

Related substances. Examine the test article using a liquid chromatographic system that meets the following requirements:

Test solution: Dissolve 10 mg in the mobile phase, and dilute to 20 mL with the mobile phase.

Reference solution (a): Dilute 1 mL of the Test Solution to 20 mL with the mobile phase, and then dilute 1 mL of this solution to 10 mL with the mobile phase.

Reference solution (b): Dissolve 10 mg of diltiazem hydrochloride CRS in the mobile phase, and dilute to 10 mL with mobile phase. Dilute 1 mL of this solution to 20 mL with Reference Solution (a).

The chromatographic system is composed of the following:

1. *Column*: A stainless steel column 0.25 m long and 4.6 mm in internal diameter, is packed with octylsilyl silica gel for chromatography R (5 μm). The temperature of the column is maintained at 30°C.

2. *Mobile Phase*: It is prepared by dissolving 0.504 g of potassium dihydrogen phosphate R in 370 mL of water R, and adjusting to pH 3 with phosphoric acid R. To this is added 80 mL of

acetonitrile R and 550 mL of methanol R. The mobile phase is eluted at a flow rate of 1.3 mL/min.

3. *Detector*: A spectrophotometer set at an analyzing wavelength of 290 nm.

Inject 20 μL of each solution and continue the chromatography in the test solution for a length of time that is at least nine times the retention time of dipyridamole. The test is not valid unless in the chromatogram obtained with reference solution (b), the resolution between the peaks corresponding respectively to diltiazem and dipyridamole is at least 2. In the chromatogram obtained with the test solution, the area of any peak (apart from the principal peak) is not greater than the area of the peak in the chromatogram obtained with reference solution (a) (i.e., 0.5%). The sum of the areas of all the peaks (apart from the principal peak) is not greater than twice the area of the peak in the chromatogram obtained with reference solution (a) (i.e., 1%). Disregard any peak with an area less than 0.1 times that of the peak in the chromatogram obtained with reference solution (a).

Chloride. To 0.250 g of the substance under test, add 10 mL of water R and shake strongly. Filter, rinse the filtrate with 5 mL of water R, and dilute to 15 mL with water R. The combined filtrates comply with the limit test for chloride (not more than 200 ppm).

Loss on drying. After drying 1.000 g in an oven at 100°C–105°C, the substance loses not more than 0.5% of volatile substances.

Sulfated ash. After testing 1 g, the substance contains not more than 0.1% sulfated ash.

Assay. Dissolve 0.400 g in 70 mL of methanol R, and titrate with 0.1 M perchloric acid, determining the end point potentiometrically. Each milliliter of 0.1 M perchloric acid is equivalent to 50.46 mg of $C_{24}H_{40}N_8O_4$.

Storage. Store in a well-closed container, protected from light.

Impurities. 2,2′-[[4,6,8-tri(piperidin-1-yl)-pyrimido-[5,4-d]-pyrimidin-2-yl]nitrilo] diethanol.

This impurity has the following structure:

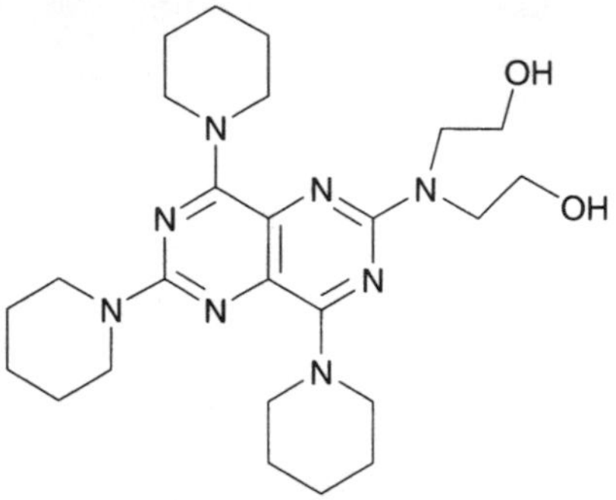

2,2′,2″,2‴,2⁗,2′′′′′-[[8-(Piperidin-1-yl)-pyrimido-[5,4-d]-pyrimidine-2,4,6-triyl]-trinitrilo]-hexaethanol. This impurity has the following structure:

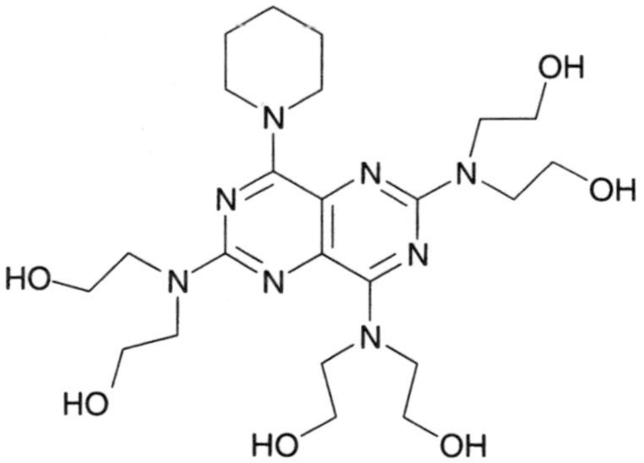

2,2′-[[2-Chloro-4,8-di-(piperidin-1-yl)-pyrimido-[5,4-d]-pyrimidin-6-yl] nitrilo] diethanol. This impurity has the following structure:

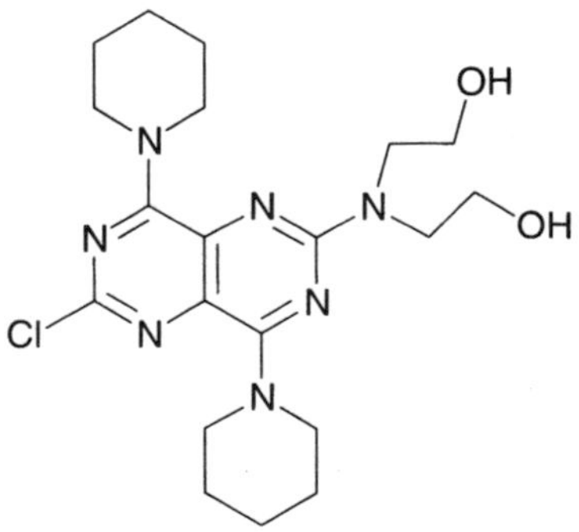

4.1.2.2 Tests for dipyridamole in formulations

Identification.

Test 1: Shake a quantity of the powdered tablets containing 50 mg of dipyridamole with 20 mL of chloroform, filter, and evaporate to dryness. The infrared absorption spectrum of the residue (see Appendix IIA of the BP) is concordant with the reference spectrum of dipyridamole (RS 108).

Test 2: To a quantity of powdered tablets containing 10 mg of dipyridamole, add 50 mL of methanol, warm slightly, shake for 15 min, and allow to cool. Add sufficient methanol to produce 100 mL, and filter. To 10 mL of the filtrate, add 1 mL of 1 M hydrochloric acid and sufficient methanol to produce 100 mL. The light absorption of the resulting solution (see Appendix II B of the BP) in the range of 220–450 nm exhibits three maxima, namely 230, 285, and 405 nm.

Related substances. Carry out the method for liquid chromatography (see Appendix III D of the BP), using the following solutions:

Solution (1): Shake a quantity of powdered tablets containing 50 mg of dipyridamole with 100 mL of mobile phase for 15 min, filter (Whatman GF/C filter is suitable), and use the filtrate.

Solution (2): Dilute 1 volume of Solution (1) to 200 volumes with mobile phase.

Solution (3): Dilute 1 volume of a 0.1% w/v solution of diltiazem hydrochloride EPCRS to 20 mL with Solution (2).

Chromatographic System: It consists of the following:

Column: A stainless steel column (25 cm × 4.6 mm), packed with octylsilyl silica gel for chromatography (5 μm) (Lichrosorb RP8 5 μ is suitable). The column is maintained at 30°C.

Mobile Phase: Dissolve 0.504 g of potassium dihydrogen orthophosphate in 370 mL of water and adjust the pH to 3 with orthophosphoric acid. To this, add 80 mL of acetonitrile and 550 mL of methanol. The mobile phase is eluted at a flow rate of 1.3 mL/min.

Detector: UV absorbance at a wavelength of 290 nm.

Procedure: Inject 20 μL of each solution. Continue the chromatography of Solution (1) for at least nine times the retention time of dipyridamole. This test is not valid unless in the chromatogram obtained with Solution (3), the resolution factor between the peaks due to diltiazem and dipyridamole is at least 2.

In the chromatogram obtained with Solution (1), the area of any secondary peak is not greater than the area of the principal peak in the chromatogram obtained with Solution (2) (i.e., 0.5%). In addition, the sum of the areas of any such peaks is not greater than twice the area of the peak in the chromatogram obtained with Solution (2) (i.e., 1%). Disregard any peaks with an area less than 0.1 times that of the peak in the chromatogram obtained with Solution (2) (0.05%).

Assay. To 10 whole tablets, add 300 mL of 1 M hydrochloric acid, heat at 40°C for 20 min with shaking, allow to cool, and add sufficient 1 M hydrochloric acid to produce 500 mL. Filter and dilute, if necessary, with 1 M hydrochloric acid to produce a solution containing 0.05% w/v of Dipyridamole. Dilute 1 volume of this solution to 50 volumes with water, and measure the absorbance of the resulting solution at the maximum 283 nm (see Appendix II B of the BP). Calculate the content of $C_{24}H_{40}N_8O_4$ from the absorbance of a solution obtained by diluting 1 volume of a 0.05% w/v solution of dipyridamole BPCRS in 1 M hydrochloric acid to 50 volumes with water.

4.2 Methods of analysis reported in the literature

4.2.1 Identification

The following two color tests have been recommended for the identification of dipyridamole [2]:

Mandelin's Test: Prepare the reagent by dissolving 0.5 g of ammonium vanadate in 1.5 mL of water and diluting to 100 mL with sulfuric acid. When dipyridamole is reacted with this reagent, the resulting solution has a violet color.

Marquis' Test: Prepare the reagent by mixing 1 volume of formaldehyde with 9 volumes of sulfuric acid. When dipyridamole is reacted with this reagent, the resulting solution has an orange color.

4.2.2 Titrimetric methods

Ganescu *et al.* used a volumetric and a spectrophotometric method for the determination of dipyridamole with Reinecke salt analogues [15]. Eight complex salts of dipyridamole, namely dipyridamole $H_2[Cr(SCN)_4(amine)_2]_2$ (amine = aniline or other aromatic heterocyclic amines), were prepared and characterized. Dipyridamole in these salts was determined through oxidation (using $KMnO_4$, $KBrO_3$ or KIO_3), through spectrophotometry, or through complexation (using EDTA). The results were subjected to statistical analysis.

Ganescu *et al.* used thiocyanato–Cr(III)-complexes in the titrimetric and spectrophotometric determination of dipyridamole [16]. For titrimetric analysis of the drug (1.36–27.2 mg), sample solutions (25–50 mL) were acidified with hydrochloric acid and treated with excess of a 3% solution of ammonium dianilinetetrakis (thiocyanato) chromium (III) in aqueous 25% ethanol. The red-violet precipitate was collected and washed, decomposed with NaOH, and the mixture again acidified. The liberated thiocyanate was titrated with potassium permanganate, potassium bromate, or potassium iodate.

Qi *et al.* studied the effect of water content in perchloric acid on the non-aqueous potentiometric titration of dipyridamole and other nitrogen-containing compounds [17]. The USP non-aqueous titrimetric method specifies a 0.1 N perchloric acid solution, with a water content in the range of 0.02–0.05%. Experiments were carried out on nitrogen-containing compounds to see if this water range could be extended. When using an increased water content, the results showed a maximum difference of 0.7% when compared with those obtained using the standard method. Relative standard deviations were less than 0.4%. The study authors recommended that perchloric acid with a water content in the range of 0.02–0.5% should be permitted for the titration of nitrogen-containing compounds.

4.2.3 Potentiometric methods

Hassan and Rizk developed potentiometric dipyridamole sensors based on lipophilic ion-pair complexes and native ionic polymer membranes [18]. The sensors are based on the use of the ion-association complexes of dipyridamole cation with tetraphenylborate and reineckate counter-anions as ion-exchange sites in plasticized PVC matrices. A plasticized native polymer (carboxylated polyvinyl chloride) can also be used. These sensors exhibit linear and near-Nernestian responses for 10 mM–1 μM dipyridamole, with a cationic slope of 53.4–55.1 mV/decade, and a working pH range of 2–5. The sensors proved to be useful for determining dipyridamole in various dosage forms. The average recovery was reported to be 98.7%, with a standard deviation of 0.6%.

Issa *et al.* described a potentiometric titration method based on the use of ion-selective electrodes [19]. The electrode response was Nernestian over the range of 6.3 μM–0.1 M, when operated in the pH range of 2–7 and at a temperature range of 25–70°C. The response time was reported to be 20–30 s. The recovery was 97.37–103.35%, with relative standard deviation of 0.06–0.5%.

4.2.4 Extraction-gravimetric method

Liu *et al.* determined dipyridamole in tablets using a modified extraction-gravimetric method combined with a surface acoustic wave resonator sensor [20]. The powdered sample (200 mg) was mixed with 0.02 M hydrochloric acid and filtered. A portion of the filtrate was mixed with 0.2 M sodium hydroxide, shaken with chloroform, and a portion (1 μL) of the organic layer was then applied to a surface acoustic wave resonator sensor. After the chloroform was evaporated by a nitrogen flow, frequency shift responses were measured after 3 min. The device comprised 62 MHz one-port resonators fabricated on Y-cut 2-propagation $LiNbO_3$ crystals with aluminum metallization, which were coated with polyvinylidene fluoride. The sensor was mounted in a measuring chamber and connected to an oscillator. The calibration graph was linear for 0.065–1.12 μg dipyridamole, and the relative standard deviation ($n = 5$) for 0.48 μg was 3.2%. The detection limit was 38 ng, and the mean recovery was 99.8%.

4.2.5 Conductimetric method

Bahbouh *et al.* used a conductimetric method for the analysis of dipyridamole [21]. Ground tablets containing the drug were dissolved in 100 mL of 67% ethanol, and a portion was transferred to a titration cell equipped with a conductivity cell. The sample solution was diluted to 50 mL with water and titrated with 5 mM phosphotungstic acid. After each addition of titrant, conductivity measurements were recorded after a 2 min interval and plotted against the amount of titrant added. The end point was determined at the curve break. Results obtained by this method were in good agreement with those obtained by the official pharmacopoeial methods.

4.2.6 Spectrophotometric methods

4.2.6.1 Infrared spectrometry

Huvenne and Lacroix described a mathematical procedure for the correlation of band intensities of the Fourier-transform infrared absorption spectra with those of the corresponding infrared transmission spectra of compounds in KBr discs [22]. The procedure was applied to spectra of flunitrazepam, dipyridamole, and lactose that were obtained through the use of a Nicolet 7199 B FTIR spectrometer with photo-acoustic detection. When the photoacoustic spectrum of a plant charcoal was used to correct the spectra for inequalities in the incident light flux before applying the procedure, the correlated band intensities were generally consistent with those obtained using infrared transmission spectra. The procedure may be useful for the direct identification of the drugs.

Umapathi *et al.* determined aspirin and dipyridamole simultaneously in bulk form, and in dosage forms, by an infrared spectrophotometric method [23]. Accurately weighed quantities of the drug (250 mg, 375 mg, 500 mg, 625 mg, 750 mg, 875 mg, and 1000 mg) were transferred quantitatively to volumetric flasks, dissolved in and diluted to 25 mL with chloroform. This procedure yielded solutions containing 10–40 mg/mL and 10–60 mg/mL of aspirin and dipyridamole, respectively. Two other mixed standard solutions of aspirin and dipyridamole were prepared containing varying concentrations (10–40 mg/mL) of aspirin and powdered tablets equivalent to 600 mg dipyridamole (aspirin) and 480, 800 or 320 mg aspirin (dipyridamole) dissolved in and diluted to 25 mL with chloroform. The solutions were filtered, and the infrared spectra recorded using a calibrated sodium chloride cavity cell (path length 0.5 mm) against a chloroform reference. The absorbance at 920 and 2935 cm^{-1} for aspirin and dipyridamole, respectively, was used for quantitation.

4.2.6.2 Ultraviolet spectrometry

El-Yazbi *et al.* reported an application of a derivative-differential ultraviolet spectrophotometric method for the determination of oxazepam or phenobarbitone in the presence of dipyridamole [24]. Tablets containing the drugs were powdered and dissolved in ethanol. For solutions of oxazepam and dipyridamole, two portions of each were diluted with 0.1 N sulfuric acid and 0.05 M sodium borate, and subjected to differential spectrophotometry with measurements being made at 283, 292, 298 and 282, and 307, and 296 nm. First derivative (ΔD_1)

and second derivative (ΔD_2) determinations were made. For solutions of dipyridamole and phenobarbitone, two portions of each were diluted with 0.1 N sulfuric acid and 0.1 N sodium hydroxide, and differential measurement were made at 253 and 309, 238 and 310, and 244 and 315 nm for ΔD_1 and ΔD_2, respectively. The calibration graphs were linear over the range of 0.2–1.6 mg %, the recoveries were similar (close to 100%), and the coefficients of variation were in the range of 0.8–1.2%.

Hu *et al.* described an ultraviolet spectrophotometric method for the analysis of dipyridamole [25]. A sample (50 mg) of the powdered tablet was dissolved and diluted to 100 mL with 0.01 M hydrochloric acid. A 2 mL portion of the solution was diluted to 100 mL with 0.01 M hydrochloric acid, and the absorbance of the final solution was measured at 283 nm (or at 403 nm for pure dipyridamole). The calibration graph was linear for upto 12 or 60 μg/mL, and suitable for high-content determinations. Results were in good agreement with those obtained by titration with bromate.

Barary *et al.* used a derivative spectrophotometric method for the determination of dipyridamole and carbocromen hydrochloride in the presence of their oxidative degradation products [26]. Dipyridamole in 0.1 M hydrochloric acid was determined by first derivative (D_1) spectrophotometry at 240 and 260 nm, and by second derivative (D_2) spectrophotometry at 246 and 268 nm. For both drugs, 0.1 M-hydrochloric acid was used as a blank. Graphs of D_1 and D_2 were linear for 4–12 μg/mL of dipyridamole. The technique effectively corrects for the presence of degradation products. This method was used successfully to determine dipyridamole in persantin tablets.

Khashaba *et al.* descried a first derivative spectrophotometric method for the analysis of dipyridamole in the presence of aspirin in a ratio of 1:4.4, as well as for buphenine hydrochloride in the presence of aescin in a ratio of 1:20 [27]. The method depends on the measurement of the first derivative amplitude (peak to zero) at 312 nm and at 250 nm for minor components dipyridamole and buphenine hydrochloride, respectively. No interference was observed from the other combined drugs (major components) at the selected wavelength. The method response obeys Beer's law in the concentration range of 2–18 μg/mL for dipyridamole, and 3–38 μg/mL for buphenine hydrochloride. Laboratory synthetic mixtures of the combined drugs in different ratios were analyzed showing good and reasonable recoveries. For dipyridamole the reported recovery range was 97.11 ± 0.09 to 100.17 ± 0.47, and for buphenine hydrochloride the reported range was 98.23 ± 0.89 to 99.53 ± 0.31. The method was also used for the assay of both drugs in single pharmaceutical dosage forms. Small relative standard deviation values (i.e., less than 2%) were noted, and the results obtained were found to be in good agreement with those obtained using official methods.

Korany and Haller described a spectrophotometric method for the determination of dipyridamole and oxazepam in two component mixtures [28]. The Δp_j method was applied to the simultaneous determination of the two drugs in two component mixture. Authentic mixtures and tablets containing the two drugs were assayed by measuring the absorbances in 0.1 N sulfuric acid and 0.05 M borax solutions, and then calculating the Δp_2 at 2 wavelength sets {272–312 nm (at 8 nm intervals) and 288–328 nm (at 8 nm intervals)}. The concentration is then determined by solving a pair of simultaneous equations. The results obtained were reasonably reproducible, exhibiting a coefficient of variation that was less than 2%. The method was compared with a ΔA modification of Vierordt's method of two-component analysis. The Δp_j method was found to yield more accurate and precise results.

4.2.6.3 Colorimetric methods

Sane *et al.* used a spectrophotometric method for the determination of dipyridamole and methdilazine hydrochloride from pharmaceutical formulations [29]. The method uses a simple technique of ion pair extraction of the two drugs with acid dyes like solochrome cyanine (SCR), solochrome dark blue (SDB), solochrome black T (SBT), and chromotrope 2B (C2B) in the presence of an acidic buffer. The colored complexes of the drugs are extracted into chloroform, and their absorbances measured at the wavelength of maximum absorption. The drug was detected at 460, 520, and 520 nm when using SCR, SDB, and SBT dyes, respectively. The percentage recovery of the SCR, SDB, and SBT dye products from tablet formulations were 100.42, 99.55, and 98.98%, respectively, and the corresponding standard deviations were 0.122, 0.171, and 0.147. Beer's law was obeyed over the respective ranges of 4–28, 4–24, and 4–32 μg/mL and the respective coefficients of variation were 0.95%, 1.18%, and 1.17%.

Mahrous *et al.* reported the use a spectrophotometric method for the determination of dipyridamole and other cardiovascular drugs using *p*-chloranilic acid [30]. Tablets, or aqueous solutions containing salts of dipyridamole and the other drugs, were rendered alkaline with aqueous ammonia and extracted with chloroform (4 × 20 mL). The extract was diluted to 100 mL with chloroform, a portion (0.4 to 2 mL) was evaporated to dryness, and the residue was dissolved in 2 mL of acetonitrile. 0.2% chloranilic acid in acetonitrile (1 mL) was added, and the mixture was diluted to 5 mL with acetonitrile. The absorbance was measured at 522 nm against a reagent blank. Recoveries were in the range 99–102%, and Beer's law held from 0.1–0.5 mg/mL for dipyridamole.

Issa *et al.* used 2,3-dichloro-5,6-dicyano-*p*-benzoquinone for the spectrophotometric determination of dipyridamole and other cardiovascular drugs [31]. Aqueous sample solutions are extracted with chloroform, portions of the organic solution are evaporated, and the residues dissolved in 2 mL of acetonitrile. After that, 2 mL of the cited reagent (0.2% solution in acetonitrile) was added, and the solutions diluted to 5 mL with acetonitrile. After 15 min, the absorbance was measured at 430 nm. Calibration graphs are linear in the range 10–50 μg/mL of dipyridamole (molar absorptivity equal to 10,300). The results obtained using this method compared well with those obtained by the official methods.

Ganescu *et al.* used a thiocyanto–Cr(III)-complex for the spectrophotometric determination of dipyridamole (1–15 mg) in 25 mL of solution [16]. The solution was acidified with 1–2 mL of 1 M hydrochloric acid, and treated with a slight excess of ammonium diimidazoletetrakis (thiocyanato) chromium(III). The precipitate was collected and washed, dissolved in acetone, and diluted to 25 mL with the same solvent. The absorbance was measured at 540 nm (molar absorptivity equal to 846).

Shoukry *et al.* used selected chromotropic acid azo dyes to develop a spectrophotometric method for the determination of dipyridamole and chlorpheniramine maleate [32]. When using *p*-chlorophenylazochromotropic and 2-(*p*-nitrophenylazo)chromotropic acids, the method is simple, accurate, and highly sensitive for the analysis of the drugs. The method consists of extracting the ion-pair complexes into methylene chloride. After reaction with dipyridamole, the *p*-chlorophenylazochromotropic acid product exhibits an absorption maximum at 544 nm, while the 2-(*p*-nitrophenylazo)chromotropic acid product exhibits a maximum at 524 nm. Dipyridamole was determined upto 3.5–3.78 mg/mL using these reagents. The molar absorptivity and Sandell sensitivity of the reaction products were calculated. Statistical treatment of the results reflects that the procedure is precise, accurate, and is easily applied for the determination of the drugs in pure form and in their pharmaceutical preparations.

4.2.6.4 Fluorimetric methods

Li described a spectrofluorimetric method for the determination of dipyridamole in plasma [33]. Three to 5 mL aliquots of chloroform are used to extract dipyridamole from 0.5 to 1 mL samples of plasma. After centrifugation, the fluorescence of the extract is measured at 465 nm (excitation at 415 nm) against a blank. The calibration graph was found to be linear over the range of 40–800 ng/mL, and the detection limit was 10 ng/mL. The coefficients of variation ($n = 5$) were 3.43 and 0.97% at 76.31 and 261.02 ng/mL, respectively.

Shao *et al.* recommended the use of a simultaneous fluorimetric method for the determination of the dissolution rate of dipyridamole and aspirin tablets [34]. The powdered tablets (equivalent to weight of one tablet) of dipyridamole and aspirin were dissolved in simulated digestive fluids at 37°C, cooled, and diluted to 1 L with simulated digestive juice. The solution was filtered, and a 1 mL portion of the filtrate was mixed with 0.1 M sodium hydroxide and then set aside at room temperature for 1 h. The solution was mixed with 8 mL of phosphoric acid buffer (pH 6.8) and fluorimetrically detected for dipyridamole at 493 nm (excitation at 418 nm). The coefficients of variation for within-day and within 5 days were 2% for both dipyridamole and aspirin.

Pulgarin *et al.* reported the use of a direct method for the determination of dipyridamole in serum [35]. Dipyridamole (50–1000 ng) was added to 2.8 mL of 1 M ammonium chloride/sodium hydroxide buffer solution of pH 10 and 1.5 mL ethanol, followed by 0.5 mL of serum. The resultant mixture was diluted to 10 mL with water. Detection was by fluorescence spectrometry at 495 nm (excitation at 412.2 nm). Calibration graphs were linear from 5–100 ng/mL dipyridamole, and the detection limit was 2.8 ng/mL. Relative standard deviations were 1.6–14.4% ($n = 3$) for repeatability, and 2.5–23.3% ($n = 3$) for reproducibility. Recoveries were in the range of 99.1–102.7%. The effect of varying ethanol content, pH, and temperature was discussed.

Umapathi *et al.* used spectrofluorimetric methods for the estimation of aspirin and dipyridamole in pure admixtures and in dosage forms [36]. The methods are described for the determination of the two drugs in pharmaceutical preparation without prior separation from each other and the matrix. A finely ground sample of the tablets was extracted with chloroform. The extract was filtered, and appropriate volumes of the filtrate were diluted with 1% acetic acid in chloroform for the determination of aspirin and pure chloroform for the determination of dipyridamole. Aspirin and dipyridamole were determined at 345 nm (excitation at 246 nm) and 475 nm (excitation at 420 nm), respectively. Linear calibration graphs were obtained for 1–12 μg/mL aspirin and 2–12 μg/mL dipyridamole. Recoveries from 60 to 100 mg aspirin samples, and 75 mg dipyridamole from tablets, were 98.9–99.75 and 98.94–100.14%, respectively. The relative standard deviations ($n = 5$) were less than 0.53%.

Steyn described and made a comparison between two spectrofluorimetric methods for the determination of dipyridamole in serum [37]. The thin-layer chromatographic-fluoridensitometric method utilizes 1 mL of plasma, which is extracted at pH 10 with diethyl ether–dichloromethane (80:20). The organic phase is evaporated to dryness, reconstituted in 250 μL dichloromethane, and 5 μL aliquots spotted on a silica gel 60 plate. The plate is developed in ethyl acetate–methanol–ammonia (85:10:5), dried, dipped in a paraffin wax solution, dried, and scanned at an excitation wavelength of 380 nm. Using a 430 nm cut-off filter, all emitted light is collected using a photomultiplier tube. Quantitation was performed by the external standard method, with peak heights being measured and a calibration graph constructed. For the spectrofluorimetric method, 1 mL of plasma is extracted at pH 10 with 8 mL of hexane-isoamyl alcohol (95:5), and the organic phase directly used for the measurement of the fluorescence response (excitation at 405 nm, and emission at 495 nm). Quantitation was accomplished by measuring the fluorescence of standards that were treated as above and constructing a calibration graph of concentration versus fluorescence intensity. Concentrations of unknowns were found by interpolation from this graph. The two methods were found to exhibit good correlation, but the spectrofluorimetric method proved to be more amenable to the analysis of a large number of samples.

Murillo-Pulgarin *et al.* used a non-linear variable-angle synchronous fluorescence method for the simultaneous determination of dipyridamole and three other drugs that have potential for abuse by athletes [38]. The drugs studied (atenolol, propranolol, amiloride, and dipyridamole) were determined simultaneously at concentrations of 10–400, 6–200, 5.6–280, and 5–100 ng/mL, respectively. Analyses were performed at 20°C in aqueous 70% ethanol at pH 7.5, adjusted with trishydroxymethyl aminomethane buffer solution. A spectrum based on a single scan was obtained by nonlinear variable-angle synchronous fluorimetry, with measurement of the absolute values of the first derivative at excitation/emission wavelengths of 228.8/300, 287.2/340, 366.4/412.8, and 288/487.2 nm for the four drugs, respectively. The detection limits were 5.2, 1.9, 2.1, and 0.1 ng/mL for the four drugs, respectively, and the relative standard deviations for the repeatability and reproducibility were 0.8–2.5% and 0.85–11.5%, respectively. When applied to dosage forms of one compound to which the others were added, recoveries were in the range of 99.8–110%.

Gong and Zhang developed a modified β-cyclodextrin based fluorosensor for dipyridamole [39]. Samples (2 mL) were injected into a carrier stream (2 mL/min) consisting of 0.1 M-sodium hydroxide–0.1 M sodium dihydrogen phosphate buffer of pH 8. The stream passed through a flow cell (25 mL) packed with silica gel (100–120 mesh) on which 2,6-*O*-diethyl-β-cyclodextrin had been immobilized, and the fluorescence of the complex with dipyridamole was measured at 468 nm (excitation at 414 nm). Between injection of samples or standard, 2 mL of aqueous 70% methanol was injected to desorb the dipyridamole from the cell. Calibration graphs were linear, with a detection limit of 0.8 nM dipyridamole. The relative standard deviation ($n = 10$) was 1.4% for 50 nM dipyridamole. No interference was observed with the 15 compounds tested (sugars, ascorbic and salicylic acids, trace metals, or starch). Urine samples were analyzed after mixing 5 mL with 5 mL of the buffer and diluting to 25 mL. Recoveries of 100–150 nM dipyridamole solutions were in the range 96–101%. Extracts of tablets and dilutions of injection formulations of dipyridamole gave results that were in agreement with the labeled values and with those obtained using the 1995 Chinese Pharmacopoeial method.

Bahbouh *et al.* used a spectrophotometric method for the determination of dipyridamole in pure drug and its dosage forms using chromotrope 2B and phosphotungestic acid [21]. Ground tablets containing 0.1–1 mg dipyridamole were dissolved in 100 mL of 1 mM hydrochloric acid. A portion of this solution was transferred to a 50 mL separating funnel, treated with 1 mL of 0.1 M of hydrochloric acid and 10 mL of 1 mM chromotrope 2B, and then extracted with two 5 mL aliquots of chloroform. The separated organic extracts were centrifuged, and the absorbance was measured at 540 nm against a reagent blank. Beer's law was obeyed from 5 to 60 μg/mL, and the recoveries were 101.9–102.8%. Results obtained using this method were in good agreement with those obtained by the official pharmacopoeial methods.

El-Yazbi *et al.* developed a spectrophotometric method for the determination of dipyridamole in ternary mixtures by the derivative ratio spectrum-zero crossing approach [40]. This method was used for the resolution of ternary mixtures with overlapping spectra, and is based on the simultaneous use of the zero-crossing and Salinas method developed by Berzas-Nivado *et al.* [41]. The method involves the use of the first derivative of the ratio spectra (obtained by dividing the absorption spectrum of the mixture by that of one of its components) and zero-crossing wavelength measurements. The concentrations of the other components were obtained from their calibration graphs. The utility of the method was illustrated by resolution of two ternary mixtures, one being dipyridamole, aspirin, and salicylic acid, and the other being dipyridamole, oxazepam, and 2-amino-5-chlorobenzophenone. The method was found to be suitable for both synthetic mixtures and commercial dosage forms. The strength of the procedure was in that it requires no solvent extraction and provides a direct estimate of degradation products in the presence of the intact drug.

Cao *et al.* used chemometrics to enhance a spectrofluorimetric resolution for the determination of dipyridamole in mixture with a amiloride and propranolol (closely overlapping drug mixtures) [42]. The mixture was determined in aqueous solution by fluorimetry, with excitation wavelengths of 220–230 nm (4 nm intervals) and emission wavelengths of 330–522 nm (4 nm intervals), at a scan rate of 1200 nm/min. A $49 \times 26 \times 9$ data array was collected, and treated by a modified parallel factors analysis algorithm with penalty diagonalization error. Linear ranges for dipyridamole, amiloride, and propranolol were 20–110, 10–150, and 9–100 ng/mL, respectively, and the corresponding correlation coefficients were better than 0.999.

Ruiz-Medina *et al.* prepared a flow-through optosener with fluorimetric transduction for the sensitive and selective determination of dipyridamole in aqueous solutions and biological fluids [43]. The method is based on a monochannel flow-injection analysis system using Sephadex QAE A-25 resin, placed into a Hellma 176-QS fluorimetric flow-through cell, as an active sorbing substrate. The native fluorescence of dipyridamole fixed on the solid sorbent is continuously monitored at wavelengths 305 nm and 490 nm for excitation and emission, respectively. After obtaining the maximum fluorescence intensity, the eluted solution (potassium dihydrogen phosphate/sodium hydroxide buffer solution, $c(T) = 0.05$ mol/L, pH 6) is allowed to reach the flow cell, the analyte is removed, and the resin support is regenerated. When a sodium hydroxide (10^{-4} mol/L)/sodium chloride (0.1 mol/L) solution is used as a carrier solution at a flow rate of 1.56 mL/min, the sensor responds linearly in the measuring range of 10–500 μg/L with a detection limit of 0.94 μg/L and a throughput of 22 samples per hour (300 μL of sample volume). The relative standard deviation for ten independent determinations (200 μg/L) is less than 0.82%. The method was satisfactorily applied to the determination of dipyridamole in pharmaceutical preparations and in human plasma.

4.2.6.5 Phosphorimetry methods

Murillo-Pulgarin *et al.* used a phosphorimetry method for the determination of dipyridamole in pharmaceutical preparations [44]. Ten tablets or capsules were powdered and homogenized, and then 0.1 g was dissolved in 0.1 M sodium dodecyl sulfate. The determination of dipyridamole was carried out in 26 mM sodium dodecyl sulfate/ 15.6 mM thallium nitrate/20 mM sodium sulfite, whose pH was adjusted to 11.5 by the addition of sodium hydroxide. After 15 min at 20°C, the phosphorescence was measured at 616 nm (after excitation at 303 nm). The calibration graph was linear from 100–1600 ng/mL, with a detection limit of 16.4 ng/mL. Relative standard deviations were in the range of 0.5–7.3%, and sample recoveries were in the range of 95–97%.

Munoz-de-la-Pena *et al.* described a micellar-stabilized room temperature phosphorescence method for the stopped-flow determination of dipyridamole in pharmaceutical preparations [45]. The pharmaceutical preparation was dissolved in and diluted with 0.1 M sodium dodecyl sulfate. A portion of the resulting solution was mixed with 9 mM sodium dodecyl sulfate, 1 mM sodium hydroxide, and 15 mM sodium sulfite in a drive syringe. A second drive syringe was filled with 1 mM sodium hydroxide, 15 mM sodium sulfite, and 60 mM thallium nitrate. The contents of the two syringes were mixed in a stopped-flow mixing chamber, and the phosphorescence intensity was recorded at 616 nm (after excitation at 303 nm). Dipyridamole was quantified by measuring the phosphorescence intensity after 10 s (method A), or by measuring the maximum slope of phosphorescence development over a time interval of 1 s (Method B). The calibration graph was linear for 50–400 ng/mL dipyridamole for both methods. The detection limits were 21.5 and 37.5 ng/mL with methods A and B, respectively. The relative standard deviations were 1.4–11.5% for method A and 2.7–21% for method B.

Salinas-Castillo *et al.* used a new, rapid, sensitive, simple, and selective phosphorimetric method for the determination of dipyridamole in pharmaceutical preparation [46]. The phosphorescence signals are a consequence of intermolecular protection when analytes are exclusively in the presence of heavy atom salts, and sodium sulfite is used as an oxygen scavenger to minimize RTP quenching. The determination was performed in 0.1 mol/L thallium(I) nitrate and 8 mmol/L sodium sulfite at a measurement temperature of 20°C. The phosphorescence intensity was measured at 635 nm, after excitation at 305 nm. A linear concentration range was obtained between 0 and 100 ng/mL, with a detection limit of 940 ng/L, an analytical sensitivity of 2.5 ng/mL, and a standard deviation of 2.7% at 60 ng/mL concentration. The method has been successfully applied to the analysis of dipyridamole in a unique Spanish commercial formulation containing 100 ng/mL per capsule. The recovery was 101.6%, with a standard deviation of 6.5%. The method has been validated using standard addition methodology.

4.2.6.6 Chemiluminescence methods

Nakushima *et al.* determined dipyridamole on the basis of its chemiluminescence intensity after injecting into a carrier stream of triethylamine in acetonitrile (1 mL/min) combined with a stream of bis-(2,4,6-trichlorophenyl)oxalate and hydrogen peroxide [47].

Sugiura *et al.* studied the effect of additives of reaction solvents in a peroxyoxalate chemiluminescence detection system for liquid chromatography [48]. The method described previously by Imai *et al.* [49] was adopted to obtain the reaction profile and optimum conditions for the chemiluminescent detection with bis[4-nitro-2-(3,6,9-trioxadecyloxycarbonyl) phenyl]oxalate and hydrogen peroxide by changing the reaction coil lengths between the post-column mixer and the detector. The addition of a phthalate ester to the basic solvent mixture for the chemiluminogenic reagents, ethyl acetate-acetonitrile (1:1), gave a considerable reduction in the baseline noise level without decreasing the fluorescent signal from the test compound, dipyridamole. A mixture of 0.25 mM bis-[4-nitro-2-(3,6,9-trioxadecyloxycarbonyl) phenyl]oxalate and 25 mM hydrogen peroxide in ethyl acetate–acetonitrile–0.25% di(2-ethylhexyl)phtlalate was optimum for detection of amol levels of fluorescent compounds. The detection limit of dipyridamole was 10 amol.

Imai *et al.* studied the catalytic effect of imidazole and other bases on the peroxyoxalate chemiluminescence reaction for HPLC [49]. The peak height of dipyridamole obtained using the eluent containing buffers was largest at pH 7, a few times less at pH 6 and pH 5, 100 times less at pH 4, and a few hundred times less at pH 3. Dipyridamole was separated on the ODS column, with a detection limit of dipyridamole equal to 40 amol.

Sugiura *et al.* also developed a chemiluminescence detection system using bis-[4-nitro-2-(3,6,9-trioxadecyloxycarbonyl) phenyl]oxalate for the sensitive determination of dipyridamole and the other fluorescent compounds separated with an acidic mobile phase [50]. Dipyridamole samples (10 fmol) and dansylated amino acids (10 fmol) were subjected to HPLC on two systems. System A involved the mixing of the cited chemiluminescent reagent and an imidazole buffer solution in a coil, resulting in the generation of active intermediates, which were then mixed with the column eluate in a two dimensional mixing device. System B involved the mixing of the chemiluminescent reagent, buffer, and column eluate directly in a rotating flow-mixing device fitted with three directional inlets (0.3, 0.3, and 0.5 mm, i.d.) followed by peroxyoxalate chemiluminescence detection. Detection limits were 10 amol and sub-fmol for dipyridamole and the dansylated amino acids, respectively.

Nie *et al.* used a flow injection chemiluminescence method for the determination of dipyridamole in Persantin tablets [51]. A flow injection analysis system equipped with a chemiluminescence detector was used. Sodium hypochlorite was used as a chemiluminescent substance, and Triton X-100 was used as an enhancer of the chemiluminescence reaction. Sample solutions (pH 4–6) was injected into 0.2% Triton X-100 and treated with 0.5% sodium hypochlorite, and the chemiluminescence intensity then measured. The calibration graph was linear from 0.04 to 10 μg/mL of Persantin tablets (as its active ingredient dipyridamole), with a detection limit of 11 ng/mL. Recoveries were quantitative and the relative standard deviation ($n = 3$) was 2.7%. The method has been used to determine Persantin in injection and tablets, and the results obtained using this method agreed with those obtained by the Chinese Pharmacopoeial method.

Zheng and Zhang developed a new electrogenerated chemiluminescence method for the determination of Persantin with flow-injection analysis [52]. A sample obtained from 20 ground tablets was dissolved in water, and a 90 μL portion of the solution was injected into a carrier stream of 8 μM luminol at 4 mL/min in a flow-injection analysis manifold. The mixed stream was merged with a reagent stream containing hypobromite (obtained online by galvanostatic electrolysis of 0.15 M potassium bromide in sodium carbonate/sodium bicarbonate buffer of pH 11) at 3 mL/min, in an electrolytic flow-cell, at a current of 200 μA. The reduced chemiluminescence due to inhibition caused by the dipyridamole in Persantin was measured. The calibration graph was linear from 0.01 to 2 mg of dipyridamole, with a detection limit of 4 μg/L. The relative standard deviation was 2%, and no interference was observed.

Yang *et al.* determined dipyridamole by flow-injection analysis method with chemiluminescence detection [53]. Portions (50 μL) of a standard solution of dipyridamole were injected into a carrier stream of 2 mM hydrogen peroxide at 1.3 mL/min in a flow-injection manifold, and merged with a confluent of two reagent streams (50 mM potassium permanganate and 2 M sulfuric acid) flowing at the same rate. The merged stream was mixed in transfer tubing (0.8 mm i.d.), and the resulting chemiluminescence intensity was measured. The calibration graph was linear from 0.2 to 80 μg/mL of dipyridamole, with a detection limit of 58 ng/mL. Most co-existing ions and substances did not interfere. The method was used in the direct analysis of dipyridamole in both tablets and injection solutions, with relative standard deviations of 1.2–1.4%.

Pascual *et al.* studied the effect of antioxidants on the luminal chemiluminescence produced by dipyridamole [54]. The method is based on the use of the reaction between luminal and horse radish peroxidase, and is proposed for the determination of dipyridamole.

4.2.7 *Electrochemical methods*

4.2.7.1 Polarography method

Tuncel *et al.* investigated the polarographic behavior of dipyridamole in a supporting electrolyte consisting of 0.2 M potassium chloride, 20% (v/v) ethanol, and buffer [55]. The optimum polarographic conditions were pH 7–8, potential rate of 4 mV/s, drop time of 1 s, and a pressure of 1000 dyne cm^{-2}. The effect of concentration on the limiting current was examined using direct current, superimposed constant amplitude pulse, superimposed amplitude pulse, and differential pulse polarographic modes at pH 7.8 and optimum conditions. Differential pulse polarography was compared with ultraviolet spectrophotometry for the determination of dipyridamole in tablets. The methods were similar in terms of simplicity, reproducibility, and time consumption, but the differential pulse polarography method was ten times more sensitive and more specific.

4.2.7.2 Voltammetric methods

Zeng *et al.* reported the measurement of trace dipyridamole by adsorptive stripping voltammetry [56]. A solution (10 mL) of dipyridamole in 0.05 M sodium hydroxide containing 10% ethanol was purged with nitrogen for 4 min before pre-concentration of dipyridamole on a static-Hg-drop electrode at -1.2 V versus silver/silver chloride. A platinum wire counter electrode was used. After quiescence for 15 s, the voltammogram was recorded by applying a linear-sweep scan to -1.6 V at 100 mV/second. The calibration graph was rectilinear from 0.005 to 1 μM dipyridamole (pre-concentration for 1 min), and a detection limit of 1 nM was attainable by pre-concentration for 5 min. For 0.5 μM dipyridamole and a pre-concentration of 15 s, the relative standard deviation was 1.7% ($n = 12$). The method was applied to injection solutions (recoveries 88–92%), tablets (recovery was 105%, and relative standard deviation of 9.4%), and urine.

Wang *et al.* determined trace amounts of dipyridamole by stripping voltammetric using a Nafion modified electrode [57]. Human serum (0.5 mL) containing dipyridamole was mixed with 0.5 mL water and 1 mL 20% TCA, and centrifuged. A portion (0.1 mL) of the supernatant was mixed with 10 mL of Britton-Robinson buffer (pH 1.7), and analyzed by anodic stripping voltammetry using a Nafion-modified vitreous C working electrode (0.126 cm^2). After accumulation for 1 min at 0 V versus silver/silver chloride (platinum counter electrode) with stirring, and a 30 s rest period, then a positive scan was initiated at 100 mV/s until $+1$ V was reached. The calibration graph for dipyridamole was linear for 1–80 nM, and the detection limit was 80 pM for 4 min accumulation. The relative standard deviation for the determination of 3.78 μg/mL of dipyridamole in serum was 4.76%, and recoveries were in the range of 96.7–104%.

Almeida *et al.* studied the oxidation of dipyridamole in acetonitrile and ethanol by voltammetric and spectroscopic methods [58]. The electrochemical oxidation of dipyridamole in acetonitrile and ethanol at high and low scan speeds was similar, and exhibited two reversible voltammetric waves corresponding to consecutive one-electron reactions. The $E_{1/2}$ values were 120 and 400 mV in acetonitrile, and 470 and 680 mV in ethanol. The release of one electron produced a radical cation which was detected by electron spin resonance spectrometry. The electrochemical oxidation of dipyridamole was accompanied by a decrease in fluorescence intensity at 470 nm, and significant changes in the ultraviolet–visible absorption spectrum. The oxidation of dipyridamole proceeded more slowly in ethanol than in acetonitrile.

Yang studied the anodic voltammetry of dipyridamole at a glassy carbon electrode [59]. Portions of a standard dipyridamole solution were mixed with 10 mM hydrochloric acid for anodic voltammetry, using a glassy carbon working electrode and measurement of the anodic oxidation peak at 0.62 V (versus saturated calomel electrode). The calibration graph was linear from 0.5 to 10 mg/L of dipyridamole, and over 20 foreign species did not interfere. The method was directly applied to the analysis of dipyridamole in tablets and urine, with recoveries of 100.5–101.2% and a relative standard deviation of 1.5–2.2%.

Ghoneim *et al.* described a cathodic adsorptive square-wave stripping voltammetric method for the assay of dipyridamole in human serum [60]. The rapid and sensitive square-wave voltammetric procedure was optimized for the determination of dipyridamole after its adsorption pre-concentration onto a hanging mercury drop electrode. The peak current of the first of the two peaks developed for this drug in Britton-Robenson buffer at pH 8 was considered for the analytical study. An accumulation potential of −1.0 V against $Ag/AgCl/KCl_s$, a pulse amplitude equal to 100 mV, a scan increment of 10 mV, and a frequency of 120 Hz were found to be the optimal experimental parameters. Dipyridamole can be determined in the concentration range of 9 nM–5 μM using accumulation times of 30–300 s. A detection limit of 40 pM was achieved after a 300 s accumulation time. Applicability to serum samples was illustrated. The average recoveries for dipyridamole spiked to serum at 0.25–4.5 μg/mL were 96–102%, with a standard deviation of 2.9%. A detection limit of 0.06 μg/mL of serum was obtained.

4.2.8 Mass spectrometry

Rodrigues Filho *et al.* studied the fragmentation of dipyridamole and several of its derivatives by electrospray ionization combined with collisional activated decomposition mass spectrometry in both positive and negative modes [61]. These compounds produce abundant mono-charged ions ($[M + H]^+$) under electrospray ionization. Interpretation of the collisional activated decomposition spectra showed that fragmentation occurs preferentially in the ethanolamine groups attached at C-2, C-4, C-6, and C-8. 2-Methoxyethanol is eliminated when ethanolamines are in positions C-2/C-6, and 2-aziridinethanol is eliminated from

C-4/C-8 ethanolamines. The fragmentation schemes were supported by deuterium labeling experiments and tandem mass spectrometry.

4.2.9 Chromatographic methods

4.2.9.1 Thin-layer chromatography

Strelyuk described a thin-layer chromatographic method for the qualitative detection of dipyridamole [62]. Data were presented on the use of various reagents for the detection by solution color change and precipitate formation in the thin-layer chromatography of dipyridamole. Thin-layer chromatography was on silica gel–calcium sulfate dihydrate, with methanol–aqueous ammonia or benzene-dioxane being used as the mobile phase with detection at 270–330 nm. The R_f values were 0.7–0.75 and 0.5–0.55 for the two solvent systems, and the detection limit was 0.08 μg.

Argekar and Kunjir reported the use of a high performance thin-layer chromatographic method for the simultaneous determination of aspirin and dipyridamole in pharmaceutical preparations [63]. Tablets containing aspirin and dipyridamole were powdered and dissolved with ultrasonication in methanol-ethanol (1:1). The solution was filtered and further diluted with methanol. The resulting solution was applied to aluminum-backed silica gel $60F_{254}$ high performance thin-layer chromatographic plates. The plates were developed with the upper layer of ethyl acetate–ethanol–13.5 M ammonia (5:1:1) to 7 cm, and the absorbance of the spots was measured by densitometry at 290 nm. Calibration graphs were linear for 60–340 ng of aspirin and 60–380 ng of dipyridamole, with detection limits of 25 and 10 ng, respectively, and average recoveries of 99.72 and 97.08%, respectively.

The following three TLC systems [2, 64] were reported for the separation of dipyridamole:

System 1:

Plate: Silica gel G, 250 μm thick, dipped in, or sprayed with, 0.1M potassium hydroxide in methanol, and dried.
Mobile phase: Methanol:strong ammonia solution (100:1.5).
Reference compounds: Diazepaim R_f 75, chlorprothexine R_f 56, codeine R_f 33, and atropine R_f 18.
Analyte R_f value: 68 [2, 64].

System 2:

Plate: Silica gel G, 250 μm thick, dipped in, or sprayed with, 0.1 M potassium hydroxide in methanol, and dried.
Mobile phase: Cyclohexane:toluene:diethylamine (75:15:10).
Reference compounds: Dipipanone R_f 66, pethidine, R_f 37, desipramine R_f 20, codeine R_f 0.6.
Analyte R_f value: 00 [2, 64].

System 3:

Plate: Silica gel G, 250 μm thick, dipped in, or sprayed with, 0.1 M potassium hydroxide in methanol, and dried.
Mobile phase: Chloroform:methanol (90:10).
Reference compounds: Meclozine R_f 79, caffeine R_f 58, dipipanone, R_f 33, desipramine R_f 11.
Analyte R_f value: 37 (acidified iodoplatinate solution, positive) [2, 64].

4.2.9.2 Gas chromatography
The following gas chromatographic system [2, 65] was recommended for the separation of dipyridamole:

Column: 2.5% SE-30 on 80–100 mesh Chromosorb G (acid washed and dimethyl dichlorsilane-treated), 2 m × 4 mm internal diameter glass column. It is essential that the support is fully deactivated.

Column temperature: Normally between 100°C and 300°C. As an approximate guide, the temperature to use is the retention index divided by 10.

Carrier gas: Nitrogen at 45 mL/min.

Reference compounds: n-alkanes, with an even number of carbon atoms.

Analyte retention index: 1640 [2, 65].

4.2.9.3 High performance liquid chromatography
The following high performance liquid chromatography (HPLC) system [2, 66] has been described for the analysis of dipyridamole:

Column: Silica (Spherisorb S5W, 5 μm 12.5 cm × 4.9 mm i.d.).

Eluent: A solution containing 1.175 g (0.01 M) of ammonium perchlorate in 1000 mL of methanol, adjusted to pH 6.7 by the addition of 1 mL of 0.1 M sodium hydroxide in methanol.

k′ Values: 0.2 [2, 66].

Zhou *et al.* developed an HPLC method for the determination of dipyridamole, aspirin, and salicylic acid [67]. The HPLC system consisted of a Lichrospher 5-C_{18} (4.6 mm × 150 mm) column and a UV detector operated at 227 nm. The mobile phase (eluted at a flow rate of 1 mL/min) was composed of disodium hydrogen phosphate solution(methanol (50:50, pH adjusted to 3 by addition of phosphoric acid). The mean recovery of dipyridamole and aspirin was 100.9–104.1% and 97.01–97.89, respectively.

Chevalier *et al.* analyzed dipyridamole directly in the blood samples by a reversed phase high performance chromatographic method with fluorimetric detection [68]. The HPLC system uses a 15 cm × 4.8 mm column packed with Lichrosorb RP_8 (10 µm) and a mobile phase consisting of 40:60 v/v acetonitrile–water. Detection was made on the basis of fluorimetric detection at 410 nm (excitation at 305 nm). Depending on the model of the detector used, the detection limit of the method was between 2×10^{-13} and 2×10^{-15} mole injected. The precision was 5% for a concentration of 180 ng/mL of dipyridamole, and the method was used for a biodisposition study of dipyridamole in dogs.

Schmid *et al.* described a rapid and sensitive high-performance liquid chromatographic method for the determination of dipyridamole in human plasma [69]. The column used was a 12.5 cm × 4.6 mm filled with Lichrosorb RP 18, 5 µm. The mobile phase consisted of methanol–0.2 M Tris HCl buffer (80:20), eluted at a flow rate of 1 mL/min. Spectrofluorimetric detection at an emission wavelength of 478 nm (excitation at 415 nm) was used for detection. The results were confirmed by re-chromatographing the eluate from the column on silica gel G thin-layer plates, which were developed using a solvent system composed of toluene–isopropanol–ethanol–ammonia (70:15:15:1) and *n*-butanol–methyl ethyl ketone (80:20). The R_f values of dipyridamole for the two solvent systems are 0.60 and 0.80, respectively. The plates were dried in a stream of cold air, and then inspected under ultraviolet light at 254 nm.

Wolfram and Bjornsson described a rapid, sensitive, and specific high performance liquid chromatographic method for the quantitative analysis of dipyridamole in plasma and whole blood [70]. The method involves a single extraction on an alkalinized sample with diethyl ether, followed by evaporation of the organic solvent and ion-pair chromatography using fluorescence detection. The HPLC system consists of a Waters model 6000 A high-pressure solvent delivery system (equipped with a Model U6K injector), and a Waters μBondapak C_{18} reversed-phase column (30 cm × 39 mm; 10 μm). The mobile phase is a mixture of methanol–water (65:35), containing 1-heptanesulphonic acid sodium salt (0.005 M) with 0.1% acetic acid. The mobile phase is eluted at a flow rate of 2 mL/min with a column input of 170 atm (2500 p.s.i.). The column is insulated with sponge rubber to minimize ambient temperature variation. The fluorescence signal was measured with a Schoeffel Model FS 970 fluorimeter. An excitation wavelength of 285 nm was selected in conjunction with a 470 nm emission filter. The lower limit of sensitivity for dipyridamole is 1 ng/mL. Concentrations of dipyridamole between 1 and 500 ng per sample are measured, with an average coefficient of variation of 4.5% for plasma and 7.4% for whole blood.

Williams *et al.* used a high performance liquid chromatographic assay method for dipyridamole monitoring in plasma [71]. The HPLC system uses a Waters model 6000 A solvent delivery pump equipped with a U6K injector, a μBondapak C_{19} column (30 cm × 39 mm; 10 μm), and a Model 440 absorbance detector. The signal from the detector was quantified using a Shimadzu data processor and an Omni-Scribe recorder. A mobile phase flow rate of 1.5 mL/min was produced by a pressure of approximately 102 atm (1500 p.s.i.). The mobile phase was 50:50 mixture of acetonitrile and 0.01 M sodium phosphate in water (adjusted to pH 7). The absorbance reading of dipyridamole in methanol was made at 280 nm.

Rosenfeld *et al.* described a high-performance liquid chromatographic method with ultraviolet detection for the determination of dipyridamole in plasma [72]. The method is sufficiently sensitive to follow the concentration of dipyridamole in man for 48 h, and is linear over the range of concentrations found in man after a standard therapeutic regimen. The HPLC system used was a Waters model 6000 liquid chromatograph equipped with a U6K injector and a model 440 dual-channel filter absorbance detector in conjunction with a Helwett-Packard 3380 A integrator. The column used was an Altex Ultrasphere 5μm reversed-phase C_{18} column (25 cm × 0.46 mm). The drug was eluted with 33% acetonitrile in 0.02 M phosphate buffer containing

0.01 M *N,N,N,N*-tetramethylethylenediamine at pH 2.9. The flow rate was 2.5 mL/min, which was generated a pressure of approximately 238 bar. The effluent was monitored at 280 nm.

Sane *et al.* described a HPLC method for the simultaneous estimation of the combined dipyridamole and aspirin dosage form [73]. Twenty tablets containing dipyridamole and aspirin were accurately weighed and powdered. Powder equivalent to 75 mg of dipyridamole was accurately weighed and 40 mg of pyrimethamine (internal standard) was added. The contents were first dissolved in 20 mL of acetonitrile, and then in about 40 mL of the mobile phase. The solution was filtered through Whatman filter paper no. 1. The residue was washed with the mobile phase. The filtrate and the washings were collected in a 100 mL standard volumetric flask, and diluted to the mark with mobile phase. A 5 mL portion was transferred into a 50 mL volumetric flask, and diluted to the mark with mobile phase. A 10 mL portion of this solution was further diluted to 10 mL, and then 20 μl of this solution was injected onto the HPLC system. The chromatographic system used a Spectra physics HPLC equipped with a SP 8810 precision isocratic pump, and a SP 8450 UV–visible detector with wavelength of detection as 254 nm. The column used was a Nucleosil C_{18} (25 cm × 4 mm, 10 μm), with a mobile phase consisting of methanol : acetonitrile : water : triethylamine (55:0:40:0.1; pH adjusted to 4 using phosphoric acid). The flow rate was 1.5 mL/min, and the sample injected at ambient temperature using a Rheodyne 7125 (20 μL loop) injector. The standard deviation, coefficient of variation, and percent recovery for dipyridamole were 0.77, 1.03, and 101.09%, respectively.

Pederson described a specific HPLC method for the determination of dipyridamole in serum [74]. The HPLC system used was a Waters model 600 liquid chromatograph equipped with a U6K injector, a μBondapak C_{18} column (30 cm × 39 mm) (10 μm), and a model 440 dual channel filter absorbance detector in conjunction with a Tarkan W + W 600 recorder. The mobile phase was a 75:25 mixture of methanol and a 0.02 M solution of sodium acetate (adjusted to pH 4 with acetic acid). The solvent flow rate of 2 mL/min was produced by an applied pressure of approximately 2000 p.s.i. Detection of the analyte was made at the UV absorption maximum of 280 nm.

Table 6 contains a summary of the conditions of other reported HPLC methods that has also been reported for the determination of dipyridamole [49, 75–84].

Table 6

HPLC Conditions of the Methods Used for the Determination of Dipyridamole Hydrochloride

Column	Mobile phase and flow rate	Detection	Remarks	Ref.
15 cm × 4 mm of TSK ODS 80 TM (5 μm)	50 mM imidazole in 50 mM phosphate buffer of pH 6-acetonitrile (1:1) [1 mL/min] and 0.5 mM bis-[4-nitro-2-(3,6,9-trioxadecyloxy-carbonyl)-phenyl]oxalate (TDPO) in acetonitrile-25 mM-H_2O_2 in ethyl acetate for eluents of pH 5–7 and a mixture of 0.8 mMTDPO and 4 mM-bis-(2,4-dinitrophenyl)oxalate for eluents at pH 3–4.	Chemiluminescence	Separation of the drug and benzydamine HCl. Detection limit for dipyridamole was 40 amol.	49
30 cm × 3.9 mm of μBondapak C_{18} (10 μm)	0.1 M Sodium acetate buffer pH 4-methanol (1:3) [1 mL/min]	Amperometric at +0.70 V vs a SCE	Detection and determination of the drug by HPLC in pharmaceuticals	75

7 cm × 4.6 mm of Ultrasphere XL ODS (3 μm)	Methanol-0.02 M ammonium acetate buffer pH 5 (13:7) [1.5 mL/min].	280 nm	Analysis of free and plasma protein-bound dipyridamole by HPLC	76
10 cm × 2 mm of Hypersil ODS (5 μm)	Aqueous 20 mM-Na_2HPO_4 (containing 50 mM-Na dodecyl sulfate and H_3PO_4 to pH 2)-acetonitrile (13:12) [0.5 mL/min]	305 nm and by fluorimetry at 468 nm (excitation at 293 nm)	Rapid isocratic HPLC assay for the drug using a micro-bore column technique. Down to 0.05 μg/mL of the drug can be detected.	77
25 cm × 4.6 mm of Phenomenex ODS 1 (5 μm) operated at 60°C.	Methanol-aqueous 200 mM sodium pentanesulfonate (7:3) containing triethylamine 2 mL/L and adjusted to pH 3 with phosphoric acid [1.5 mL/min).	288 nm	Stability indicating analysis used to analyse the raw material and the capsule formulations.	78
30 cm × 3.9 mm of μBondapak C_{18} (10 μm)	0.1 M sodium acetate buffer of pH 4/methanol (9:11) [1 mL/min].	Coulometric at +0.65 V	Sensitive isocratic HPLC for the assay of the drug in plasma	79

(continued)

Table 6 (continued)

Column	Mobile phase and flow rate	Detection	Remarks	Ref.
TSK gel ODS-80-TM	5 mM phosphate buffer pH 5-acetonitrile (3:2) [1 mL/min] and 5 mM bis [4-nitro-2-(3,6,9-trioxadecyl oxycarbonyl)phenyl] -oxalate TDPO and 50 mM-acetonitrile [1.3 mL/min]	Chemiluminescent reagent.	Separates the drug and dansylated valine. Detection limit under the optimum conditions is 20 amol for dipyridamole.	80
30 cm × 3.9 mm of μBondapak C_{18} (10 μm)	Sodium acetate-acetic acid of pH 5.1 ± 0.1 in methanol (70:130) [1 mL/min].	276 nm	Stability indicating HPLC method. Analysis of the drug in injection	81
30 cm × 3.9 mm of Waters μBondapak C_{18} (10 μm)	Aqueous sodium acetate pH 5–36% acetic acid–650 mL methanol [1 mL/min].	276 nm	Stability indicating HPLC method for analysis of the degradation products in dipyridamole injection.	82

25 cm × 4 mm of LiChrosorb RP-18	15 mM-NaH_2PO_4 buffer (pH 7.6) containing 60–90% of acetonitrile [2 mL/min]	280 nm and by fluorimetry at 470 nm (excitation at 285 nm)	Purity evaluation of the drug by HPLC. Three impurities were separated from the drug.	83
15 cm × 4.6 mm of TSK ODS 80 TM (5 μm) operated at 40°C	50 mM imidazole buffer (pH 6)-acetonitrile (1:1). Eluate was mixed with 0.25 mM-bis-[(4-nitro-2-(3,6,9-trioxadecyloxy-carbonyl)phenyl]oxalate and 12.5 mM-H_2O_2 solution in acetonitrile–ethyl acetate (1:1)	Chemiluminescence	Analysis in rat plasma by HPLC of dipyridamole and benzydamine.	84

5. STABILITY

The stability of freshly prepared standard solutions of dipyridamole exposed to daylight showed considerable variation [74]. The rate of degradation seemed to follow first-order kinetics, and the rate constant was obviously dependent on the light intensity in the laboratory. The maximum degradation rate was found in the solutions with TRIS-buffered water, in which the half-time of dipyridamole varied from about 3–30 h. The corresponding half-time in ethanol was 12–240 h. Samples prepared with deproteinized serum were stable for several days. No difference between the reagent tubes was found, and there was no evidence of adsorption on the glassware or plastic tubes. These findings strongly indicate that all samples and standards should be protected from light [74].

6. PHARMACOKINETICS AND METABOLISM

Dipyridamole is completely absorbed from the gastrointestinal tract, with peak plasma concentration occurring about 75 min after oral administration. Dipyridamole is highly bound to plasma proteins. A terminal half-life of 10–12 h has been reported. Dipyridamole is metabolized in the liver, and is mainly excreted as glucuronide conjugates in the bile. Excretion may be delayed by enterohepatic recirculation. A small amount is excreted in urine [2, 5, 85, 86].

7. ACKNOWLEDGMENT

The authors would like to thank Mr. Tanvir A. Butt for his secretarial assistance in preparing this manuscript.

8. REFERENCES

1. ***Index Nominum 2000, International Drug Directory***, 17th edn., Swiss Pharmaceutical Society, eds., Medpharm GmbH-Scientific Publishers Stuttgart 2000, p. 360 (2000).

2. ***Clarke's Isolation and Identification of Drugs***, 2nd edn., A.C. Moffat, ed., The Pharmaceutical Press, London, p. 562 (1986).

3. ***The Merck Index***, 12th edn., S. Budavari, ed., Merck and Co., NJ, p. 567 (1996).

4. ***The British Pharmacopoeia (1993)***, Her Majesty's Stationary Office, London (1993), British Pharmacopoeia CD-ROM, Volume 1, p. 887 (2000).

5. ***Martindale, The Complete Drug Reference***, 33rd edn., Sean C. Sweetman, ed., Pharmaceutical Press, p. 877 (2002).
6. M.P. Rivey, M.R. Alexander, and J.W. Taylor, *Drug Intell. Clin. Pharm.*, ***18***, 869 (1984).
7. G.A. FitzGerald, *N. Engl. J. Med.*, ***316***, 1247 (1987).
8. C.R. Gibbs and G.Y. Lip, *Br. J. Clin. Pharmacol.*, ***45***, 323 (1998).
9. A. Eldor, I. Voldavsky, Z. Fuks, T.H. Muller, and W.G. Eisert, *Thromb. Haemost*, ***56***, 333 (1986).
10. E. Weber, T.A. Haas, T.H. Muller, W.G. Eisert, J. Hirsh, M. Richardson, and M.R. Buchanan, *Thromb. Res.*, ***57***, 383 (1990).
11. W.G. Eisert and T.H. Muller, *Thromb. Res. Suppl.*, ***12***, 65 (1990).
12. F.G. Fischer, J. Roch, and A. Kottler, **U.S. Patent 3, 031**, 450 (1962).
13. D. Lednicer and L.A. Mitscher, ***The Organic Chemistry of Drug Synthesis***, Volume 1, John Wiley and Sons, New York, p. 428 (1977).
14. ***United States Pharmacopoeia 24.*** The United States Pharmaceutical Convension, Rockville, MD (2000) p. 590; **USP 23**, pp. 539–540 (1995).
15. I. Ganescu, M. Preda, I. Papa, and A. Vladoianu, *Zentralbl Pharm Pharmakother Laboratoriumsdiagn*, ***127***, 577 (1988).
16. I. Ganescu, M. Preda, and I. Papa, *Arch. Pharm.*, ***324***, 321 (1991).
17. X.S. Qi, R.B. Miller, Y.H. Namiki, J. Zhang, and R. Jocobus, *J. Pharm. Biomed. Anal.*, ***16***, 413 (1997).
18. S.S.M. Hassan and N.M.H. Rizk, *Anal. Lett.*, ***33***, 1037 (2000).
19. Y.M. Issa, M.S. Rizk, A.F. Shoukry, and R.M. El-Nashar, *Electroanal.*, ***9***, 24 (1997).
20. D.Z. Liu, R.H. Wang, L.H. Nie, and S.Z. Yao, *J. Pharm. Biomed. Anal.*, ***14***, 1471 (1996).
21. M.S. Bahbouh, A.A. Salem, and Y.M. Issa, *Mikrochim. Acta*, ***128***, 57 (1998).
22. J.P. Huvenne and B. Lacroix, *Spectrochim. Acta*, ***44A***, 109 (1988).
23. P. Umapathi, P. Parimoo, and S.K. Thomas, *Indian Drugs*, ***31***, 489 (1994).
24. F.A. El-Yazbi, M.A. Korany, and M. Bedair, *J. Pharm. Belg.*, ***40***, 244 (1985).
25. J. Hu, H. Sun, and Y. Xu, *Huaxue Shijie*, ***28***, 19 (1987).

26. M.H. Barary, M.A.H. El-Sayed, M.H. Abdel-Hay, and S.M. Mohamed, *Anal. Lett.*, ***22***, 1643 (1989).
27. P.Y. Khashaba, H.F. Askal, and O.H. Abdelmageed, *Bollettino Chimico Farmaceutico*, ***137***, 298 (1998).
28. M.A. Korany and R. Haller, *J. Assoc. Off. Anal. Chem.*, ***65***, 144 (1982).
29. R.T. Sane, V.G. Nayak, N.R. Naik, and D.D. Gupte, *Indian Drugs*, ***20***, 334 (1983).
30. M.S. Mahrous, A.S. Issa, M.A. Abdel-Salam, and N. Soliman, *Anal. Lett.*, ***19***, 901 (1986).
31. A.S. Issa, M.S. Mahrous, M. Abdel-Salam, and N. Soliman, *Talanta*, ***34***, 670 (1987).
32. A.F. Shoukry, N.T. Abdel-Ghani, Y.M. Issa, and O.A. Wahdan, *Anal. Lett.*, ***34***, 1689 (2001).
33. B. Li, *Yaoxue Tangbao*, ***20***, 731 (1985).
34. Z. Shao, Z. Pan, and Z. Gan, *Zhongguo Yaoxue Zazhi*, ***26***, 222 (1991).
35. J.A.M. Pulgarin, A.A. Molina, and P.F. Lopez, *Anal. Biochem.*, ***245***, 8 (1997).
36. P. Umapathi, P. Parimoo, S.K. Thomas, and V. Agarwal, *J. Pharm. Biomed. Anal.*, ***15***, 1703 (1997).
37. J.M. Steyn, *J. Chromatogr. Biomed. Appl.*, ***6***, 487 (1979).
38. J.A. Murillo Pulgarin, A. Alanon Molina, and P. Fernandez Lopez, *Anal. Chim. Acta*, ***370***, 9 (1998).
39. Z.L. Gong and Z.J. Zhang, *Fresenius' J. Anal. Chem.*, ***360***, 138 (1998).
40. F.A. El-Yazbi, H.H. Abdine, R.A. Shaalan, and E.A. Korany, *Spectrosc. Lett.*, ***31***, 1403 (1998).
41. J.J. Berzas Nevado, C.C. Guiberteau, and F. Salinas, *Talanta*, ***39***, 547 (1992).
42. Y.Z. Cao, C.Y. Mo, J.G. Long, H. Chen, H.L. Wu, and R.Q. Yu, *China Anal. Sci.*, ***18***, 333 (2002).
43. A. Ruiz-Medina, M.L. Fernandez de Cordova, and A. Molina-Diaz, *Eur J. Pharm. Sci.*, ***13***, 385 (2001).
44. J.A. Murillo-Pulgarin, A.A. Molina, and P.F. Lapez, *Analyst*, ***122***, 253 (1997).
45. A. Munoz-de-la-Pena, A.E. Mansilla, J.A.M. Pulgarin, A.A. Molina, and P.F. Lapez, *Talanta*, ***48***, 1061 (1999).
46. A. Salinas-Castillo, A. Segura Carretero, and A. Fernandez-Gutierrez, *Anal. Bioanal. Chem.*, ***376***, 1111 (2003).
47. K. Nakushima, K. Maki, S. Akiyama, and K. Imai, *Biomed. Chromatogr.*, ***4***, 105 (1990).

48. M. Sugiura, S. Kanda, and K. Imai, *Anal. Chim. Acta*, ***266***, 225 (1992).
49. K. Imai, A. Nishitani, Y. Tsukamoto, W.H. Wang, S. Kanda, K. Hayakawa, and M. Miyazaki, *Biomed. Chromatogr.*, ***4***, 100 (1990).
50. M. Sugiura, S. Kanda, and K. Imai, *Biomed. Chromatogr.*, *7*, 149 (1993).
51. F. Nie, L.D. Zhang, M.L. Feng, and J.R. Lu, *Fenxi Hauxue*, ***25***, 879 (1997).
52. X.W. Zheng and Z.J. Zhang, *Fenxi Hauxue*, ***27***, 145 (1999).
53. M.L. Yang, L.Q. Li, M.L. Feng, J.R. Lu, and Z.J. Zhang, *Fenxi Hauxue*, ***28***, 161 (2000).
54. C. Pascual and C.J. Romay, *Biolumin. Chemilumin.*, *7*, 123 (1992).
55. M. Tuncel, Y. Yazan, D. Dogrukol, and Z. Atkosar, *Anal. Lett.*, ***24***, 1837 (1991).
56. X.Q. Zeng, S.C. Lin, and N.F. Hu, *Talanta*, ***40***, 1183 (1993).
57. Z.H. Wang, H.Z. Zhang, and S.P. Zhou, *Talanta*, ***44***, 621 (1997).
58. L.E. Almeida, M. Castilho, L.H. Mazo, and M. Tabak, *Anal. Chim. Acta*, ***375***, 223 (1998).
59. Y.F. Yang, *Fenxi Hauxue*, ***29***, 28 (2001).
60. M.M. Ghoneim, A. Tawfik, and A. Radi, *Anal. Bioanal. Chem.*, ***374***, 289 (2002).
61. E. Rodrigues Filho, A.M.P. Almeida, and M. Tabak, *J. Mass Spectrom.*, ***38***, 540 (2003).
62. A.N. Strelyuk, *Farmatsiya*, ***36***, 76 (1986).
63. A.P. Argekar and S.S. Kunjir, *J. Planar Chromatogr. Mod-TLC*, ***9***, 65 (1996).
64. A.H. Stead, R. Gill, T. Wright, J.B. Gibbs, and A.C. Moffat, *Analyst*, ***107***, 1106 (1982).
65. R.E. Ardrey and A.C. Moffat, *J. Chromatogr.*, ***220***, 195 (1981).
66. I. Jane, A. McKinnon, and R.J. Flanagan, *J. Chromatogr.*, ***323***, 191 (1985).
67. L. Zhou, Q.N. Ping, and L. Yang, *Yaowu Fenxi Zazhi*, ***23***, 199 (2003).
68. C. Chevalier, P. Rohrbach, and M. Caude, *Feuill Biol.*, ***20***, 129 (1979).
69. J. Schmid, K. Beschke, W. Roth, G. Bozler, and F.W. Koss, *J. Chromatogr. Biomed. Appl.*, ***5***, 239 (1979).
70. K.M. Wolfram and T.D. Bjornsson, *J. Chromatogr. Biomed. Appl.*, ***9***, 57 (1980).

71. C. Williams II, C.S. Huang, R. Erb, and M.A. Gonzalez, *J. Chromatogr. Biomed. Appl.*, ***14***, 225 (1981).
72. J. Rosenfeld, D. Devereaux, M.R. Buchanan, and A.G.G. Turpie, *J. Chromatogr. Biomed. Appl.*, ***20***, 216 (1982).
73. R.T. Sane, J.K. Ghadge, A.B. Jani, A.J. Vaidya, and S.S. Kotwal, *Indian Drug*, ***29***, 240 (1992).
74. A.K. Pedersen, *J. Chromatogr.*, ***162***, 98 (1979).
75. C. Deballon and M. Guernet, *J. Pharm. Biomed. Anal.*, ***6***, 1045 (1988).
76. M. Barberi, J.L. Merlin, and B. Weber, *J. Chromatogr. Biomed. Appl.*, ***103***, 511 (1991).
77. A.R. Zoest, S. Wanwimolruk, J.E. Watson, and C.T. Hung, *J. Liq. Chromatogr.*, ***14***, 1967 (1991).
78. J.H. Bridle and M.T. Brimble, *Drug Dev. Ind. Pharm.*, ***19***, 371 (1993).
79. M. Barberi-Heyob, J.L. Merlin, L. Pons, M. Calco, and B. Weber, *J. Liq. Chromatogr.*, ***17***, 1837 (1994).
80. R. Gohda, K. Kimoto, T. Santa, T. Fukushima, H. Homma, and K. Imai, *Anal. Sci.*, ***12***, 713 (1996).
81. J. Zhang, R.B. Miller, S. Russell, and R. Jacobus, *J. Liq. Chromatogr. Relat. Technol.*, ***20***, 2109 (1997).
82. J. Zhang, R.B. Miller, and R. Jacobus, *Chromatographia*, ***44***, 247 (1997).
83. F. Fontani, G.P. Finardi, G. Targa, G.P. Besana, and M. Ligorati, *J. Chromatogr.*, ***280***, 181 (1983).
84. A. Nishitani, Y. Tsukamoto, S. Kanda, and K. Imai, *Anal. Chim. Acta*, ***251***, 247 (1991).
85. C. Mahony, K.M. Wolfram, D.M. Cocchetto, and T.D. Bjornsson, *Clin. Pharmacol. Ther.*, ***31***, 330 (1982).
86. C. Mahony, J.L. Cox, and T.D. Bjornsson, *J. Clin. Pharmacol.*, ***23***, 123 (1983).

Mefenamic Acid: Analytical Profile

Hadi Poerwono[1], Retno Widyowati[1], Hajime Kubo[2], Kimio Higashiyama[2] and Gunawan Indrayanto[1]

[1]*Faculty of Pharmacy, Airlangga University*
Jl. Dharmawangsa Dalam
Surabaya 60286, Indonesia

[2]*Institute of Medicinal Chemistry Hoshi University*
4-41, Ebara 2-chome, Shinagawa-ku
Tokyo 142-8501, Japan

PROFILES OF DRUG SUBSTANCES,
EXCIPIENTS, AND RELATED
METHODOLOGY – VOLUME 31
DOI: 10.1016/S0000-0000(00)00000-0

CONTENTS

1. **Introduction** 283
2. **Compendial Methods of Analysis** 284
    2.1 Identification 284
        2.1.1 Bulk drug substance 284
        2.1.2 Drug substance in pharmaceutical preparations 286
    2.2 Physical methods of analysis 287
        2.2.1 Loss on drying 287
        2.2.2 Residue on ignition 288
        2.2.3 Sulfated ash 288
        2.2.4 Light absorption 288
    2.3 Impurity analyses 288
        2.3.1 Heavy metals 288
        2.3.2 Copper 288
        2.3.3 Chromatographic purity 289
        2.3.4 Related substances 289
        2.3.5 2,3-Dimethylaniline 290
    2.4 Assay methods 291
        2.4.1 Titration methods 291
        2.4.2 High performance liquid chromatography 291
3. **Titrimetric Methods of Analysis** 291
    3.1 Non-aqueous titration methods 291
    3.2 Potentiometric titration methods 292
4. **Electrochemical Methods of Analysis** 293
    4.1 Polarography 293
    4.2 Conductimetry 293
5. **Spectroscopic Methods of Analysis** 293
    5.1 Spectrophotometry 293
    5.2 Colorimetry 296
    5.3 Fluorimetry 298
    5.4 Chemiluminescence 301
    5.5 Proton magnetic resonance spectrometry 302
    5.6 Mass spectrometry 302
    5.7 Atomic absorption spectrometry 302
6. **Chromatographic Methods of Analysis** 303
    6.1 Thin-layer chromatography 303
    6.2 Column chromatography 304
    6.3 Gas chromatography 305

6.4 High performance liquid chromatography 309
6.5 Supercritical fluid chromatography 313
6.6 Capillary electrophoresis 325
6.7 Capillary isatachophoresis 328
7. **Determination in Body Fluids and Tissues** 329
7.1 Spectroscopic methods 329
7.2 Chromatographic methods 330
8. **References** 330

1. INTRODUCTION

Mefenamic acid has the following structure:

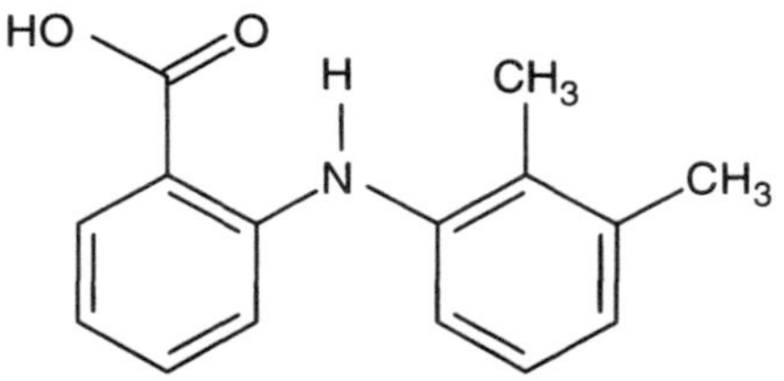

HO O H CH3 N CH3

The elemental analysis of mefenamic acid calculates as follows:

Carbon	74.67%
Hydrogen	6.27%
Nitrogen	5.81%
Oxygen	13.26%

The empirical formula of mefenamic acid is $C_{15}H_{15}NO_2$, and its formula weight is calculated to be 241.285.

According to USP 26 [1] and the Indonesian Pharmacopoeia [4], mefenamic acid is specified to contain not less than 98.0% and not more than 102.0% of $C_{15}H_{15}NO_2$, calculated on the dried basis. The substance is to be preserved in tight, light-resistant containers. The British Pharmacopoeia 2002 [2], European Pharmacopoeia [5] and Indian Pharmacopoeia [6] describe that mefenamic acid contains not less than

99.0%, and not more than the equivalent of 100.5% of *N*-2,3-xylylanthranilic acid, calculated with reference to the dried substance. The Pharmacopoeia of the People's Republic of China [7] describes that mefenamic acid is 2-[(2,3-dimethylphenyl)amino]benzoic acid, and contains not less than 99.0% $C_{15}H_{15}NO_2$, calculated on the dried basis.

According to USP 26 [1], Indonesian Pharmacopoeia [4], Indian Pharmacopoeia [6], and Pharmacopoeia of the People's Republic of China [7], mefenamic acid capsules are specified to contain not less than 90.0% and not more than 110.0% of the labeled amount of $C_{15}H_{15}NO_2$. According to the British Pharmacopoeia 2002 [3], mefenamic acid capsules and tablets are specified to contain not less than 95.0% and not more than 105.0 percent of the stated amount of the substance.

2. COMPENDIAL METHODS OF ANALYSIS

2.1 Identification

2.1.1 Bulk drug substance

According to the United States Pharmacopoeia 26 [1] and Indonesian Pharmacopoeia [4], mefenamic acid is identified by examination using infrared absorption spectrophotometry (method ⟨197 K⟩), comparing with the spectrum obtained with USP Mefenamic Acid reference standard (Test A). In addition to the infrared absorption test, the USP 26 [1] includes examination using Liquid Chromatography, where the retention time of the major peak in the chromatogram of the Assay preparation corresponds to that in the chromatogram of the Standard preparation, as obtained in the Assay (Test B). Test B according to the Indonesia Pharmacopoeia [4] uses a Thin Layer Chromatography test, where the R_f value of the main spot in the chromatogram of the Test solution corresponds to that in the chromatogram of the Standard solution.

Identification testing in the British Pharmacopoeia 2002 [2] and European Pharmacopoeia [5] is arranged in two tiers, classified as the (a) first identification by test B, and (b) the second identification by tests A, C and D.

Test A: About 20 mg of the acid sample is dissolved in a mixture of 1 volume of 1 M hydrochloric acid and 99 volumes of methanol, and diluted to 100 mL with the same mixture of solvent. About 5 mL of the solution is diluted to 50 mL in a mixture of 1 volume of 1 M hydrochloric acid and 99 volumes of methanol. The resulting solution is examined between 250 nm and 380 nm, whereas the solution shows two absorption maxima at 279 nm and 350 nm. The ratio of the absorbance measured at the maximum at 279 nm to that measured at 350 nm is 1.1 to 1.3.

Test B: Examination of the acid sample by infrared absorption spectrophotometry, comparing with the spectrum obtained with mefenamic acid reference standard. Examine the substances prepared as KBr pellets. If the spectra obtained show differences, dissolve the substance to be examined and the reference substance separately in alcohol, evaporate to dryness, and record new spectra using the residue.

Test C: About 25 mg of the acid sample is dissolved in 15 mL of methylene chloride. The solution is examined in ultraviolet light at 365 nm to exhibit a strong greenish-yellow fluorescence. Carefully add 0.5 mL of a saturated solution of trichloroacetic acid dropwise and examine under ultraviolet light at 365 nm. The solution does not exhibit fluorescence.

Test D: About 5 mg of the acid sample is dissolved in 2 mL of sulfuric acid, to which is added 0.05 mL of potassium dichromate solution. An intense blue color is produced, turning rapidly to brownish-green.

Identification testing according to the Indian Pharmacopoeia [6] is performed by means of three tests (denoted as tests A, B, and C).

Test A: The infrared absorption spectrum is concordant with the reference spectrum of mefenamic acid or with the spectrum obtained from mefenamic acid reference standard. If the spectra are not concordant, dissolve a sufficient quantity of the substances in the minimum volume of ethanol (95%), evaporate to dryness, and prepare new spectra of the residues.

Test B: About 25 mg of the acid sample is dissolved in 15 mL of chloroform. The solution is examined under ultraviolet light at 365 nm to exhibit a strong greenish-yellow fluorescence. Carefully add 0.5 mL of a saturated solution of trichloroacetic acid dropwise and examine under ultraviolet light at 365 nm. The solution does not exhibit fluorescence.

Test C: About 5 mg of the acid sample is dissolved in 2 mL of sulfuric acid, to which is added 0.05 mL of 0.0167 M potassium dichromate solution. An intense blue color is produced, which fades rapidly to brownish-green.

According to the Pharmacopoeia of the People's Republic of China [7], mefenamic acid is identified by the performance of four tests (denoted as Tests 1, 2, 3, and 4).

Test 1: About 25 mg of the acid sample is dissolved in 15 mL of chloroform. The solution is examined under ultraviolet light at 254 nm to exhibit a strong green fluorescence.

Test 2: About 5 mg of the acid sample is dissolved in 2 mL of sulfuric acid, to which is added 0.05 mL of 0.5% potassium dichromate solution. A deep blue color is produced which fades immediately to brownish-green.

Test 3: The light absorption of a 20 μg/mL solution in a mixture of hydrochloric acid (1 mol/L) and methanol (1:99), exhibits two absorption maxima, at 279 nm and 350 nm. The absorbance of these is about 0.69~0.74 and 0.59~0.6, respectively.

Test 4: The infrared absorption spectrum is concordant with the spectrum of mefenamic acid reference standard.

2.1.2 Drug substance in pharmaceutical preparations

Identification test of mefenamic acid in capsule preparations according to the United States Pharmacopoeia 26 [1] and Indonesian Pharmacopoeia [4] is performed by the execution of two tests (denoted as Tests A and B)

Test A: A portion of capsule contents (equivalent to about 250 mg of mefenamic acid) is placed in a 250 mL volumetric flask, to which is added about 100 mL of a mixture of chloroform and methanol (3:1) with vigorous shaking. It is then diluted with a mixture of chloroform and methanol (3:1) to volume, mixed, and filtered and the filtrate is obtained. This solution responds to the Thin-Layer Chromatographic Identification Test ⟨201⟩ when eluted with a solvent system consisting of a mixture of chloroform, ethyl acetate, and glacial acetic acid (75:25:1). The Ordinary Impurities general test ⟨466⟩ visualization technique is used.

Test B: Examination using Liquid Chromatography: the retention time of the major peak in the chromatogram of the Assay preparation corresponds to that in the chromatogram of the Standard preparation, obtained as directed in the Assay.

According to the British Pharmacopoeia 2002 [2] and Indian Pharmacopoeia [6], mefenamic acid in capsule and tablet preparations are identified by examination using infrared absorption spectrophotometry as the following procedure. Extract a quantity of the capsule contents (or powdered tablets) containing 0.25 g of mefenamic acid with two 30 mL quantities of ether. Wash the combined extracts with water, evaporate to dryness on a water bath, and dry the residue at 105°C. Dissolve a sufficient quantity in the minimum volume of absolute ethanol, and evaporate to dryness on a water bath. The infrared absorption spectrum is concordant with the reference spectrum of mefenamic acid.

According to the Pharmacopoeia of the People's Republic of China [7], the method of identification testing of mefenamic acid in capsule and tablet preparations is based on light absorption spectrophotometry. A specific quantity of the powdered contents of capsules (or powdered tablets), equivalent to 0.25 g mefenamic acid, is dissolved in a mixture of 10 mL of 0.1 mL/L hydrochloric acid/methanol (1:99), shaken, and filtered. Then some quantity of the filtrate is diluted with the above-mixed solution to produce a solution having a concentration of about 20 μg/mL. The absorption spectrum of the solution exhibits maxima at 279 nm and 350 nm.

2.2 Physical methods of analysis

2.2.1 Loss on drying

When the acid sample is dried at 105°C for 4 h according to the USP 26 general procedure ⟨731⟩, or by the procedure of Indonesian and Indian

Pharmacopoeias, it loses not more than 1.0% of its weight [1, 4, 6]. According to the British and European Pharmacopoeias, as well as the Pharmacopoeia of the People's Republic of China, it loses not more than 0.5% determined on 1.000 g by drying at 100–105°C to a constant weight [2, 5, 7].

2.2.2 Residue on ignition

When examined according to the USP general procedure ⟨281⟩ as well as the procedure of Indonesian Pharmacopoeia and Pharmacopoeia of the People's Republic of China, the ash obtained for an acid sample may not exceed 0.1% [1, 4, 7].

2.2.3 Sulfated ash

When examined according to the British Pharmacopoeia 2002 general procedure ⟨2.4.14⟩, as well as the procedure of the European and Indian Pharmacopoeias, the sulfated ash obtained for an acid sample may not exceed 0.1%, determined on 1.0 g [2, 5, 6].

2.2.4 Light absorption

The absorbance of a 0.002% w/v solution (in a mixture of 99 volumes of methanol and 1 volume of 1 M hydrochloric acid) is 0.69–0.74 at the first maximum of 279 nm, and 0.56–0.60 at the second maximum of 360 nm [6].

2.3 Impurity analyses

2.3.1 Heavy metals

The heavy metal content of mefenamic acid is determined using the USP 26 general procedure ⟨231⟩, as well as the procedure of Indonesian Pharmacopoeia [1, 4]. The limit on heavy metals in the substance is 0.002%.

2.3.2 Copper

The copper content of mefenamic acid is determined using atomic absorption spectrometry, according to the British Pharmacopoeia 2002 general procedure ⟨2.2.23, Method I⟩, as well as the procedure of the European Pharmacopoeia [2,5]. The substance may not contain more than 10 ppm of Cu.

Indian Pharmacopoeia determines the copper content using a color intensity method, according to the following procedure. Moisten 1 g of the substance with sulfuric acid and ignite until all the carbon is removed. Add 10 mL of 1 M sulfuric acid to the residue and allow to stand for 10 min. Transfer to a separating funnel using 20 mL of water and add 10 mL of a solution containing 20% w/v diammonium hydrogen citrate and 5% w/v of disodium edetate. Add 0.2 mL of thymol blue solution and neutralize with 5 M ammonia. Add 10 mL of sodium diethyldithiocarbamate solution and 15 mL of carbon tetrachloride, shake, and allow to separate. The yellow color of the carbon tetrachloride layer is not more intense than that produced by treating 2 mL of copper standard solution (10 ppm Cu) in the same manner beginning at the directions, "Transfer to a separating funnel using 20 mL of water, as described above" (20 ppm) [6].

According to the Pharmacopoeia of the People's Republic of China, the above separated carbon tetrachloride layer is measured by the light absorption spectrophotometry at 435 nm. The absorbance of that solution is not more than 0.35 [7].

2.3.3 Chromatographic purity

Chromatographic purity according to the USP 26 is determined using Liquid Chromatography. The impurity content of the substance is not more than 0.1% for any individual impurity, and not more than 0.5% for total impurities [1].

2.3.4 Related substances

According to the British Pharmacopoeia 2002 and the European and Indian Pharmacopoeias, the content of related substances is examined by thin-layer chromatography using TLC silica gel GF_{254} plates. A test solution is prepared by dissolving 0.125 g of the substance to be examined in a mixture of 1 volume of methanol and 3 volumes of methylene chloride, and diluting to 5 mL with the same mixture of solvents. Reference solution (a) is prepared by diluting 1 mL of the test solution to 50 mL with a mixture of 1 volume of methanol and 3 volumes of methylene chloride, then diluting 1 mL of the solution to 10 mL with the same mixture of solvents. Reference solution (b) is prepared by dissolving 5 mg of flufenamic acid and 5 mg of mefenamic acid reference standard in a mixture of 1 volume of methanol and 3 volumes of methylene chloride, and diluting to 10 mL with the same mixture of solvents. Apply 20 μL of each solution to the TLC silica gel GF_{254} plate. Develop over a path of 15 cm using a mixture of 90 volumes of toluene, 25 volumes of dioxane, and 1 volume of glacial acetic acid. Dry the plate in a current of warm air, and expose the plate to iodine vapor for 5 min, and examine in ultraviolet light (254 nm). Any spot in the chromatogram obtained with the test solution, apart from the principal spot, is not more intense than the spot in the chromatogram obtained with reference solution (a) (0.2%). The test is not valid unless the chromatogram obtained with reference solution (b) shows two clearly separated spots [2, 5, 6].

A similar procedure is mentioned in the Pharmacopoeia of the People's Republic of China that uses a mixture of 3 volumes of isobutanol and 1 volume of concentrated ammonia solution as the mobile phase [7].

2.3.5 2,3-Dimethylaniline

The 2,3-dimetylaniline content is determined by color intensity method, according to the British Pharmacopoeia 2002 and the European Pharmacopoeia. Solution (a) is prepared by dissolving 0.250 g of the substance to be examined in a mixture consisting of 1 volume of methanol and 3 volumes of methylene chloride, and diluting to 10 mL with the same mixture of solvents. This solution is used to prepare the test solution. Solution (b) is prepared by dissolving 50 mg of 2,3-dimethylaniline in a mixture of 1 volume of methanol and 3 volumes of methylene chloride, and diluting to 100 mL with the same mixture of solvents. Then 1 mL of the solution is diluted to 100 mL with the same mixture of solvents. This solution is used to prepare the standard. Using three flat-bottomed tubes, place 2 mL of solution (a) in the first, 1 mL of solution (b) in the second, and 1 mL of a mixture of 1 volume of methanol and 3 volumes of methylene chloride in the third. A blank tube is prepared by making 2 mL of a mixture of 1 volume of methanol and 3 volumes of methylene chloride. To each tube add 1 mL of a freshly prepared 10 g/L solution of dimethylamino-benzaldehyde in methanol and 2 mL of glacial acetic acid. Allow to stand at room temperature for 10 min. The intensity of the yellow color of the test solution is between that of the blank and that of the standard (100 ppm) [2, 5].

According to the Indian Pharmacopoeia, 2,3-dimethylaniline is examined by TLC using a silica gel G plate, and a mixture of 90 volumes of toluene, 25 volumes of dioxane, and 1 volume of 18 M ammonia as the mobile phase. Apply 40 μL of each of the two solutions in a mixture of 3 volumes of chloroform and 1 volume of methanol containing (1) 2.5% w/v of the substance being examined separately to the plate, and (2) 0.00025% w/v of 2,3-dimethylaniline. After removal of the plate, dry it in a current of warm air. Spray the dried plate with ethanolic sulfuric acid (20%), heat at

105°C for 30 min, and immediately expose to nitrous fumes in a closed glass chamber for 15 min (the nitrous fumes may be generated by adding dilute sulfuric acid dropwise to a solution containing 10% w/v of sodium nitrite and 3% w/v of potassium iodide). Place the plate in a current of warm air for 15 min, and spray with a 0.5% w/v solution of *N*-(1-naphthyl)ethylenediamine dihydrochloride in ethanol (95%). If necessary, allow to dry, and repeat the spraying. Any spot corresponding to 2,3-dimethylaniline in the chromatogram obtained with solution (1) is not more intense than the spot in the chromatogram obtained with solution (2) [6].

2.4 Assay methods

2.4.1 Titration methods

According to the British Pharmacopoeia 2002, the European and Indian Pharmacopoeias, and the Pharmacopoeia of the People's Republic of China, the assay for mefenamic acid is performed by a titration method. About 0.200 g of the substance to be assayed is dissolved with the aid of ultrasound in 100 mL of warm ethanol that has been previously neutralized to the phenol red endpoint. About 0.1 mL of phenol red solution is added and titrated with 0.1 M sodium hydroxide. Each milliliter of 0.1 M NaOH is equivalent to 24.13 mg of mefenamic acid [2, 5–7].

2.4.2 High performance liquid chromatography

According to the USP 26 and the Indonesian Pharmacopoeia, the assay for mefenamic acid is performed by a liquid chromatography method. A buffer solution is prepared as a 50 mM solution of monobasic ammonium phosphate, adjusted with 3 M ammonium hydroxide to a pH of 5.0. The system uses a filtered and degassed mixture of acetonitrile, buffer solution, and tetrahydrofuran (23:20:7) as the mobile phase. The liquid chromatograph is equipped with a 254 nm detector and a 4.6 mm × 25 cm column that contains packing L1. The flow rate is about 1 mL/min. The column efficiency is not less than 8200 theoretical plates, the tailing factor for the analyte peak is not more than 1.6, and the relative standard deviation for replicate injections is not more than 1.0% [1, 4].

3. TITRIMETRIC METHODS OF ANALYSIS

3.1 Non-aqueous titration methods

A quantity of the mixed contents of 20 mefenamic acid capsules containing 0.5 g of the substance, or a quantity of the powdered tablets (after weighing and powdering of 20 mefenamic acid tablets) containing 0.5 g of the substance, is dissolved in 100 mL of warm absolute ethanol previously neutralized to the phenol red endpoint. The resulting solution of each preparation is titrated with 0.1 M sodium hydroxide using phenol red solution as the indicator. Each mL of 0.1 M NaOH is equivalent to 24.13 mg of mefenamic acid [2, 6, 7].

Based on the fact that mefenamic acid reacts quantitatively with *N*-bromosuccinimide (NBS) in an acidic medium, Hassib *et al.* [8] reported the titrimetric determination of mefenamic acid. A solution of mefenamic acid in acetic acid (0.02% m/V) was prepared. A volume of this solution containing 1.5–3.5 mg of mefenamic acid was allowed to react with 40 mL of 0.05 M NBS solution for 10 min in the dark (temperature of 25 ± 2°C). Potassium iodide solution (10 mL) was added, and the solution was titrated with 0.01 N sodium thiosulfate to the starch end-point. A blank determination was carried out. For the determination of mefenamic acid in the capsules, the contents of 20 Ponstan® capsules were mixed thoroughly and an accurately weighed portion of the mixed powder (nominally containing 25 mg of mefenamic acid) was extracted with diethyl ether (3 × 10 mL). The residue remaining after evaporation of diethyl ether was dissolved in acetic acid, and the procedure was continued as described for the determination of mefenamic acid.

Issa *et al.* [9] used various metal ions for the volumetric determination of mefenamic acid. Mefenamic acid was precipitated from its neutral alcoholic solution by a standard solution of either silver nitrate, mercurous acetate, or potassium aluminum sulfate. In the argentimetric procedure, residual Ag(I) was titrated with standard NH_4SCN. With $Hg(OAc)_2$ or potash alum, the residual metal was determined by adding EDTA and conducting back titration of excess of EDTA with standard $Pb(NO_3)_2$ using xylenol orange indicator. The applied methods were used for the determination in bulk drug substance, and in its formulations.

3.2 Potentiometric titration methods

Çakirer *et al.* [10] reported a potentiometric titration method in nonaqueous media for the determination of some commonly used antiinflammatory agents. The direct potentiometric titration of mefenamic acid, fenbufen, and ibuprofen, and the indirect potentiometric titration of diclofenac sodium, were carried out in acetonitrile using tetrabutylammonium hydroxide solution in 2-propanol/methanol as the titrant. The titration was performed at 25°C under an atmosphere of nitrogen. This method gave highly precise recoveries, having a relative standard deviation of not more than 1.0% for all antiinflammatory agents studied.

4. ELECTROCHEMICAL METHODS OF ANALYSIS

4.1 Polarography

Song *et al.* [11] developed a polarographic method for the determination of mefenamic acid in tablets, which was based on rapid nitrosation of mefenamic acid with sodium nitrite in acetic acid, and subsequent measurement of the *N*-nitroso derivative of mefenamic acid by linear-sweep polarography. The method is simple, sensitive, and specific, and was characterized by a detection limit of 2×10^{-7} mol/L.

4.2 Conductimetry

Aly and Belal reported the conductimetric determination of some non-steroidal antiinflammatory drugs in their dosage forms [12].

5. SPECTROSCOPIC METHODS OF ANALYSIS

5.1 Spectrophotometry

On the basis of spectral changes induced by changing the solvent medium from HCl (0.01 M) to NaOH (0.01 M), Hassan and Shaaban [13] developed a coefficient-difference spectrophotometric method of analysis for mefenamic acid using the six-points quadratic order of the orthogonal polynomials. The optimum wavelength range was 326–386 nm, measured at 12 nm intervals. The mean recovery was 100.1 ± 0.89 % over a concentration range of 0.6–2.6 mg/100 mL ($p = 0.05$). Trials to adapt single wavelength difference technique as well as the cubic order of the orthogonal polynomials were also performed; their results were discussed on a statistical basis. The developed procedures have been applied to the analysis of some randomly collected market preparations, and the results proved suitability of the method for its application in routine analysis.

Shinkuma *et al.* [14] reported a spectrophotometric assay of different commercial mefenamic acid capsules. The contents of 20 capsules of each brand were carefully removed and accurately weighed. After finding the mean weight of the capsule contents, the contents were mixed uniformly and used as the brand sample. A weighed amount of this powder, equivalent to about 150 mg of mefenamic acid, was dissolved in 200 mL of hydrochloric acid–methanol solution (0.84 mL of concentrated hydrochloric acid/1000 mL methanol). After filtering and appropriate dilution, the sample was assayed spectrophotometrically at 350 nm using a double-beam spectrophotometer. Beer–Lambert law was obeyed over the concentration range used in this work.

Das *et al.* [15] established a spectrophotometric procedure for the simultaneous determination of mefenamic acid and paracetamol in mixtures. Using 0.01 M methanolic hydrochloric acid as a solvent, the absorbance of the mixture is measured at 248, 279, and 351 nm. The concentration of each component can be calculated by solving two equations using two wavelengths, either 248 and 279 nm, or 248 and 351 nm. The amounts of mefenamic acid and paracetamol were calculated using either of two sets of equations:

$$\text{Set A: Mass (mg) of mefenamic acid} = 332.2A_{351}$$

$$\text{Mass (mg) of paracetamol} = 113.04A_{248} - 84.4A_{351}$$

$$\text{Set B: Mass (mg) of mefenamic acid} = 314.8A_{279} - 53.0A_{248}$$

$$\text{Mass (mg) of paracetamol} = 126.6A_{248} - 79.9A_{279}$$

The method was applied for the determination of these drugs in commercial tablets with recoveries in the ranges of 99.1–101.7% (set A), and 96.7–100.75% (set B) for mefenamic acid and 95.9–99.5% (set A) and 96.4–99.6% (set B) for paracetamol.

Gangwal and Sharma [16] described a simultaneous spectrophotometric method for the determination of mefenamic acid and paracetamol in their combined dosage forms based on the native UV absorbance maxima of mefenamic acid and paracetamol in 0.02 M NaOH. Mefenamic acid exhibits two absorption maxima at 285 nm and 333 nm, while paracetamol has one absorbance maxima at 257 nm. In a separate study, the same group [17] also reported a spectrophotometric procedure for mefenamic acid and paracetamol in two component tablet formulations. The method is based on the two-wavelength method of calculation. The difference in absorbances at 217 nm and 285 nm was used for determination of mefenamic acid, and the difference in absorbances at 257 nm and 308.8 nm was used for the determination of paracetamol. Beer's law is obeyed by both the drugs within the concentration ranges employed for analysis. The method has been statistically validated, and was found to be satisfactory.

Parimoo *et al.* [18] established a UV spectrophotometric method for the simultaneous determination of mefenamic acid and paracetamol in combination preparations. The determination was carried out in two different solvents, namely methanol and 0.1 N NaOH. The wavelength maximum for mefenamic acid was found at 284 nm, and 248 nm for paracetamol. The wavelength maximum for mefenamic acid in NaOH was found at 219 nm, and 256 nm for paracetamol. There were no interferences in the analyte estimations.

A direct and simple first derivative spectrophotometric method has been developed by Toral *et al.* [19] for the determination of mefenamic acid and paracetamol in pharmaceutical formulations. A methanolic hydrochloric acid solution was used as a solvent for extracting the drugs from the formulations, and subsequently the samples were directly evaluated by derivative spectrophotometry. Simultaneous determination of both the drugs can be carried out using the zero-crossing and the graphical methods, and the methods do not require simultaneous equations to be solved. The calibration graphs were linear in the ranges from 1.8×10^{-6} to 1.6×10^{-4} M of mefenamic acid, and from 4.1×10^{-6} to 1.4×10^{-4} M of paracetamol. The ingredients commonly found in commercial pharmaceutical formulations did not interfere, so the proposed method was applied to the determination of these drugs in tablets.

Dinç *et al.* [20] developed four new methods for the simultaneous determination of mefenamic acid and paracetamol in their combination formulations. In the ratio spectra derivative method, analytical signals were measured at the wavelengths corresponding to either maxima or minima for both the drugs in the first derivative spectra of the ratio spectra obtained by dividing the standard spectrum of one of the two drugs in a mixture of 0.1 M NaOH and methanol (1:9). In three chemometric methods {classical least-squares, inverse least-squares, and principal component regression (PCR)}, the conduct of procedures was randomly selected by using the different mixture compositions containing two drugs in a mixture of 0.1 M NaOH and methanol (1:9). The absorbance data was obtained by the measurement at 13 points in the wavelength range of 235–355 nm. Chemometric calibration was constructed from the absorbance data and the training set for the prediction of the amount of mefenamic acid and paracetamol in samples. In the PCR method, the covariance matrix corresponding to the absorbance data was calculated for the basis vectors and matrix containing the new coordinates. The obtained calibration curve was used to determine the drugs in their mixtures. Linearity range in all the methods was found over the range of 2–10 μg/mL for mefenamic acid and 4–20 μg/mL for paracetamol. Mean recoveries were found to be satisfactory (greater than 99%). The procedures do not require any separation step, and were successfully applied to a pharmaceutical formulation (tablet).

5.2 Colorimetry

Deveaux *et al.* [21] reported that mefenamic acid, being an N-aryl derivative of anthranilic acid, can be characterized by two color reactions. The color reactions, negative with N-aryl derivatives of aminonicotinic acid, are associated with the diphenylamine structure. For the first color reaction, add to a test tube approximately 0.5 g of oxalic acid dihydrate and at least 1 mg of the test material. Place the tube into an oil bath at 180–200°C for 4–5 min. After cooling, dissolve the residue in 10 mL of 95% ethanol to obtain a stable, intense blue solution (absorption maximum at 586–590 nm). To use the reaction for capsule formulations, extract the active ingredient with acetone and filter prior to the assay. For the second color reaction, add mefenamic acid (100–800 μg) in 1 mL of HOAc–H_2SO_4 (98:2, v/v), 5 mL of HOAc–HCl (50:50, v/v) and 1 mL of 0.10% (w/w) aqueous levulose. Heat the mixture for 25 min at 100°C, and after cooling, measure the absorbance at 597 nm.

Vinnikova [22, 23] described the spectrophotometric determination of mefenamic acid at 490 nm after conversion to its colored complex with Fast Red B salt at pH 6.60 (phosphate buffer). The method was applied for the determination of free and bound mefenamic acid, and found to be useful for studying the blood plasma protein binding, absorption, distribution, metabolism, and excretion of mefenamic acid.

Sastry and Rao [24] reported the determination of mefenamic acid and some NSAIDs in pharmaceuticals by a spectrophotometric method. The method is based on the formation of a blue complex with Folin-Ciocalteu reagent, and subsequent measurement of the absorbance at 750–760 nm. The method was found to be simple, sensitive, accurate, and rapid, with recoveries of 98.0–102.3% and relative standard deviations of 1.04–1.66%.

Sastry *et al.* [25] also described a fairly sensitive spectrophotometric method for the determination of mefenamic acid based on the formation of a chloroform-soluble, colored ion-association complex between mefenamic acid and methylene violet at pH 7.6. The absorbance of the separated chloroform layer was measured at 540 nm. The molar absorptivity is 2.29×10^4 L/mol $\times$ cm, and the limit of detection is 2–16 μg mefenamic acid/mL. The method was applied successfully to the determination of mefenamic acid in bulk samples and in pharmaceutical preparations, with recoveries of 99.2–99.9%.

Another report by Sastry's group [26] described the determination of mefenamic acid based on the reaction of the mefenamic acid with *p-N,N*-dimethylphenylenediamine in the presence of $S_2O_8^{2-}$ or Cr(VI), whereby an intensely colored product having maximum absorbance at 740 nm is developed. The molar absorptivity is 4.15×10^4 L/mol cm. The reaction is sensitive enough to permit the determination of 0.25–4.0 μg mefenamic acid/mL, and was successfully applied to the determination of mefenamic acid in bulk samples and in pharmaceutical preparations, with recoveries of 99.4–99.8%.

Mahrous *et al.* [27] described the colorimetric determination of mefenamic acid with potassium ferricyanide in NaOH medium. The orange product is measured at 464 nm, and the molar absorptivity is 1.9×10^3 L/mol cm. The method was applied successfully to the determination of mefenamic acid in capsules. Garcia *et al.* [28] reported a flow injection spectrophotometric method for the determination of mefenamic acid in bulk samples and pharmaceuticals, also based on the reaction of mefenamic acid with potassium ferricyanide in NaOH media. The absorbance of the product obtained was measured at 465 nm, and the corresponding calibration graph was linear over the range of 1.00–100 mg/L, with a limit of detection of 0.18 mg/L.

Hassib *et al.* [29] determined mefenamic acid by ferric hydroxamate complex formation and measurement of the absorbance of the colored complex at 530 nm. The method is applicable to mefenamic acid amounts varying from 0.5 to 74.5 mg/25 mL, and Ponstan® capsules were successfully determined using this method.

Khier *et al.* [30] determined mefenamic acid after complexation with copper(II) amine sulfate. The complex is extracted with chloroform and treated with diethyldithiocarbamate solution, whereupon another copper (II) complex (wavelength maximum of 430 nm) is formed. Beer's law is followed over the mefenamic acid concentration range of 6–48 μg/mL. The method was applied successfully to the determination of mefenamic acid in bulk samples and in pharmaceutical preparations, with recoveries of 98.0–101.0%.

Abdel-Hay *et al.* [31] described the determination of mefenamic acid based on its reaction in ethanolic medium with 2-nitrophenylhydrazine in the presence of dicyclohexylcarbodiimide to give an acid hydrazine, which showed intense violet color (maximum absorption at around 550 nm). The effect of reagent concentration (2-nitrophenylhydrazine-HCl, pyridine, and dicyclohexylcarbodiimide), heating temperature, and heating time were studied to optimize the reaction conditions. The method was successfully applied to the determination of mefenamic acid in pure and dosage forms, with a relative standard deviation less than 2%.

Bojarowicz and Zommer-Urbanska [32] reported that the ferric complexes of mefenamic acid (stoichiometries of 1:1, 1:2, 1:3, and 1:4, depending on the amount of the ligand used) were formed in 90% aqueous methanol at pH 2.8. All showed absorbance at 540–550 nm.

A simple photometric method in the visible region was described by Babu *et al.* [33] for the determination of mefenamic acid in its pharmaceutical dosage forms. The method is based on the reaction of the drug with 4-aminophenazone and potassium ferricyanide to yield a reddish-green colored chromophore, which exhibits an absorption maximum at 590 nm. The chromophore is stable for 40 min, and Beer's law is obeyed over the concentration range of 0.5–4 μg/mL.

Amin *et al.* [34] reported the determination of mefenamic acid based on the formation of the blue complex obtained from the reaction of methylene blue with mefenamic acid. The method was applied for the analysis of pharmaceutical preparations containing mefenamic acid, and found to be simple, precise and reproducible.

The nitrosation conditions of aniline and diphenylamine derivatives (paracetamol, phenacetin, mefenamic acid, and diclofenac sodium), with subsequent formation of colored products in alkaline media, were studied by Shormanov *et al.* [35]. Suitable procedures for quantitative determination of analytes in bulk drug substance and tablets were developed.

5.3 Fluorimetry

The influence of sample pretreatment by alkaline saponification and oxidation by $K_2Cr_2O_7$ in acetic acid, and the influence of the solvent and addition of haloacetic acids (e.g., trichloroacetic acid required for fluorescence), on the fluorescence intensity of mefenamic acid and other *N*-arylanthranilic acids was studied by Dell and Kutschbach [36] with reference to their fluorimetric determination in urine. Hydrolysis is necessary for determining esters and amides of mefenamic acid. Polar solvents decrease the fluorescence intensity, and solvents with a negligible dielectric constant (especially CCl_4, $Cl_2C=CCl_2$, and decalin) were required to render the solution fluorescent. Among the haloacetic acids studied, only acids with pK values less than 1 were found to be suitable (i.e., Cl_3CCO_2H and F_3CCO_2H). Aryl substitution affected the position of the fluorescence maximum but not the intensity. Anthranilic acid was found to fluoresce in CCl_4–Cl_3CCO_2H, whereas its meta and para analogs and Ph_2NH were not.

Deveaux *et al.* [21] explained the fluorescent reaction of mefenamic acid by the formation of a substituted acridone after dissolving mefenamic acid in concentrated H_2SO_4 and heating for 10 min at 100°C. The acridone exhibits an intense green fluorescence when excited by white light, and blue when excited by ultraviolet light.

In contrast to the above Deveaux method, Mehta and Schulman [37] described that the native fluorescence of mefenamic, flufenamic and meclofenamic acids is more useful for determination of these drugs compared to the fluorescence of the derivatives-substituted acridones and benzoxazines of these drugs by treatment with H_2SO_4 and HCHO, respectively. The determination of mefenamic acid at trace level by fluorescence spectrometry was also reported by Miller *et al.* [38].

Huang and Xu [39] developed a fluorimetric method for analysis of some acid drugs by using pyronine to form a drug-dye complex. In Na_2HPO_4 solution, mefenamic acid reacts with pyronine to form a complex which is soluble in $CHCl_3$. Pyronine fluoresces strongly at an excitation wavelength of 542 nm, and emits at an emission wavelength of 566 nm. The intensity of pyronine in the complex is proportional to the amount of mefenamic acid in the complex. The method sensitivity was found to be 10–50 ng/mL, and good recovery and precision were obtained.

Based on the formation of the complex of the compound with Al(III) in an ethanolic medium, a sensitive and rapid flow injection fluorimetric method was proposed by Albero *et al.* [40] for the determination of mefenamic acid. The calibration graph resulting from the measurement of the fluorescence after excitation of the complex with mefenamic acid at 355 nm (emission wavelength of 454 nm) was linear over the range of

0.30–16.1 μg/mL. The method has been applied to the determination of the substance in pharmaceutical preparations.

Ioannou *et al.* [41] reported the use of terbium sensitized fluorescence to develop a sensitive and simple fluorimetric method for the determination of the anthranilic acid derivative, mefenamic acid. The method makes use of radiative energy transfer from anthranilate to Tb(III) in alkaline methanolic solutions. Optimum conditions for the formation of the anthranilate-Tb(III) complex were investigated. Under optimized conditions, the detection limit was 1.4×10^{-8} mol/L, and the range of application was 2.5×10^{-8} to 5.0×10^{-5} mol/L. The method was successfully applied to the determination of mefenamic acid in serum after extraction of the sample with ethyl acetate, evaporation of the organic layer under a stream of nitrogen at 40°C, and reconstitution of the residue with alkaline methanolic terbium solution prior to instrumental measurement. The mean recovery from serum samples spiked with mefenamic acid (3.0×10^{-6}, 9.0×10^{-6}, 3.0×10^{-5} mol/L) was 101 ± 5%. The within-run precision (RSD) for the method for the two serum samples varied from 2 to 8%, and the day-to-day precision for two concentration levels varied from 2 to 13%.

The use of second-derivative synchronous fluorescence spectrometry was reported by Ruiz *et al.* [42] to develop a simple, rapid and sensitive fluorimetric method for the determination of binary mixtures of the nonsteroidal antiinflammatory drugs flufenamic (FFA), meclofenamic (MCFA) and mefenamic (MFA) acids in serum and in pharmaceutical formulations. The method is based on the intrinsic fluorescence of these compounds in chloroform. A differential wavelength of 105 nm was used for the resolution of FFA–MFA and MFA–MCFA mixtures, whereas the FFA–MCFA mixture was determined at a differential wavelength of 40 nm. Serum samples were treated with trichloroacetic acid to remove the proteins, and the analytes were extracted in chloroform prior to determination. Pharmaceutical preparations were analyzed without prior separation steps.

Capitán-Vallvey *et al.* [43] has developed a spectrofluorimetric method for the quantitative determination of flufenamic, mefenamic and meclofenamic acids in mixtures by recording emission fluorescence spectra between 370 and 550 nm with an excitation wavelength of 352 nm. The excitation-emission spectra of these compounds are deeply overlapped which does not allow their direct determination without previous separation. The proposed method applies partial least squares multivariate calibration to the resolution of this mixture using a set of wavelengths previously selected by Kohonen artificial neural networks. The linear calibration graphs used to construct the calibration matrix were selected in the ranges from 0.25 to 1.00 μg/mL for flufenamic and meclofenamic acids, and from 1.00 to 4.00 μg/mL for mefenamic acid. A cross-validation procedure was used to select the number of factors. The selected calibration model has been applied to the determination of these compounds in the synthetic mixtures and pharmaceutical formulations.

Sabry [44] reported that the fluorescence properties of flufenamic and mefenamic acids were obtained in aqueous and binary aqueous-organic solvents with or without α- and β-cyclodextrins (CDs). The spectral behavior of the model compounds in aqueous micellar systems of ionic and non-ionic surfactants were investigated. In aqueous acidic solution, the investigated analytes were better incorporated in CDs (flufenamic acid) and micelles (flufenamic and mefenamic acids). The luminescence emission from the analytes were found to be greatly enhanced by surfactants. The fluorescence of the analyte–surfactant system was quenched by increasing amounts of organic solvents. The changes in the dissociation constant of flufenamic and mefenamic acids in solutions of anionic and nonionic surfactants (sodium dodecyl sulfate and Triton X-100, respectively) were studied spectrofluorimetrically. The fluorescence intensity enhancements in CDs media relative to aqueous solution ranged from 1.3 to 1.6. Sensitive micellar-enhanced spectrofluorimetric methods for the determination of flufenamic and mefenamic acids were developed, with ranges of application down to ng/mL. The linear dynamic ranges and relative standard deviations were reported. The methods were applied satisfactorily to the determination of both the compounds in pharmaceutical preparations with recoveries of 98–101%.

5.4 Chemiluminescence

Aly *et al.* [45] has investigated a novel chemiluminescent method using flow injection for the rapid and sensitive determination of flufenamic and mefenamic acids. The method is based on a tris(2,2′-bipyridyl)ruthenium (III) chemiluminescence reaction. $Ru(bipy)_3^{3+}$ is chemically generated by mixing two streams containing solution of tris(2,2′-bipyridyl)ruthenium (II) and acidic cerium (IV) sulfate. After selecting the best operating parameters calibration graphs were obtained over the concentration ranges 0.07–6.0 and 0.05–6.0 μg/mL for flufenamic and mefenamic acids, respectively. The limits of detection ($s/n = 3$) were 3.6×10^{-9} M flufenamic acid and 2.1×10^{-7} M mefenamic acid. The method was successfully applied to the determination of these compounds in pharmaceutical formulations and biological fluids.

5.5 Proton magnetic resonance spectrometry

Mansour *et al.* [46] developed a simple, rapid efficient quantitative proton magnetic resonance method for the determination of mefenamic acid in Ponstan® capsules. The method is based on a comparison between the sum integrals of the two methyl singlets of mefenamic acid and that integral of the sharp singlet of *p*-methoxybenzylidenemalononitrile (internal standard). The method gave accurate and reproducible results.

5.6 Mass spectrometry

De Kanel *et al.* [47] described a method for the simultaneous analysis of 14 nonsteroidal antiinflammatory drugs in human serum using negative electrospray ionization-tandem mass spectrometry. After addition of internal standard and protein precipitation using acetonitrile, samples were transferred to autosampler vials for direct analysis without chromatography. Injection of an air bubble with the sample and a multiple reaction monitoring method using argon collision-induced dissociation of analyte $(M–H)^-$ ions permitted integration of the product ion peak areas to produce reproducible quantitative data over the range of concentrations expected in serum during routine use of these drugs. The method permitted the analysis of 30 samples per hour. Two hundred and fifty consecutive analyses did not adversely affect instrument sensitivity.

5.7 Atomic absorption spectrometry

Salem *et al.* [48] reported simple and accurate methods for the quantitative determination of flufenamic, mefenamic and tranexamic acids utilizing precipitation reactions with cobalt, cadmium and manganese. The acidic drugs were precipitated from their neutral alcoholic solutions with cobalt sulfate, cadmium nitrate or manganese chloride standard solutions followed by direct determination of the ions in the precipitate or indirect determination of the ions in the filtrate by atomic absorption spectroscopy (AAS). The optimum conditions for precipitation were carefully studied. The molar ratio of the reactants was ascertained. Statistical analysis of the results compared to the results of the official methods revealed equal precision and accuracy. The suggested procedures were applied for determining flufenamic, mefenamic and tranexamic acids in pharmaceutical preparations as well and proved validity.

The similar method using nickel (II) instead of cobalt, cadmium and manganese was also investigated by Salem *et al.* [49] for the quantitative determination of flufenamic, mefenamic and tranexamic acids, furosemide, diclofenac sodium and thiaprofenic acid. Statistical analysis of the results compared to assays used in pharmacopoeias and the A_{max} methods revealed equal precision and accuracy. Furthermore, the assays were also applied for the determination of these drugs in pharmaceutical preparations.

Alpdogan and Sungur [50] developed an indirect atomic absorption spectroscopy method for the determination of mefenamic and flufenamic acids, and diclofenac sodium, based on the complexation with copper (II) amine sulfate. The complex was extracted into chloroform, and the concentrations of substances were determined indirectly by AAS measurement of copper after re-extraction into 0.3 N nitric acid solution. The developed method was applied to the assay of the substances in commercial tablet formulations. The results were statistically compared with those obtained by HPLC method by *t*- and *F* tests at 95% confidence level. Calculated *t* and *F* values were both lower than the table values.

Khalil *et al.* [51] described the microquantitative determination of mefenamic acid based on the reaction of mefenamic acid with a silver nitrate solution in a neutral alcoholic medium. The formed precipitation is quantitatively determined directly or indirectly through the silver content of the precipitation formed or the residual unreacted silver ions in the filtrate by atomic absorption spectrophotometry. The results obtained in both the procedures either in their pure form or in their pharmaceutical formulations are accurate and precise. The stoichiometric relationship of the reaction was studied using Job's continuous variation method, and it was found to be (1:1) drug:Ag^+ for the mefenamic acid.

6. CHROMATOGRAPHIC METHODS OF ANALYSIS

6.1 Thin-layer chromatography

The detection of mefenamic acid and its metabolites {*N*-(2-methyl-3-carboxyphenyl)anthranilic acid and *N*-(2-methyl-3-hydroxymethyl-phenyl)anthranilic acid} in urine by thin layer chromatography was reported by Demetriou and Osborne [52]. Plates were prepared from a slurry of silica gel G (30 g) in water (65 mL). Layers of thickness 250 μ were prepared and dried at 110°C for 30 min. For the extraction procedure, urine (10 mL) was treated with sodium hydroxide (10 M, 1 μL) for 20 min at room temperature. The sample was then acidified with concentrated hydrochloric acid, and extracted with petroleum ether (b.p. 40–60°C; 10 mL) for 10 min on a mechanical shaker. The organic layer was separated, and placed in a 10 mL conical tube, then the solvent was removed. The extract was reconstituted with methanol (100 μL) and a 30-μL aliquot of the solution was used for spotting, along with 10-μL aliquot of the reference solution (mefenamic acid in methanol, 1 μg/1 μL). The solvent system used for the development of the chromatogram was toluene–acetic acid (9:1) providing R_f value of 0.75 for mefenamic acid. Sodium nitrite (1% in 1% sulfuric acid) was used as the detection reagent to give a green color of mefenamic acid.

By *et al.* [53] described the identification and chromatographic determination of some undeclared medicinal ingredients of traditional oriental medicine. The qualitative examination of indomethacin, mefenamic acid, diazepam and hydrochlorothiazide was performed on silica gel using solvent system of ethyl acetate (system A) and ethyl acetate–methanol (4:1, v/v) (system B). The spots were identified using short-wave UV light to give R_f values of 0.42 and 0.50 for mefenamic acid in solvent system A and B, respectively. The utilization of the TLC method was also reported by Dharmananda [54] for the investigation of imported Chinese herb products adulterated with drugs.

Argekar and Sawant [55] developed simultaneous determination of paracetamol and mefenamic acid in tablets by high performance thin layer chromatography method (HPTLC). The procedure was performed on silica gel using toluene–acetone–methanol (8:1:1) as the solvent system. Detection and quantification of the substance was assayed by densitometry at 263 nm. The linearity ranges were 120–360 ng/μL; and the RSD were between 0.41 and 0.78%. Limit of detection and limit of quantification were found to be 1.80 and 5.5 ng/μL, respectively. Recoveries were in the range of 95.50–103.60%.

6.2 Column chromatography

Adams [56] reported ion-pairs partition chromatography of mefenamic acid with tetraalkylammonium cations as an effort to develop an analytical method from the extraction data. Ion pairs of mefenamic acid with Me_4NOH, Et_4NOH, Pr_4NOH and Bu_4NOH were formed and extracted from aqueous solutions with $CHCl_3$ or mixtures of $CHCl_3$ and isooctane. The ion-pair extraction constant (K_{QA}) increased by a factor of $\sim$500,000 from the methyl to butyl homologs, and the plot of K_{QA} vs. C number was linear. The distribution ratio (D) of the ion-pairs between aqueous and organic phases was affected by solvent polarity, which was varied by adding up to 60% isooctane to the $CHCl_3$. Chromatography requires D values <0.01 for retention and >0.06 for elution. For chromatography, Celite 545 is mixed with Pr_4NOH (the best ion-pairing reagent), placed in a column, covered with an alkaline solution of ion-paired with excess Pr_4NOH and containing more Celite 545. The column is washed with 33% $CHCl_3$ in isooctane to reduce the distribution ratio to <0.01, and pure $CHCl_3$ is used to elute the ion-pair, which is collected in HCl–MeOH to break the ion-pair. Mefenamic acid is then detected by spectrometry in $CHCl_3$ at 353 nm. The average recovery from standard solutions was 99.01%, and results with capsule solutions were reproducible.

6.3 Gas chromatography

Bland *et al.* [57] reported the determination of mefenamic acid in blood and urine of the horse. To 2.0 mL of biological fluids were added 1.0 mL of an aqueous solution of saturated KH_2PO_4 and 2.0 mL of CH_2Cl_2. The drug was extracted into CH_2Cl_2 phase, transferred to a clean culture tube, and evaporated to dryness under a stream of nitrogen. To the residue was added 50 μL of pentafluoropropionic anhydride and 50 μL of ethyl acetate saturated with $ZnCl_2$ which acts as a Lewis acid. The tube was capped and placed in a water bath at 60°C for 5 min to complete the derivatization. Following derivatization, 1.0 mL of cyclohexane was added to the residue and the solvent was washed vigorously with 8.0 mL of a saturated sodium tetraborate solution. After centrifugation, the derivative in the cyclohexane phase was then analyzed using gas-liquid chromatograph which was a Beckman GC-5, equipped with a Beckman nonradioactive electron capture detector. The column was a 4 ft, 2 mm i.d. glass column packed with 3% OV-1 on 80/100 mesh Chromosorb GHP provided by Supelco, Inc. The instrument was operated isothermally at an oven temperature of 190°C. The samples were injected onto the column, at an inlet temperature of 215°C. Detector compartment temperature was 290°C. The carrier gas (He) flow was 40 mL/min. The drug derivative eluted in 3 min 43 s under these conditions. As little as 50 ng/mL of mefenamic acid can be quantitated with this procedure.

Gas-liquid chromatographic determination of mefenamic acid in human serum was established by Dusci and Hackett [58]. A 2 mL sample of serum was acidified with 0.5 mL of 1 M hydrochloric acid and extracted with 15 mL of CH_2Cl_2 by shaking for 5 min in a glass tube. The tube was centrifuged and the aqueous phase aspirated from the surface. The organic layer was filtered through a solvent-washed Whatman No. 54 filter paper. To a 12 mL aliquot of the solvent was added 50 μL of 1000 mg/L solution of 5-(*p*-tolyl)-5-phenylhydantoin in ethanol as the internal standard, and the mixture taken to dryness at 50°C under a stream of nitrogen. To the residue were added 80 μL *N*,*N*-dimethylacetamide, 10 μL tetramethylammonium hydroxide and 20 μL 1-iodobutane. The contents of the tube were mixed after each addition and left to stand for 5 min. The sample was centrifuged and an aliquot of 2.0 μL of the supernatant injected to a Pye GCD gas chromatograph equipped with a flame ionization detector. The column was a 1.5 m × 4 mm i.d. glass tube packed with 3% SP 2250 DA on Supelcoport 100–200 mesh. The instrument settings were as follows; injection temperature 285°C; detector temperature 290°C; carrier gas flow-rate, 60 mL/min; hydrogen flow-rate, 240 mL/min. Under these conditions the retention time of mefenamic acid and the internal standard were 1.8 min and 4.1 min, respectively. The ratio of the peak height of mefenamic acid to that of the internal standard was linear over a range of 0.05–0.8 μg of the drug.

Singh *et al.* [59] described a gas chromatographic method for the simultaneous analysis of flunixin, naproxen, ethacrynic acid, indomethacin, phenylbutazone, mefenamic acid and thiosalicylic acid in horse plasma and urine. For sample preparation from plasma, a 1-mL plasma sample containing 0.1, 0.5, 1.0, 2, 2.5, 5.0 and 10 μg of various drugs was mixed with 1 mL of 0.1 M hydrochloric acid and 10 mL of dichloromethane. The mixture was rotoracked for 10 min and centrifuged (1500 *g*) for 15 min, and the aqueous phase was aspirated from the surface. The organic layer was collected into another tube and dried at 45°C under a stream of nitrogen. The dried residue was derivatized by mixing with 10 μL of bis(trimethylsilyl)trifluoroacetamide (BSTFA) before analysis. Mefenamic acid and indomethacin was used as the internal standard for the quantitative analysis. For sample preparation from urine, 1–2 mL of urine (control or NaOH-hydrolyzed) were acidified to pH 3 with saturated phosphate buffer and extracted with 10 mL of dichloromethane. The dichloromethane layer was separated by centrifugation at 10,000 *g*, and dried at 45°C under nitrogen. The dried residue was derivatized by adding 10 μL of BSTFA to the sample, and 1 μL of the derivatized residue was injected into the GC–MS system.

GC–MS analysis was performed by using an Econocap capillary column, SE-54 (30 m × 0.25 mm i.d.). The oven temperature was programmed at 20°C/min from an initial temperature of 150°C to a final temperature of 280°C; the run time was 15 min. The injector temperature was 250°C and the injection mode was splitless. For selected-ion monitoring (SIM), three ions were selected for each drug. The recovery of each drug from plasma was ca. 95%, and the assay demonstrated good precision. The accuracy of the assay was best at the 10.0 μg/mL level.

Giachetti *et al.* [60] compared the performance of mass selective detector (MSD), electron capture detector (ECD) and nitrogen–phosphorus detector (NPD) of gas chromatography systems in the assay of six non-steroidal antiinflammatory drugs in the plasma samples. As a practical test, six NSAIDs (mefenamic, flufenamic, meclofenamic and niflumic acids, diclofenac and clonixin) added to plasma samples were detected and quantified. The analyses were carried out after solvent extraction from an acidic medium and subsequent methylation. The linearity of response was tested for all the detection systems in the range of 1–25 ng/mL. Precision and accuracy were detected at 1, 5 and 10 ng/mL. The minimum quantifiable level for the six drugs was about 1 ng/mL with each of the three detection systems.

Gonzáles *et al.* [61] described a gas chromatographic–mass spectrometric (GC–MS) procedure for the detection of seventeen nonsteroidal anti-inflammatory drugs (including mefenamic acid) in equine plasma and urine samples. The extraction of the compounds from the biological matrix was performed at acidic pH (2–3) with diethyl ether. Ethereal extracts were washed with a saturated solution of sodium hydrogencarbonate (urine) or treated with a solid mixture of sodium carbonate and sodium hydrogencarbonate (plasma). The ethereal extracts were dried and derivatized by incubation at 60°C with methyl iodide in acetone in the presence of solid potassium carbonate. Mono- or bismethyl derivatives of the NSAIDs were obtained. After derivatization kinetic studies, 90 min was the incubation time finally chosen for screening purposes for adequate methylation of all the compounds under study. For individual confirmation analyses, shorter incubation times can be used. The chromatographic analysis of the derivatives was accomplished by a Model 5890A Series II gas chromatograph coupled to a Model 5970 electron impact mass-selective detector via a direct capillary interface (Hewlett-Packard). The column was a 25 m × 0.2 mm i.d. fused silica cross-linked methylsilicone with a 0.11-μm film thickness (Hewlett-Packard). Helium was used as the carrier gas at 0.65 mL/min. The injection port and detector temperatures were 280°C. The oven temperature was increased from 100 to 200°C at 25°C/min and then to 300°C at 15°C/min, with a final hold time of 2.3 min; the total run time was 13 min. The injection volume was 2 μL and a splitting ratio of 10:1 was used. In general, extraction recoveries ranged from 23.3 to 100% in plasma and from 37.5 to 83.8% in urine samples. Detection limits from less than 5 to 25 ng/mL were obtained for both plasma and urine samples using selected-ion monitoring. The procedure was applied to the screening and confirmation of NSAIDs in routine doping control of equine samples.

Maurer *et al.* [62] developed a gas chromatographic–mass spectrometric (GC–MS) screening procedure for the detection of NSAIDs in urine as part of a systematic toxicological analysis procedure for acidic drugs and poisons after extractive methylation. The compounds were separated by capillary GC and identified by computerized MS in the full-scan mode. Using mass chromatography with the ions m/z 119, 135, 139, 152, 165, 229, 244, 266, 272, and 326, the possible presence of NSAIDs and their metabolites could be indicated. The identity of positive signals in such mass chromatograms was confirmed by comparison of the peaks underlying full mass spectra with the reference spectra recorded during this study. The method allowed the detection of therapeutic concentrations of acemetacin, acetaminophen (paracetamol), acetylsalicylic acid, diclofenac, diflunisal, etodolac, fenbufen, fenoprofen, flufenamic acid, flurbiprofen, ibuprofen, indomethacin, kebuzone, ketoprofen, lonazolac, meclofenamic acid, mefenamic acid, mofebutazone, naproxen, niflumic acid, phenylbutazone, suxibuzone, tiaprofenic acid, tolfenamic acid, and tolmetin in urine samples. The overall recoveries of the different NSAIDs ranged between 50 and 80% with coefficients of variation of less than 15% ($n = 5$), and the limit of detection of the different NSAIDs were between 10 and 50 ng/mL ($s/n = 3$) in full-scan mode. Extractive methylation has also been used for plasma analysis.

Takeda *et al.* [63] described a systematic analysis of acid, neutral and basic drugs in horse plasma by a combination of solid-phase extraction, nonaqueous partitioning and gas-chromatography–mass spectrometry. Sample preparation method started by using a 6 mL column packed with 1 g of Bond Elut Certify for the extraction of 4 mL of horse plasma. Fractionation is performed with 6 mL of $CHCl_3$–Me_2CO (8:2) and 5 mL of 1% TEA–MeOH according to its property. Simple and effective clean-up based on nonaqueous partitioning is adopted to remove co-eluted contaminants in both acid and basic fractions. Two kinds of

1-(*N*,*N*-diisopropylamino)-*n*-alkanes are co-injected with the samples into the GC–MS system for the calculation of the retention index. Total recoveries of 107 drugs (including mefenamic acid) are examined. This procedure achieves sufficient recoveries and clean extracts for GC–MS. This method is able to detect ng/mL drug levels in horse plasma.

Lo *et al.* [64] reported acidic and neutral drugs screen in blood with quantitation using capillary gas chromatography–ion flame detector. Blood previously acidified with aqueous saturated ammonium chloride solution was extracted with ethyl acetate. The dried extract was subjected to acetonitrile–hexane partition. The acetonitrile portion was analyzed for the presence of acidic and neutral drugs by GC–FID (25 m narrow-bore × 0.25 mm i.d. HP-5 column with 0.33 μm film thickness). The protocol was found to be suitable for both clinical toxicology (including emergency toxicology) and postmortem toxicology. Quantitation was facilitated by incorporating calibration blood standards in each assay batch. The five drugs most commonly encountered in clinical blood specimens (1150 cases) were: paracetamol (47.4% of the cases), chlormezanone (6.6%), theophylline (1.74%), naproxen (1.65%) and mefenamic acid (1.56%).

Arrizabalaga *et al.* [65] reported the determination of the composition of a fraudulent drug by coupled chromatographic methods (GC–MS, GC–FTIR, HPLC–DAD). The Chinese pills, *Chuifong Toukuwan*, as traditional herbal medicine claimed to have a natural composition, was studied by these appropriate methods. The analysis of crude and derivatized extracts of the pills showed that they comprise a mixture of synthetic drugs (mefenamic acid, indomethacin, hydrochlorothiazide, and diazepam) and some herbal compounds. A similar procedure was also reported by Fraser *et al.* [66] for the determination of undeclared prescription drugs in methanol–ether extracts of Black Pearl products, a type of Chinese herbal preparation generally marketed to American consumers suffering from arthritis and back pain. They detected the following prescription drugs: diazepam, diclofenac, hydrochlorothiazide, indomethacin and mefenamic acid.

6.4 High performance liquid chromatography

As mentioned above, a compendial assay using high performance liquid chromatography was described by the USP 26 and the Indonesian Pharmacopoeia [1, 4].

The separation of samples that contain more than 15–20 analytes ($n > 15$–20) is typically difficult and usually requires gradient solution. Considering this problem, Dolan *et al.* [67, 68] reported reversed-phase liquid chromatographic (RP–LC) separation of complex samples by optimizing temperature and gradient time. They examined the reversed-phase liquid chromatographic separation of 24 analytes with $8 \leq n \leq 48$ (n = the sum of analytes) as a function of temperature T and gradient time t_G. The required peak capacity was determined for each sample, after selecting T and t_G for optimum selectivity and maximum sample resolution. A comparison of these results with estimates of the maximum possible peak capacity in reversed-phase gradient elution was used to quantify the maximum value of "n" for some required sample resolution (when T and t_G have been optimized). These results were also compared with literature studies of similar isocratic separations as a function of ternary-solvent mobile phase composition, where the proportions of methanol, tetrahydrofuran and water were varied simultaneously. This turn provides information on the relative effectiveness of these two different method development procedures (optimization of T and t_G vs. %methanol and %tetrahydrofuran) for changing selectivity and achieving maximum resolution. The chromatographic conditions: 0–100% acetonitrile in buffer gradient in 40 min, temperature 50°C, 25 × 0.46 cm column, 2.0 mL/min. This observation is surprising, since the ability of either T or t_G to vary band spacing and maximize separation is much less than observed for changes in solvent type (e.g., varying the ratio of methanol to tetrahydrofuran). An alternative approach is to carry out two separations with different conditions (T, t_G) in each run. The combination of results from these two runs then allows the total analysis of the sample, providing that every sample component is adequately resolved in one run or the other. One of the compounds having the poorest-resolved peak was mefenamic acid ($R_s = 1.7$, $t_G = 103$ min, $T = 74$°C).

Ahrer *et al.* [69] developed methods for the determination of drug residues in water based on the combination of liquid chromatography (LC) or capillary electrophoresis (CE) with mass spectrometry (MS). A 2 mM ammonium acetate at pH 5.5 and a methanol gradient was used for the HPLC–MS allowing the separation of a number of drugs such as paracetamol, clofibric acid, penicillin V, naproxen, bezafibrate, carbamazepin, diclofenac, ibuprofen, and mefenamic acid. Apart from the analytical separation technique, water samples have to be pretreated in order to get rid of the matrix components and to enrich the analytes; the usual way to accomplish this aim is to perform a solid-phase extraction

(SPE) step employing suitable stationary phase (reversed-phase materials) and conditions. Using the optimized SPE procedure, a 500 mL volume of water sample was brought to pH 2 with concentrated hydrochloric acid and passed through an SPE cartridge (6 mL) packed with 500 mg Bondesil ODS 40 μm (Varian) conditioned with acetone, methanol and water, pH 2 (one cartridge volume each). The flow-rate was adjusted to approximately 10 mL/min. After the cartridges had been allowed to dry for 30 min, the drugs were eluted using an overall volume of 2 mL methanol. The extract was brought to dryness in a nitrogen stream (purity 4.6) and finally redissolved in 50 μL methanol and diluted with 300 μL water which was then used for injection (100 μL injection volume). Liquid chromatography separations were performed on a HP 1100 HPLC system equipped with a HP 1050 autosampler (Agilent) employing a YMC ODS-AM column 250 × 2 mm i.d. (YMC). Using a flow-rate of 150 μL/min the optimized mobile phase was a ternary gradient of a 2 mM ammonium acetate solution adjusted to pH 5.5 with 1 M acetic acid (A), methanol (B) and water (C) in the following form: 0 min: A–B–C (10:15:75); 3 min: A–B–C (10:15:75); 10 min: A–B–C (10:50:40); 20 min: A–B–C (10:90:0). MS detection was performed on a quadrupole system HP 5989B (Agilent) equipped with a radiofrequency-only hexapole (Analytica of Branford) using either a pneumatically assisted electrospray ionization (ESI) interface HP 59987A (Agilent) or an atmospheric pressure chemical ionization (APCI) interface (Analytica of Branford) for the combination with LC instrument. HPLC–MS is a powerful technique for the determination of drug residues in water offering both a high selectivity and good detection limits between 0.05 and 1 μg/L in the solution injected; with a sample pretreatment procedure based on SPE before LC separation, detection limits in the sample of about 1 ng/L and below can be achieved. The applicability of the technique could be demonstrated for the analysis of several river water samples; concentrations between approximately 2 and 130 ng/L of bezafibrate, carbamazepine, diclofenac, and mefenamic acid were found.

Ku *et al.* [70] used HPLC with photodiode array detector (DAD) for the identification of adulterants in rheumatic and analgesic Chinese medicine. A study was performed by using a gradient elution with acetonitrile and 0.1% acetic acid solution on a C-18 column. The method was applied to simultaneously screen seventeen synthetic drugs adulterated in traditional Chinese medicine within half an hour. The drugs belonged to six pharmacological categories, namely antipyretic analgesics, glucocorticoids, diuretics, CNS stimulants, muscle relaxants and sedatives. They included acetaminophen, aminopyrine, bucetin, ethoxybenzamide, indomethacin, ketoprofen, mefenamic acid, phenylbutazone, piroxicam, salicylamide, dexamethazone, prednisolone, chlormezanone, chlorzoxazone, hydrochlorothiazide, diazepam, and caffeine. This method can provide higher resolution and greater efficiency than thin-layer chromatography for screening adulterated synthetic drugs.

A reversed-phase HPLC post-column ion-pair extraction system was developed by Kim and Stewart [71, 72] for the analysis of carboxylic acid drugs and their salts (sodium formate, sodium acetate, 3-bromopropionic acid, 6-aminocaproic acid, 11-bromoundecanoic acid, 1-heptanesulfonic acid, *p*-nitrophenylacetic acid, sodium benzoate, sodium salicylate, valproic acid, probenecid, naproxen, ketoprofen, ibuprofen, mefenamic acid, flufenamic acid, and cefuroxime sodium) using α-(3,4-dimethoxyphenyl)-4′-trimethylammoniummethylcinnamonitrile methosulfate (DTM) as a fluorescent ion-pair reagent. The HPLC system is composed of three pumps; an Alcott Chromatography Model 760 HPLC pump for the mobile phase, a LDC/Milton Roy mini-pump VS with pulse dampener (SSI Model LP-21) for the ion-pair reagent, and a Kratos Spectraflow 400 solvent delivery system with pulse dampener for the extraction solvent; a Rheodyne Model 7125 injector equipped with a 50 μL sample loop, and two mixing tees (Alltech). The detector was a HP Model 1046A programmable Fluorescence Detector (Hewlett-Packard) set at an excitation of 355 nm and an emission of 460 nm (also equipped with a 408 nm emission cut-off filter). The analytical column was a 5 μm RP-8 column (Spheri-5, 10 cm × 4.6 mm i.d.). The extraction coils were three dimensional knitted teflon coils (1.8 m × 0.3 mm i.d., LDC/Milton Roy) knitted using a four pin "Strickliesel". The mobile phase was prepared as an absolute methanol–pH 7 aqueous phosphate buffer mixture (30:70, v/v) and the flow rate was set at 1 mL/min. An ion-pair reagent concentration of 1.12×10^{-4} M (5 mg/100 mL of mobile phase) was the most appropriate in order to minimize the contribution of background noise in the system. A 30% *n*-pentanol in chloroform was used as the extracting solvent. Sodium salicylate, ketoprofen, ibuprofen, probenecid and valproic acid were used as model compounds to evaluate the ion-pair extraction system. The method was applied to pharmaceutical dosage forms containing ketoprofen and valproic acid. Other acidic compounds evaluated using the ion-pair reagent showed that lipophilic acids produced more extractable ion-pairs and higher sensitivities than hydrophilic acids. Mefenamic acid exhibited the capacity factor (k') of 7.5039 and the limit of detection of 100 ng and 2 μg/mL ($s/n = 2$).

Shinozuka *et al.* [91] developed a sensitive method for the determination of four anthranilic acid derivatives (diclofenac sodium, aluminium flufenamate, mefenamic and tolfenamic acids) by HPLC procedure. The four drugs were converted into methylphthalimide (MPI) derivatives in a constant yield by reaction with *N*-chloromethylphthalimide at 60°C for 30 min. The production of the MPI derivatives were confirmed by mass spectrometry. The MPI derivatives of the four drugs were separated by HPLC using a C-18 bonded phase LiChrospher RP-18 column (250 × 4 mm i.d.) with acetonitrile–water (80:20, v/v) as mobile phase. The flow rate was 0.8 mL/min. The UV absorbance was measured at 282 nm. The calibration curves of the MPI derivatives of the drugs were linear from 1.0 to 5.0 μg/mL. The detection limits of the four drugs were 0.5–5 ng. The extraction procedure for the four anthranilic acid derivatives added in the plasma and urine was performed by using Extrelut 1 column. Yields of column extraction of 100 μL of plasma and urine samples (containing 0.5 μg of anthranilic acid derivatives) with 6 mL of ethyl acetate were 84–106%.

Other methods for the HPLC analysis of mefenamic acid, as described by several groups, are presented in Table 1.

6.5 Supercritical fluid chromatography

Jagota and Stewart [102] investigated super critical fluid chromatography (SFC) for the separation of nonsteroidal antiinflammatory agents. Chromatography was performed on a Lee Scientific Model 600D supercritical fluid chromatograph equipped with a pump, oven and flame-ionization detector and controlled by a Dell computer (ACI 600D, software version 2.2). SFC was performed on three different stationary phases: a 5 m × 100 μm i.d. SB-methyl-100 (200 μm o.d. and 0.25 μm flim thickness), 10 m × 50 μm i.d. SB-biphenyl-30 and a SB-cyanopropyl-50 (both 195 μm o.d. and 0.25 μm film thickness). Solutions of each NSAID were prepared by accurately weighing 5 mg of each drug and dissolving in 5 mL of absolute methanol to give a final concentration of approximately 1 mg/mL. Chromatographic parameters were pump program, multilinear pressure program: 7 min hold at an initial pressure of 100 atm, then 25 atm/min ramp to 250 atm, followed by a 4.0 atm/min ramp to 290 atm; oven program, isothermal at 130°C; injection type, time split set at 200 ms, injection ratio approximately 20:1 giving an injection volume of approximately 25 mL; detector, flame ionization at 375°C; mobile phase, supercritical fluid chromatography grade carbon dioxide. Baseline separation of a flufenamic acid, mefenamic acid, fenbufen and

Table 1

HPLC Methods for Mefenamic Acid

Column	Mobile phase	Detection	Detection limit	Samples/ matrices	Ref.
Daisopak ODS column (SP-120-5-ODS-BP, 250 × 4.6 mm i.d., 5 μm)	Acetonitrile–acetate buffer (pH 3.5; 0.1 M)–methanol (35:40:25, v/v/v), 1 mL/min	UV, 220 nm	11.5–75 ng/mL	Human plasma	73
Lichrospher 60 RP select B column (250 × 4 mm i.d., 5 μm) and Lichrospher 60 RP select B guard column (4 × 4 mm i.d., 5 μm)	0.01 M Acetic acid pH 3 in methanol (35:65, v/v), 1 mL/min, 35°C	Photodiode-array at 240, 254, 284 nm	0.016 μg/mL at 240 nm, 0.027 μg/mL at 254 nm	Bull plasma	74
Discovery RP-Amide C16, 15 cm × 4.6 mm, 5 μm particles	Acetonitrile–25 M KH_2PO_4, pH 3.0 (40:60), 1 mL/min, 30°C	UV, 230 nm	ND[a]	Solution	75

Wakosil ODS 5 C_{18}column (150 × 4.6 mm i.d., 5 μm)	1.9 g Tetra-pentylammonium bromide in 1 L of a mixture of acetic acid–sodium acetate buffer solution, pH 5.0, and acetonitrile (11:9, v/v), 1.0 mL/min, 40°C	UV, 230 nm	2 ng/mL	Human urine, pharmaceutical preparations	76
ODS II Inertsil (250 × 4.6 mm, 5 μm) column connected to pre-column ODS II Inertsil (8 mm, 5 μm)	Acetic acid–acetonitrile (80:20, 2 min; 36:64, 25 min; 80:20, 5 min), 1 mL/min, 40°C	Photodiode-array 240, 278, 290 nm	0.05–64 ng/mL	Plasma	77
Inertsil ODS-2 (150 × 4.6 mm, i.d., 5 μm) column	50 mM Phosphate buffer–acetonitrile (58:42, v/v) at pH 5.0, 0.9 mL/min, ambient temperature	UV, 230 nm	0.05 μg/mL	Human urine	78
LiChrosorb-RP 18 column (250 × 4 mm i.d., 7 μm)	6 mM Phosphoric acid–acetonitrile (50:50), 2.0 mL/min, 50°C	UV, 205 nm	0.05 μg/mL	Human plasma	79

(*continued*)

Table 1 (continued)

Column	Mobile phase	Detection	Detection limit	Samples/ matrices	Ref.
Hypersil ODS (200 mm × 2.1 mm i.d.) column	ND[a]	Diode-array detection	ND[a]	Blood	64
Lichrosorb RP 18 column (250 × 4 mm i.d., 7 μm)	Acetonitrile–water–0.1 M phosphoric acid (55:42:3,v/v), 2.0 mL/min, 45°C	UV, 280 nm	ND[a]	Urine	80
TSK Gel ODS-80 TM column (250 × 4.6 mm i.d., 5 μm)	10 mM Ammonium dihydrogen phosphate–methanol (98:2, v/v) at pH 5.0, 1.0 mL/min, ambient temperature	UV, 254 nm	0.05 μg/mL	Urine	81
Nucleosil C_{18} analytical column (5 μm)	Methanol–water (77:23, v/v), 0.8 mL/min, room temperature	UV, 280 nm	0.1 μg/mL	Rat plasma	82

Hypersil SI (100 × 2.0 mm i.d., 5 μm) column	Hexane–5% water in iso-propanol (98:2, 70:30, 0:100, v/v), 0.4 mL/min, 45°C	Diode-array UV, 280 nm	ND[a]	Equine urine	83
Lichrosorb RP 18 column (250 × 4.6 mm, 7 μm)	Methanol–water (30:70,35:65,40:60, 45:55,50:50, 55:45,60:40, 65:35,70:30,75:25,80:20, v/v) containing 1% acetic acid, 1.0 mL/min, room temp.	UV, 230, nm	ND[a]	Pharmaceuti-cal prepara-tions	84
Vydac stainless-steel analytical column (250 × 4.6 mm i.d.), C_{18} bonded-phase silica, 5 μm	10 mM Phosphoric acid–acetonitrile (40:60, v/v), pH 2.6, 0.9 mL/min, ambient temperature	UV, 280 nm	0.08 μg/ml	Human plasma	85

(continued)

Table 1 (continued)

Column	Mobile phase	Detection	Detection limit	Samples/ matrices	Ref.
Lichrosorb RP 18 (250 × 4 mm i.d., 10 μm) column	0.065 M Ammonium acetate–methanol (25:75, 30:70), 0.8 mL/min, 22°C	UV, 282 nm	0.7 ng	Blood serum, urine, tablets, capsules, suppositories, suspensions, ointment	86
Lichrosorb RP 18 (250 × 4 mm i.d., μm, 10 μm) column	0.05 M Ammonium acetate–methanol (78:22, 72:28), 0.81 mL/min, 22°C	UV, 285 nm	1 pg	Blood serum, Urine, tablets, suppositories, suspensions	87
Cyano column (25 × 0.4 cm, 10 μm)	Water–acetonitrile–methanol–17 M acetic acid (69:15:15:1, v/v)	UV, 290 nm	0.05 μg/mL	Human plasma	88

Wakosil 5 C 18 (150 × 4.6 mm)	Acetonitrile–water–triethylamine (60:40:0.1, v/v/v) adjusted with H_3PO_3 to a pH of 3.0, 0.8 mL/min, 40°C	UV, 280 nm	ND[a]	Pharmaceutical preparation	89
YMC ODS (70 × 4.6 mm i.d., 5 μm)	20 mM pH 5.0 Phosphate buffer–acetonitrile (65:35, v/v) 1.0 mL/min, 40°C	UV, 219 nm	0.1 ng/mL	Human serum	90
YMC ODS (70 × 4.6 mm i.d., 5 μm)	20 mM pH 3.5 Phosphate buffer–acetonitrile (55:45, v/v) 1.0 mL/min, 40°C	UV, 219 nm	0.1 ng/mL	Human serum	90
TSK Gel ODS 80 TM (150 × 4.6 mm i.d., 5 μm)	20 mM pH 6.0 Phosphate buffer–acetonitrile (55:45, v/v) 1.0 mL/min, 40°C	UV, 219 nm	0.1 ng/mL	Human serum	90

(continued)

Table 1 (continued)

Column	Mobile phase	Detection	Detection limit	Samples/ matrices	Ref.
Supelcosil LC 8 (7.5 cm × 4.6 mm i.d., 3 μm)	0.05 M Phosphoric acid–acetonitrile (55:45, v/v), 1.0 mL/min	Diode-array, 209–402 nm	100–500 ng/mL	Plasma, urine	59
Supelcosil LC 8 (7.5 cm × 4.6 mm i.d., 3 μm)	0.05 M Phosphoric acid–acetonitrile (55:45, v/v), 1.0 mL/min	UV, 235 nm	50–150 ng/mL	Plasma, urine	59
Supelcosil LC 8 (7.5 cm × 4.6 mm i.d., 3 μm)	0.05 M Phosphoric acid–acetonitrile (55:45, v/v), 1.0 mL/min	Fluorescence detection (λexc = 235 nm, λem = 405 nm	10 ng/mL	Plasma, urine	59
10 μm μBondapak phenyl column (300 × 3.9 mm, 10 μm)	Methanol–glacial acetic acid–water (85:2:15, v/v), 1 mL/min	Photodiode array, 278 nm	ND[a]	Pharmaceutical preparations	92

Nova-Pak C 18 column (150 × 3.9 mm, 5 μm)	Acetonitrile–THF–water–glacial acetic acid (15:40:45:2, v/v), 1 mL/min	Photodiode array, 278 nm	ND[a]	Pharmaceutical preparations	92
Shimpack CLC-ODS (15 cm × 6.0 mm)	1000 mL of 85% Methanol containing 0.05 M sodium perchlorat and 5.7 mL of glacial acetic acid, 0.6 mL/min, 35°C	Electrochemical detection	0.4 pg	Human serum	93
Hypersil ODS (25 × 0.4 cm i.d., 5 μm) column	Acetonitrile–sodium acetate buffer pH 4.2 (60:40, v/v), 2.5 mL/min	UV, 280 nm	2 mg/mL	Plasma/serum	94
LiChrosorb RP 18 column (250 × 4 mm i.d., 7 μm)	6.5 mM Phosphoric acid–acetonitrile (45:55, v/v), 2.0 mL/min, 45°C	UV, 280 nm	0.1 mg/L	Plasma	95

(*continued*)

Table 1 (continued)

Column	Mobile phase	Detection	Detection limit	Samples/ matrices	Ref.
Octyl coated silica of RP-8 Spheri-10 column (250 × 4.6 mm, 10 μm)	Acetonitrile–0.05 M potassium dihydrogen phosphate pH 3.3 (15:85, 50:50, 70:30), 1 mL/min	UV, 240 nm	ND[a]	Black pearl (black herbal pill)	53
Zorbax C 8 column (25 cm × 4.6 mm i.d.)	Methanol–water–acetic acid (70:30:1), 1.4 mL/min, 35°C	UV, 280 nm	ND[a]	Human serum albumin	96
Hypersil ODS stainless-steel column (16 cm × 5 mm i.d., 5 μm)	0.1 M Potassium dihydrogen phosphate–formic acid–isopropanol (1000:1:17, v/v), 2 mL/min	UV, 240 nm	ND[a]	Blood	97

Hypersil ODS stainless-steel column (16 cm × 5 mm i.d., 5 μm)	0.1 M Potassium dihydrogen phosphate–formic acid–isopropanol (1000:1:176, v/v), 1.5 mL/min	UV, 240 nm	ND[a]	Blood	97
Hypersil ODS stainless-steel column (16 cm × 5 mm i.d., 5 μm)	0.1 M Potassium dihydrogen phosphate–formic acid–isopropanol (1000:1:540, v/v), 1.5 mL/min	UV, 240 nm	ND[a]	Blood	97
ODS silica stainless-steel column (150 × 1 mm i.d., 3 μm)	Methanol–1% acetic acid, 30 μL/min	UV, 254 nm	ND[a]	Pharmaceutical preparations	98
ODS silica stainless-steel column (150 × 1 mm i.d., 3 μm)	Methanol–1% ammonia, 30 μL/min	UV, 254 nm	ND[a]	Pharmaceutical preparations	98
Hibar RP 18 column (250 × 4 mm i.d., 5 μm) and short pre column Pherisorb RP 8 (30 × 4 mm i.d., 30–40 μm)	Acetonitrile–0.05 M phosphate buffer pH 4.5 (from 25–55% acetonitrile), 0.8 mL/min, 35°C	UV, 254 nm	ND[a]	Urine	99

(*continued*)

Table 1 (continued)

Column	**Mobile phase**	**Detection**	**Detection limit**	**Samples/ matrices**	**Ref.**
Cyanopropylsilane stainless steel column (30 cm × 4 mm i.d., 10 μm)	Water–acetonitrile–acetic acid (60:30:10, v/v/v), 1.0 mL/min	UV, 254 nm	1 μg/mL	Plasma	100
Zorbax SB-C 8 (50 × 2.1 mm, 5 μm)	0.1% TFA in water–acetonitrile (60:40), 0.4 mL/min, 35°C	ND[a]	ND[a]	Solution	101

[a]ND = no data available

indomethacin mixture was achieved on the SB-biphenyl-30 column using a pressure gradient. A mixture containing flufenamic acid, mefenamic acid, acetylsalicylic acid, ketoprofen and fenbufen, and another mixture containing ibuprofen, fenoprofen, naproxen ketoprofen and tolmetin were well separated on the SB-cyanopropyl-50 column using pressure gradients. Typical analysis time for a mixture of NSAIDs on the biphenyl or cyanopropyl column was approximately 20–25 min. Application of the method using the biphenyl column to the determination of NSAIDs present in selected dosage forms was performed.

6.6 Capillary electrophoresis

Ku *et al.* [103] developed a micellar electrokinetic capillary chromatography (MEKC) to identify adulterants in rheumatic and analgesic traditional Chinese medicine. This chromatographic technique was carried out using a buffer of (pH 9.0) 0.02 M sodium tetraborate and dihydrogenphosphate (containing 0.15 M sodium lauryl sulfate). This method was applied to simultaneously analyze fourteen synthetic chemical drugs adulterated in traditional Chinese medicine within 25 min. They represent five pharmacological categories, namley antipyretic analgesics, glucocorticoids, CNS stimulants, muscle relaxants and sedatives, and include acetaminophen, aminopyrine, bucetin, diclofenac sodium, ethoxybenzamide, indomethacin, ketoprofen, mefenamic acid, phenylbutazone, dexamethazone, prednisolone, chlorzoxazone, diazepam and caffeine. The recovery of synthetic chemical drugs added to *Dwu-Hwo-JIh-Seng-Tang* ranged from 92.1 to 102.5%. This method was used to assess two traditional Chinese medicines sold by the distributors of Chinese natural drugs collected by the consumer centers of local health bureaus from October 1991 to June 1992. MEKC was found to be an efficient and sensitive chromatographic technique to measure adulterants in rheumatic and analgesic traditional Chinese medicine.

The possibility of separating flufenamic, meclofenamic and mefenamic acids by capillary electrophoresis using β-cyclodextrin was studied by Pérez-Ruiz *et al.* [104]. A capillary electrophoresis P/ACE System 5500 (Beckman Instruments) equipped with a diode-array detector, an automatic injector, a fluid-cooled cartridge and a System Gold data station were used in this study. Electrophoresis was performed in a 57 cm × 75 μm i.d. (50 cm to the detector) fused-silica capillary tube (Beckman Instruments). The high voltage power supply was set to 12 kV (normal polarity, equivalent to a field strength of 210 V/cm) resulting in a typical current of 90–92 μA. Sample introduction was made at the anode side using the pressure option (0.5 p.s.i. ≈ 33.3 mbar) for 4–10 s. The detection of analytes was performed at 285 nm. The best approach involved combining a suitable pH of the carrier electrolyte (pH 12.0) with the host-guest complexation effects of β-cyclodextrin. A running buffer consisting of 30 mM phosphate buffer (pH 12.0), 2 mM β-CD and 10% (v/v) acetonitrile was found to provide a very efficient and stable electrophoresis system for the analysis of fenamic acids by capillary zone electrophoresis (CZE). Responses were linear from 0.4 to 40 μg/mL for the three drugs with detection limits of about 0.3 ng/mL. Intra- and inter-day precision values of about 1–2% R.S.D. ($n = 11$) and 3–4% R.S.D. ($n = 30$), respectively, were obtained. The method is highly robust and no breakdowns of the current or capillary blockings were observed for several weeks. The general applicability of this rapid CZE procedure (migration less than 12 min) is demonstrated for several practical samples, including serum, urine, and pharmaceuticals.

Cherkaoui and Veuthey [105] described the simultaneous separation of nine nonsteroidal antiinflammatory drugs by nonaqueous capillary electrophoresis (NACE). NACE was found to be a good alternative for the analysis of pharmaceutical drugs and their metabolites, which are difficult to separate in aqueous media. CE data were generated in a Hewlett-Packard CE system equipped with an on-column diode array detector, an auto sampler and a power supply able to deliver upto 30 kV. The total capillary length (Composite Metal Services) was 48.5 cm, while the length of the detector was 40 cm, with a 50 μm internal diameter. An alignment interface, containing an optical slit matched to the internal diameter, was used and the detection wavelength was set at 200 nm with a bandwidth of 10 nm. All experiments were performed in cationic mode (anode at the inlet and cathode at the outlet). The capillary was maintained in a thermostat at 20°C, unless otherwise stated. A constant voltage of 30kV, with an initial ramping of 500 V/s, was applied during analysis. Sample injection (8 nL injection volume) was achieved using the pressure mode for 10 s at 25 mbar. The selected electrophoretic medium, a methanol–acetonitrile (40:60, v/v) mixture containing 20 mM ammonium acetate, yielded the best compromise in terms of analysis time, selectivity and separation efficiency for the simultaneous separation of baclofen, indoprofen, ibuprofen, fenoprofen, indomethacin, ketoprofen, suprofen, diclofenac, and mefenamic acid.

Ahrer *et al.* [69] developed methods for the determination of drug residues in water based on a combination of liquid chromatography (LC) or capillary electrophoresis (CE) with mass spectrometry (MS). A 20 mM ammonium acetate solution adjusted to pH 5.1 with acetic acid was employed as the carrier electrolyte for the separation of clofibric acid, naproxen, bezafibrate, diclofenac, ibuprofen, and mefenamic acid by CE–MS. For the determination of the drug residues by CE–MS the water samples were pretreated by liquid–liquid extraction (LLE) prior to solid-phase extraction (SPE). A 500 mL volume of water sample was brought to pH 2 with concentrated hydrochloric acid; after the addition of 50 g sodium sulfate the sample was extracted twice with 25 mL of a mixture of hexane–MTBE (methyl *tert*-butyl ether) (1:1, v/v). The organic phase was re-extracted twice with 50 mL 2 mM sodium hydroxide solution, brought to pH 2 with hydrochloric acid and diluted to 250 mL with water, pH 2. The extract was passed through an SPE cartridge (6 mL) packed with 500 mg Bondesil ODS 40 μm (Varian) conditioned with acetone, methanol, and water, pH 2 (one cartridge volume each). The flow-rate was adjusted to approximately 10 mL/min. After the cartridges had been allowed to dry for 30 min, the drugs were eluted using an overall volume of 2 mL methanol. The extract was brought to dryness in a nitrogen stream (purity 4.6) and finally redissolved in 50 μL of a mixture of methanol–carrier electrolyte (80:20, v/v) prior to injection (5 kPa, 0.3 min). A Crystal 310 CE instrument (Thermo CE) was employed for CE–MS experiments. Fused-silica capillaries of 50 μm i.d. were obtained from Polymicro Technologies; new capillaries with a length of 70 cm were conditioned by flushing with 0.1 M NaOH followed by water (each for 10 min). Finally, the conditioning procedure was completed by flushing with the CE carrier electrolyte consisting of 20 mM ammonium acetate, pH 5.1 for 5 min; prior to each run the capillary was flushed with carrier electrolyte for 3 min. MS detection was performed on a quadrupole system HP 5989B (Agilent) equipped with a radiofrequency-only hexapole (Analytica of Branford) using a pneumatically assisted electrospray ionization interface HP 59987A (Agilent). A CE probe was used for CE–MS experiments. The sheat-liquid for CE–MS consisted of 2-propanol–water (80:20, v/v) containing 0.1% (v/v) acetic acid for the positive detection mode or 0.1% (v/v) triethylamine for the negative mode and was delivered by a syringe pump (Model 22; Harvard Apparatus) at a flow rate of 4 μL/min. The drying gas flow-rate was 1.4 L/min at 150°C. CE–MS proved to be a useful alternative to HPLC–MS, although the sample pretreatment had to be expanded by an LLE prior to the SPE. Because of rather high standard deviations of the recoveries due to the combination of three extraction steps, a method based on standard addition is recommended for quantitation. As expected the detection limits for the CE–MS method are poorer than the detection limits obtained by HPLC–MS namely between 27 and 93 μg/L for standard injections which resulted in detection limits in the samples between 4.8 and 19 ng/L. Nevertheless, CE–MS may be a useful technique for the confirmation of questionable results obtained by HPLC–MS.

Regarding the work described above, Ahrer and Buchberger [106] noted that in contrast to the technique of gas chromatography (GC) coupled with MS, CE–MS has an advantage in that derivatization of nonvolatile analytes can be avoided. HPLC–MS yields better detection limits, but CE–MS is a valuable complementary alternative, which can be used when separation selectivity of HPLC is not satisfactory or when results from HPLC must be confirmed by a different technique. Therefore, they developed a nonaqueous capillary electrophoresis (NACE) for the separation of clofibric acid, penicillin V, diclofenac, mefenamic acid, bezafibrate, naproxen and ibuprofen using carrier electrolyte consisting of 20 mM ammonium acetate, 60 mM acetic acid in methanol–acetonitrile (40:60). Capillary electrophoresis separation was performed using an HP^{3D} instrument (Agilent Technologies). Mass spectrometry detection has been done on an HP 5989 MS equipped with an HP 59897A electrospray interface (Agilent Technologies). The sprayer needle for CE–MS coupling is a grounded stainless steel capillary with a coaxial CE capillary inside. Between the CE capillary and the steel capillary is a sheath flow that forms the electrical contact and provides a stable spray. The CE capillary was 50 μm × 70 cm fused silica. A constant voltage of −30 kV was applied during analysis. Sample injection (10 mg/L each) was achieved using the pressure mode for 18 s at 50 mbar. The capillary was flushed with hexadimethrin bromide prior to the first use to obtain a permanent reverse of the display electroosmotic flow (EOF). In contrast to the aqueous CE method, the separation of basic analytes could not be performed using these conditions, but the acidic compounds could be detected with quantitation limits between 10 and 160 μg/L.

6.7 Capillary isatachophoresis

Polásek *et al.* [107] used anionic capillary isatachophoresis (ITP) with conductimetric detection for determining selected nonsteroidal antiinflammatory and analgesic drugs of the phenamate group, namely tolfenamic, flufenamic, mefenamic and niflumic acids. Isatachophoresis analyses were carried out with the use of a computer-controlled EA 100 ITP analyzer (VILLA s.r.o.) operated in the single-column mode. The analyzer was equipped with a 30-μL sampling valve, a 160 mm × 0.3 mm i.d. analytical capillary made of fluorinated ethylene-propylene (FEP) copolymer and a conductivity detector. Initially the pKa values (proton lost) of tolfenamic, flufenamic, mefenamic, and niflumic acids were determined as 5.11, 4.91, 5.39, and 4.31, respectively, by the UV spectrophotometry in aqueous 50% (w/w) methanol. The optimized ITP electrolyte system consisted of 10 mM HCl + 20 mM imidazole (pH 7.1) as the leading electrolyte and 10 mM 5,5′-diethylbarbituric acid (pH 7.5) as the terminating electrolyte. The driving and detection currents were 100 μA (for 450 s) and 30 μA, respectively (a single analysis took about 20 min). Under such conditions the effective mobilities of tolfenamic, flufenamic, mefenamic and niflumic acids varied between 23.6 and 24.6 $m^2 V^{-1} s^{-1}$ (evaluated with orotic acid as the mobility standard). The calibration graphs relating the ITP zone length to the concentration of the analytes were rectilinear ($r = 0.9987$–0.9999) in the range 10–100 mg/L of the drug standard. The RSDs were 0.96–1.55% ($n = 6$) when determining 50 mg/L of the analytes in pure test solutions. The method has been applied to assay the phenamates in six commercial mass-produced pharmaceutical preparations (Mobilisin® gel and ointment, Lysalgo® capsules, Nifluril® cream, Niflugel® gel, and Clotam® capsules). According to the validation procedure based on the standard addition technique the recoveries were 98.4–104.3% of the drug and the RSD values were 1.25–3.32% ($n = 6$).

7. DETERMINATION IN BODY FLUIDS AND TISSUES

7.1 Spectroscopic methods

Chemiluminescence can be used to determine mefenamic acid in biological fluids according to the procedure described earlier by Aly [45] (see Section 5.4). For plasma, the extraction process was carried out with ethyl acetate after acidification with acetic acid. For spiked plasma the calibration graphs were obtained over the concentration ranges 0.07–6.0 and 0.05–6.0 μg/mL for flufenamic and mefenamic acids, respectively. The respective mean percentage recoveries were 99.57 ± 0.56 and 99.17 ± 0.98 for flufenamic and mefenamic acids, respectively. In the case of urine, the extraction process was carried out with diethyl ether after acidification with hydrochloric acid. For spiked urine, the chemiluminescence intensity (I, mV) was linearly related to flufenamic and mefenamic acids concentrations over the range 0.5–6.0 μg/mL. The respective mean percentage recoveries were 99.94 ± 0.41 and 100.10 ± 0.57 for flufenamic and mefenamic acids, respectively.

The simultaneous analysis of 14 nonsteroidal antiinflammatory drugs in human serum using negative electrospray ionization-tandem mass spectrometry without chromatography was described by De Kanel [47] (see Section 5.6).

7.2 Chromatographic methods

A simple detection of mefenamic acid and its metabolites in urine by thin layer chromatography was reported earlier by Demetriou and Osborne [52] (see Section 6.1).

As described in the Section 6.3, the gas chromatography procedure has been used to determine mefenamic acid in plasma and urine of human and horse [57–64].

Similarly, the high-performance liquid chromatography is shown to be the most applied procedure for the determination of mefenamic acid in biological fluids (see Table 1).

Determination of flufenamic, meclofenamic and mefenamic acids in human urine and serum samples using capillary zone electrophoresis (CZE) was described by Pérez-Ruiz [104] (see Section 6.6). Analysis of drug in urine samples by CZE is always a delicate problem. If urine is directly injected into the capillary, proteins and the other biomolecules in the urine matrix adsorb to the wall of the capillary and thus quickly harm the column's performance. In addition, appropriate conditions must be found where no other components comigrate with the analytes. Experiments with different urine dilution ratios have shown that a 10 fold dilution of urine was suitable for the analysis because of flufenamic, meclofenamic and mefenamic acids were kept free of adverse matrix effects. The serum was treated with acetonitrile to separate proteins. After centrifugation (3 min at 1000 *g*), the liquid supernatant was treated with 0.1 mL of 0.01 M sodium hydroxide, filtered through a 0.45 μm filter and diluted with demineralized water to appropriate volume. In the concentration range studied (3–20 and 3–30 μg/mL for urine and serum samples, respectively), the calibration curves ware linear and the migration times were reproducible. Separation of the fenamic acids under the best conditions for human urine and serum samples gave results with mean recovery in the range of 98.6–102.1% and 98.0–102.7% for mefenamic acid in urine and serum samples, respectively.

8. REFERENCES

1. ***United States Pharmacopoeia 26***, Unites States Pharmacopoeial Convention, Inc., Rockville, MD, pp. 1141–1142 (2003).
2. ***British Pharmacopoeia 2002***, Volume 1, The Stationery Office, London, UK, pp. 1105–1107 (2002).
3. ***British Pharmacopoeia 2002***, Volume 2, The Stationery Office, London, UK, pp. 2296–2297 (2002).
4. ***Farmakope Indonesia*** (Indonesian Pharmacopoeia), Edisi IV, Departemen Kesehatan Republic Indonesia, Jakarta, pp. 43–44 (1995).
5. ***European Pharmacopoeia***, 4th edn., Council of Europe, Strasbourg, France, pp. 1534–1535, (2001).
6. ***Indian Pharmacopoeia 1996***, Volume I (A-O), Published by the Controller of Publications, New Delhi, pp. 459–461 (1996).
7. ***Pharmacopoeia of the Peoples' Republic of China (English Edition 1997)***, Volume II, Chemical Industry Press, Beijing, p. 363–364 (1997).
8. S.T. Hassib, H.M. Safwat, and R.I. El-Bagry, *Analyst*, ***111***, 45–48 (1986).
9. A.S. Issa, Y.A. Beltagy, M.G. Kassem, and H.G. Daabes, *Egypt. J. Pharm. Sci.*, ***28***, 59–66 (1987).
10. O. Çakirer, E. Kiliç, O. Atakol, and A. Kenar, *J. Pharm. Biomed. Anal.*, ***20***, 19–26 (1999).
11. J. Song, W. Guo, and Y. Wan, *Zhongguo Yiyao Gongye Zazhi*, ***24***, 28–30 (1993).
12. F.A. Aly and F. Belal, *Pharmazie*, ***49***, 454–455 (1994).
13. S.M. Hassan and S.A.M. Shaaban, *Pharmazie*, ***39***, 691–693 (1984).
14. D. Shinkuma, T. Hamaguchi, Y. Yamanaka, and N. Mizuno, *Int. J. Pharm.*, ***21***, 187–200 (1984).
15. S. Das, S.C. Sharna, S.K. Talwar, and P.D. Sethi, *Analyst*, ***114***, 101–103 (1989).
16. S. Gangwal and A.K. Sharma, *Indian Drugs*, ***33***, 163–166 (1996).
17. S. Gangwal and A.K. Sharma, *Indian J. Pharm. Sci.*, ***58***, 216–218 (1996).
18. P. Parimoo, A. Bharathi, and K. Padma, *Indian Drugs*, ***33***, 290–292 (1996).
19. M.I. Toral, P. Richter, E. Araya, and S. Fuentes, *Anal. Lett.*, ***29***, 2679–2689 (1996).

20. E. Dinç, C. Yücesoy, and F. Onur, *J. Pharm. Biomed. Anal.*, ***28***, 1091–1100 (2002).
21. G. Deveaux, P. Mesnard, and A.M. Brisson, *Ann. Pharm. Fr.*, ***27***, 239–248 (1969).
22. A.V. Vinnikova, *Farm. Zh.*, ***3***, 74–75 (1979).
23. A.V. Vinnikova, *Farm. Zh.*, ***1***, 32–36 (1980).
24. C.S.P. Sastry and A.R.M. Rao, *J. Pharmacol. Methods*, ***19***, 117–125 (1988).
25. C.S.P. Sastry, A.S.R.P. Tipirneni, and M.V. Suryanarayana, *Analyst*, ***114***, 513–515 (1989).
26. C.S.P. Sastry, A.S.R.P. Tipirneni, and M.V. Suryanarayana, *Microchem. J.*, ***39***, 277–282 (1989).
27. M.S. Mahrous, M.M. Abdel-Khalek, and M.E. Abdel-Hamid, *Talanta*, ***32***, 651–653 (1985).
28. M.S. Garcia, C. Sanchez-Pedreno, M.I. Albero, and C. Garcia, *Mikrochim. Acta*, ***136***, 67–71 (2001).
29. S.T. Hassib, H.M. Safwat, and R.I. El-Bagry, *Egypt. J. Pharm. Sci.*, ***28***, 203–214 (1987).
30. A.A. Khier, M. El-Sadek, and M. Baraka, *Analyst*, ***112***, 1399–1403 (1987).
31. M.H. Abdel-Hay, M.A. Korany, M.M. Bedair, and A.A. Gazy, *Anal. Lett.*, ***23***, 281–294 (1990).
32. H. Bojarowicz and S. Zommer-Urbanska, *Acta Pol. Pharm.*, ***50***, 285–290 (1993).
33. S.K. Babu, P.U. Shankar, and G.M. Kumar, *East. Pharm.*, ***38***, 129–130 (1995).
34. M.A. Amin, F.M. Salama, S.E. Barakat, O.I. Abd El-Sattar, M.W.I. Nassar, and H.H.M. Abou Seade, *Egypt. J. Pharm. Sci.*, ***37***, 157–174 (1996).
35. V.K. Shormanov, O.S. Solov'eva, and O.A. Yatsenko, *Farmatsiya (Moskow)*, ***46***, 30–33 (1997).
36. H.D. Dell and B. Kutschbach, *Fresenius Z. Anal. Chem.*, ***262***, 356–361 (1972).
37. A.C. Metha and S.G. Schulman, *Talanta*, ***20***, 702–703 (1973).
38. J.N. Miller, D.L. Phillipps, D.T. Burns, and J.W. Bridges, *Talanta*, ***25***, 46–49 (1978).
39. R.H. Huang and X. Xu, *Yaoxue Xuebao*, ***24***, 37–42 (1989).
40. M.I. Albero, C. Sanchez-Pedreno, and M.S. Garcia, *J. Pharm. Biomed. Anal.*, ***13***, 1113–1117 (1995).
41. P.C. Ioannou, N.V. Rusakova, D.A. Andrikopoulou, K.M. Glynou, and G.M. Tzompanaki, *Analyst*, ***123***, 2839–2843 (1998).

42. T.P. Ruiz, C.M. Lozano, V. Tomas, and J. Carpena, *Talanta*, ***47***, 537–545 (1998).
43. L.F. Capitán-Vallvey, N. Navas, M. del Ohmo, V. Consonni, and R. Todeschini, *Talanta*, ***52***, 1069–1079 (2000).
44. S.M. Sabry, *Anal. Chim. Acta*, ***367***, 41–53 (1998).
45. F.A. Aly, S.A. Al-Tamimi, and A.A. Alwarthan, *Anal. Chim. Acta*, ***416***, 87–96 (2000).
46. O.A. Mansour, M.F. Metwally, S.M. Sakr, and M.I. Al-Ashmawi, *Spectrosc. Lett.*, ***23***, 801–806 (1990).
47. J. De Kanel, W.E. Vickery, and F.X. Diamond, *J. Am. Soc. Mass Spectrom.*, ***9***, 255–257 (1998).
48. H. Salem, M. El-Maamli, and A. Shalaby, *Sci. Pharm.*, ***68***, 343–356 (2000).
49. H. Salem, K. Kelani, and A. Shalaby, *Sci. Pharm.*, ***69***, 189–202 (2001).
50. G. Alpdogan and S. Sungur, *Anal. Lett.*, ***32***, 2799–2808 (1999).
51. H.M. Khalil, M.M. El-Henawee, and M.M. Baraka, *Zagazig J. Pharm. Sci.*, ***4***, 40–43 (1995).
52. B. Demetriou and B.G. Osborne, *J. Chromatogr.*, ***90***, 405–407 (1974).
53. A. By, J.C. Ethier, G. Lauriault, M. LeBelle, B.A. Lodge, C. Savard, W.-W. Sy, and W.L. Wilson, *J. Chromatogr.*, ***469***, 406–411 (1989).
54. S. Dharmananda, Drugs in Imported Chinese Herb Products, http://www.itmonline.org/arts/drugherb.htm (August 1, 2002).
55. A.P. Argekar and J.G. Sawant, *J. Planar Chromatogr.*, ***12***, 361–364 (1999).
56. W.M. Adams, *J. Assoc. Off. Anal. Chem.*, ***66***, 1178–1181 (1983).
57. S.A. Bland, J.W. Blake, and R.S. Ray, *J. Chromatogr. Sci*, ***14***, 201–203 (1976).
58. L.J. Dusci and L.P. Hackett, *J. Chromatogr.*, ***161***, 340–342 (1978).
59. A.K. Singh, Y. Jang, U. Mishra, and K. Granley, *J. Chromatogr.*, ***568***, 351–361 (1991).
60. C. Giachetti, A. Assandri, G. Zanolo, and A. Tenconi, *Chromatographia*, ***39***, 162–169 (1994).
61. G. Gonzáles, R. Ventura, A.K. Smith, R. De-la-Torre, and J. Segura, *J. Chromatogr. A*, ***719***, 251–264 (1996).
62. H.H. Maurer, F.X. Tauvel, and T. Kraemer, *J. Anal. Toxicol.*, ***25***, 237–245 (2001).
63. A. Takeda, H. Tanaka, T. Shinohara, and I. Ohtake, *J. Chromatogr. B*, ***758***, 235–248 (2001).

64. D.S.T. Lo, T.C. Chao, S.E. Ng-Ong, Y.J. Yao, and T.H. Koh, *Forensic Sci. Int.*, ***90***, 205–214 (1997).
65. P. Arrizabalaga, A. Kamatari, and J.C. Landry, *Chimia*, ***46***, 161–163 (1992).
66. D.B. Fraser, J. Tomlinson, J. Turner, and R.D. Satzger, *J. Food Drug Anal.*, ***5***, 329–336 (1997).
67. J.W. Dolan, L.R. Snyder, N.M. Djordjevic, D.W. Hill, and T.J. Waeghe, *J. Chromatogr. A*, ***857***, 1–20 (1999).
68. J.W. Dolan, L.R. Snyder, N.M. Djordjevic, D.W. Hill, and T.J. Waeghe, *J. Chromatogr. A*, ***857***, 21–39 (1999).
69. W. Ahrer, E. Scherwenk, and W. Buchberger, *J. Chromatogr. A*, ***910***, 69–78 (2001).
70. Y.-R. Ku, M.-J. Tsai, and K.-C. Wen, *J. Food Drug Anal.*, ***3***, 51–56 (1995).
71. M. Kim and J.T. Stewart, *J. Liq. Chromatogr.*, ***13***, 213–237 (1990).
72. M. Kim and J.T. Stewart, *Anal. Sci. Technol.*, ***4***, 273–284 (1991).
73. Y. Sun, K. Takaba, H. Kido, M.N. Nakashima, and K. Nakashima, *J. Pharm. Biomed. Anal.*, ***30***, 1611–1619 (2003).
74. M.I. Gonzáles Martín, C.I. Sánchez Gonzáles, A. Jiménez Hernández, M.D. García Cachán, M.J. Castro de Cabo, and A.L. Garzón Cuadrado, *J. Chromatogr. B*, ***769***, 119–216 (2002).
75. ***Chromatography: Products for Analysis and Purification***, Sigma-Aldrich, Singapore, p. 500 (2001).
76. E. Mikami, T. Goto, T. Ohno, H. Matsumoto, K. Inagaki, H. Ishihara, and M. Nishida, *J. Chromatogr. B*, ***744***, 81–89 (2000).
77. P. Gowik, B. Jülicher, and S. Uhlig, *J. Chromatogr. B*, ***716***, 221–232 (1998).
78. T. Hirai, S. Matsumoto, and I. Kishi, *J. Chromatogr. B*, ***692***, 375–388 (1997).
79. J. Sato, T. Amizuka, Y. Niida, M. Umetsu, and K. Ito, *J. Chromatogr. B*, ***692***, 241–244 (1997).
80. J. Sato, N. Kudo, E. Owada, K. Ito, Y. Niida, M. Umetsu, T. Kikuta, and K. Ito, *Biol. Pharm. Bull.*, ***20***, 443–445 (1997).
81. T. Hirai, Usability of a normal solid-phase extraction as the sample clean-up procedure for urinary drug analysis by high-performance liquid chromatography, CCAB 97: Mini Review, http://neo.pharm.hiroshima-u.ac.jp/ccab/1st/mini_review/mr002/hirai.html (August 1, 2002).
82. D. Cerretani, L. Micheli, A.I. Fiaschi, and G. Giorgi, *J. Chromatogr. B*, ***678***, 365–368 (1996).

83. S.M.R. Stanley, N.A. Owens, and J.P. Rodgers, *J. Chromatogr. B*, ***667***, 95–103 (1995).
84. E. Nivaud-guernet, M. Guernet, D. Ivanovic, and M. Medenica, *J. Liq. Chromatogr.*, ***17***, 2343–2357 (1994).
85. I. Niopas and K. Mamzoridi, *J. Chromatogr. B*, ***656***, 447–450 (1994).
86. I.N. Papadoyannis, A.C. Zotou, and V.F. Samanidou, *J. Liq. Chromatogr.*, ***15***, 1923–1945 (1992).
87. I.N. Papadoyannis, V.F. Samanidou, and G.D. Panopoulou, *J. Liq. Chromatogr.*, ***15***, 3065–3086 (1992).
88. J.M. Poirier, M. Lebot, and G. Cheymol, *Ther. Drug Monit.*, ***14***, 322–326 (1992).
89. E. Mikami, S. Yamada, Y. Fujii, N. Kawamura, and J. Hayakawa, *Iyakuhin Kenkyu*, ***23***, 491–496 (1992).
90. K. Yamashita, M. Motohashi, and T. Yashiki, *J. Chromatogr.*, ***570***, 329–338 (1991).
91. T. Shinozuka, S. Takei, N. Kuroda, K. Kurihara, and J. Yanagida, *Eisei Kagaku*, ***37***, 461–466 (1991).
92. N. Maron and G. Wright, *J. Pharm. Biomed. Anal.*, ***8***, 101–105 (1990).
93. K. Shimada, M. Nakajima, H. Wakabayashi, and S. Yamato, *Bunseki Kagaku*, ***38***, 632–635 (1989).
94. P.J. Streete, *J. Chromatogr.*, ***495***, 179–193 (1989).
95. J. Sato, E. Owada, K. Ito, Y. Niida, A. Wakamatsu, and M. Umetsu, *J. Chromatogr.*, ***493***, 239–243 (1989).
96. F.J. Diana, K. Veronich, and A.L. Kapoor, *J. Pharm. Sci.*, ***78***, 195–199 (1989).
97. H.M. Stevens and R. Gill, *J. Chromatogr.*, ***370***, 39–47 (1986).
98. Y. Matsushima, Y. Nagata, M. Niyomura, K. Takakusagi, and N. Takai, *J. Chromatogr.*, ***332***, 269–273 (1985).
99. H.J. Battista, G. Wehinger, and R. Henn, *J. Chromatogr.*, ***345***, 77–89 (1985).
100. C.K. Lin, C.S. Lee, and J.H. Perrin, *J. Pharm. Sci.*, ***69***, 95–97 (1980).
101. Mac-Nod HPLC Column Companion, Section 10: Injection Solvent Effects, http://www.mac-nod.com/cc/cc-10.html (August 1, 2002).
102. N.K. Jagota and J.T. Stewart, *J. Chromatogr.*, ***604***, 255–260 (1992).
103. Y.-R. Ku, M.-J. Tsai, and K.-C. Wen, *J. Food Drug Anal.*, ***3***, 185–192 (1995).

104. T. Pérez-Ruiz, C. Martínez-Lozano, A. Sanz, and E. Bravo, *J. Chromatogr. B*, ***708***, 249–256 (1998).

105. S. Cherkaoui and J.-L. Veuthey, *J. Chromatogr. A*, ***874***, 121–129 (2000).

106. W. Ahrer and W. Buchberger, Capillary Electrophoresis Combined with Mass Spectrometry for Environmental Analytical Chemistry, Agilent Technologies, http://lifesciences.chem.agilent.com/Scripts/Generic.asp?1Page = 3128&peak = 1&indcol = Y&prodcol = N (August 1, 2002).

107. M. Polásek, M. Pospísilová, and M. Urbánek, *J. Pharm. Biomed. Anal.*, ***23***, 135–142 (2000).

Nimodipine: Physical Profile

Mohammed A. Al-Omar

Department of Pharmaceutical Chemistry
College of Pharmacy, King Saud University
P.O. Box 2457, Riyadh-11451
Kingdom of Saudi Arabia

PROFILES OF DRUG SUBSTANCES,
EXCIPIENTS, AND RELATED
METHODOLOGY – VOLUME 31
DOI: 10.1016/S0000-0000(00)00000-0

CONTENTS

1. **General Information** 338
   1.1 Nomenclature 338
      1.1.1 Systematic chemical names 338
      1.1.2 Nonproprietary names 339
      1.1.3 Proprietary names 339
      1.1.4 Synonyms 339
   1.2 Formulae 339
      1.2.1 Empirical formula, molecular weight, CAS number 339
      1.2.2 Structural formula 339
   1.3 Elemental analysis 340
   1.4 Appearance 340
2. **Physical Characteristics** 340
   2.1 Solution pH 340
   2.2 Solubility characteristics 340
   2.3 Optical activity 340
   2.4 X-ray powder diffraction pattern 340
   2.5 Thermal methods of analysis 340
      2.5.1 Melting behavior 340
      2.5.2 Differential scanning calorimetry 340
   2.6 Spectroscopy 341
      2.6.1 Ultraviolet spectroscopy 341
      2.6.2 Vibrational spectroscopy 341
      2.6.3 Nuclear magnetic resonance spectrometry 345
   2.7 Mass spectrometry 349
3. **Stability and Storage** 353
4. **References** 354

1. GENERAL INFORMATION

1.1 Nomenclature

1.1.1 Systematic chemical names [1, 2]

1,4-Dihydro-2,6-dimethyl-4-(3-nitrophenyl)-3,5-pyridinedicarboxylic acid 2-methoxyethyl 1-methylethyl ester.

2-Methoxyethyl 1,4-dihydro-5-(isopropoxycarbonyl)-2,6-dimethyl-4-(3-nitrophenyl)-3-pyridinecarboxylate.

Isopropyl 2-methoxyethyl 1,4-dihydro-2,6-dimethyl-4-(3-nitrophenyl)-3,5-pyridinedicarboxylate.

2,6-Dimethyl-4-(3′-nitrophenyl)-1,4-dihydropyridine-3,5-dicarboxylic acid-3-β-methoxyethyl, 5-isopropyl diester.

2,6-Dimethyl-4-(m-nitrophenyl)-1,4-dihydropyridine-3,5-dicarboxylic acid-3-β-methoxy ethyl-5-α-methylethyl diester.

5-(Isopropoxycarbonyl)-3-(2-methoxyethoxycarbonyl)-4-(3-nitrophenyl)-2,6-dimethyl-1,4-dihydropyridine.

1.1.2 Nonproprietary names
Nimodipine.

1.1.3 Proprietary names [2–4]
Admon, Nimotop, Periplum, Nimodipina, Norton, Noodipinia, Oxigen, Brainox, Modina, Sobrepina, Trinalion, Brainal, Calnit, Kenesil, Modus, Remontal, Vasotop, Nivas, Cebrofort, Acival, Eugerial.

1.1.4 Synonyms [3]
Bay-e-9736

1.2 Formulae

1.2.1 Empirical formula, molecular weight, CAS number [3, 4]
$C_{21}H_{26}N_2O_7$ 418.459 [66085-59-4].

1.2.2 Structural formula

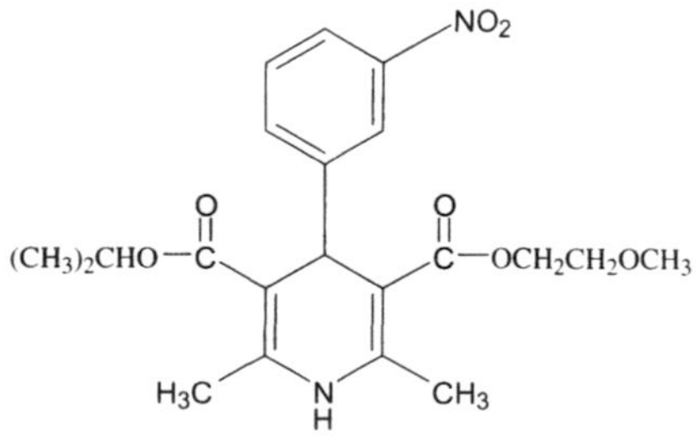

1.3 Elemental analysis

C 60.28% H 6.26% N 6.69% O 26.76%.

1.4 Appearance [1, 3]

Nimodipine is obtained as a light yellow to yellow crystalline powder.

2. PHYSICAL CHARACTERISTICS

2.1 Solution pH

The pH of a 1% (w/v) solution is 6–6.8 [3].

2.2 Solubility characteristics

Nimodipine is practically insoluble in water, freely soluble in ethyl acetate, and sparingly soluble in absolute alcohol [3].

2.3 Optical activity

The specific rotation of the (+)-form is $[\alpha]^{20}$ +7.9 ($c = 0.439$ in dioxane), and the specific rotation of the $(\overset{D}{-})$-form is $[\alpha]^{20}$ −7.9 ($c = 0.374$ in dioxane) [2]. When tested according to the British Pharmacopoeia, the angle of optical rotation, determined on solution S, is −0.10° to +0.10°.

2.4 X-ray powder diffraction pattern

The X-ray powder diffraction pattern of nimodipine is shown in Figure 1 and was obtained using a Simons XRD-5000 diffractometer. Table 1 shows the values of the scattering angles (units of degrees 2θ), the interplanar *d*-spacings (Å), and the relative peak intensities.

2.5 Thermal methods of analysis

2.5.1 Melting behavior

Nimodipine melts at about 125°C [2].

2.5.2 Differential scanning calorimetry

The differential scanning calorimetry thermogram of nimodipine was obtained using a DuPont TA-9900 thermal analyzer attached to a DuPont Data Unit. The thermogram shown in Figure 2 was obtained at a heating rate of 10°C/min, and was run from 40 to 400°C. The compound was found to melt at 130.9°C.

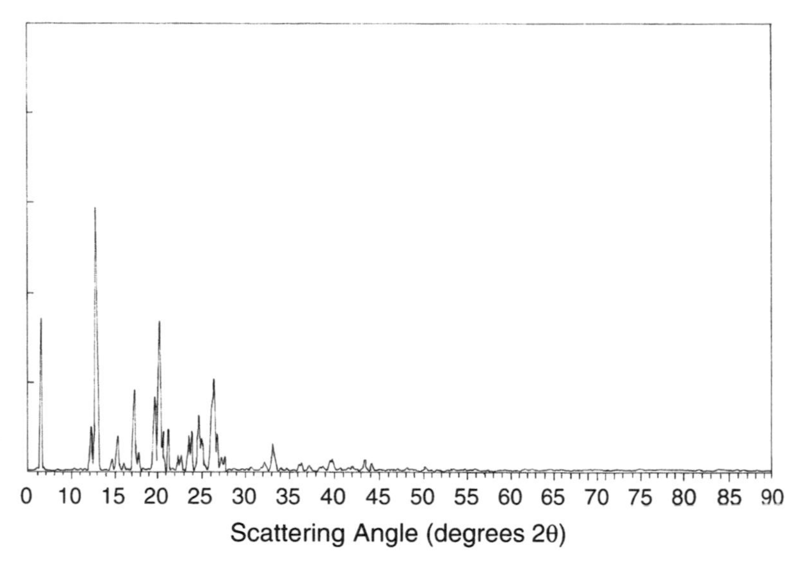

Figure 1. X-ray powder diffraction pattern of nimodipine.

2.6 Spectroscopy

2.6.1 Ultraviolet spectroscopy

The UV spectrum of nimodipine (10 μg/μL) in ethanol is shown in Figure 3, and was recorded using a Shimadzu UV–visible Spectrophotometer 1601 PC. Nimodipine exhibited two maxima at 355 and 236.6 nm, which are characterized by the following values:

λ_{max}(nm)	A [1%, 1 cm]	Molar absorptivity (L mol^{-1} cm^{-1})
355.0	162	6.77×10^3
236.6	655	2.74×10^4

2.6.2 Vibrational spectroscopy

The infrared absorption spectrum of nimodipine shown in Figure 4 was obtained in a KBr pellet, using a Perkin Elmer infrared spectrophotometer. The principal peaks were observed at 3379, 3271, 3216, 3087, 2977, 1698, 1644, 1576, 1529, 1498, and 1360 cm^{-1}. Assignments for the infrared absorption bands of nimodipine are given in Table 2.

Table 1

Crystallographic Data From the X-Ray Powder Diffraction Pattern of Minodipine

Scattering angle (degrees 2θ)	*d*-spacing (Å)	Relative intensity (%)	Scattering angle (degrees 2θ)	*d*-spacing (Å)	Relative intensity (%)
6.485	13.6180	58.19	10.303	8.5786	0.89
11.063	7.9911	1.04	12.285	7.1990	16.84
12.856	6.8801	100.00	13.480	6.5632	0.22
14.672	6.0326	4.23	15.307	5.7838	13.85
15.942	5.5547	2.61	17.286	5.1257	31.10
17.752	4.9921	6.50	18.260	4.8544	1.18
19.681	4.5071	28.46	20.226	4.3867	57.13
20.674	4.2927	15.23	21.217	4.1841	15.85
21.676	4.0965	1.04	22.257	3.9909	5.31
22.621	3.9275	5.14	23.543	3.7758	13.48
23.916	3.7177	15.32	24.671	3.6055	21.27
25.065	3.5497	11.62	25.440	3.4983	1.99
26.200	3.3985	27.42	26.379	3.3758	35.41
26.743	3.3307	14.10	27.266	3.2681	4.76
27.617	3.2273	5.39	30.683	2.9114	1.61
32.082	2.7876	3.42	33.086	2.7052	10.16
34.013	2.6336	1.29	34.647	2.5869	1.22
36.349	2.4695	3.07	37.270	2.4106	2.00
39.803	2.2628	4.79	41.968	2.1510	1.81
43.375	2.0844	4.10	44.089	2.0523	2.69
47.080	1.9286	1.10	48.154	1.8881	1.17
50.182	1.8165	1.60	62.550	1.4837	0.92

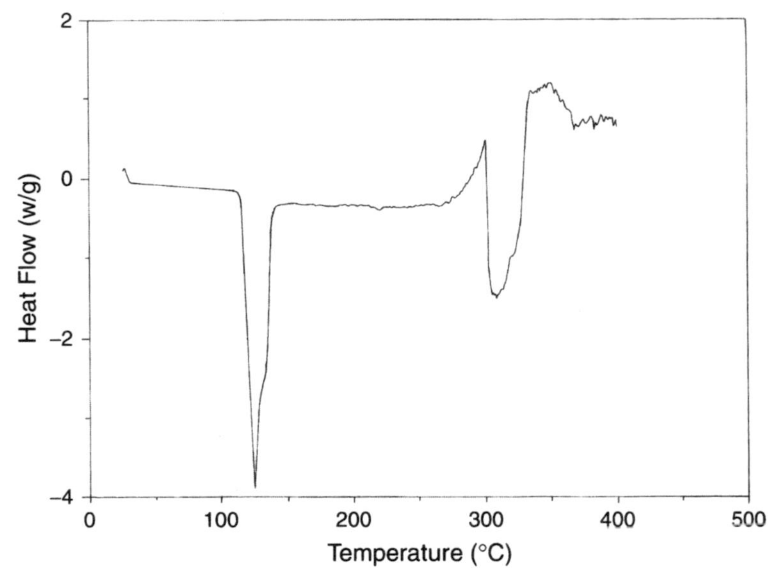

Figure 2. Differential scanning calorimetry thermogram of nimodipine.

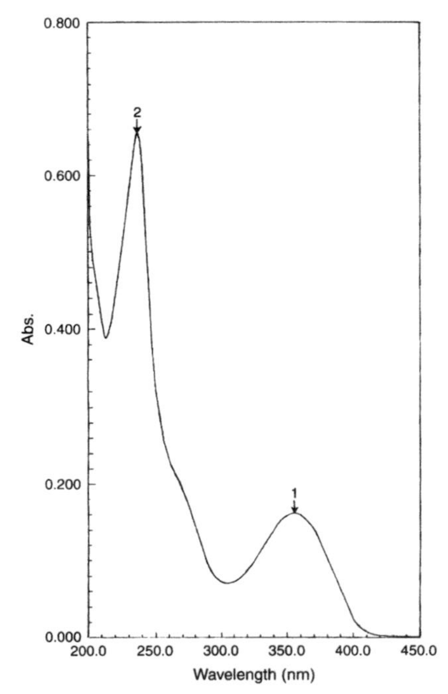

Figure 3. The ultraviolet absorption spectrum of nimodipine in ethanol.

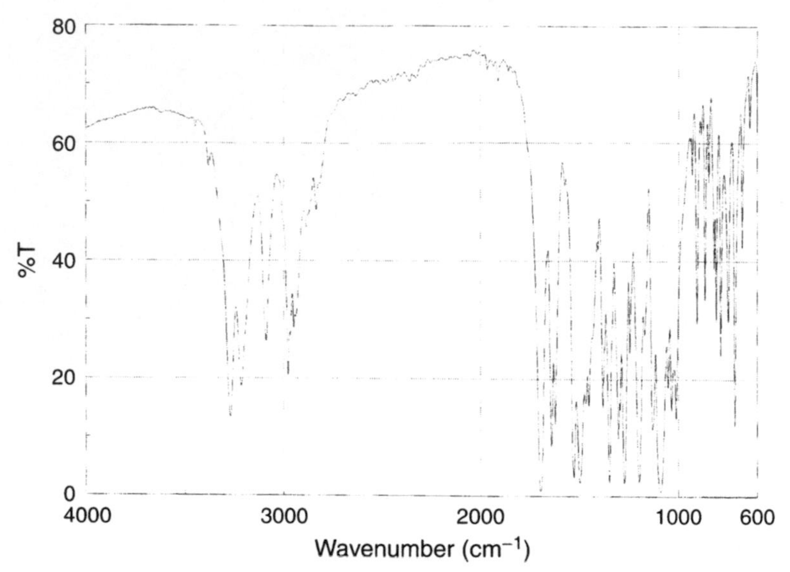

Figure 4. The infrared absorption spectrum of nimodipine obtained in a KBr pellet.

Table 2

Assignments for the Infrared Absorption Bands of Nimodipine

Frequency (cm^{-1})	Assignments
3379, 3271	N–H stretch
3216, 3087	Aromatic C–H stretch
2977	Aliphatic C–H stretch
1698	C=O stretch
1644, 1576	N–H bending
1529, 1498, 1360	C=C ring stretch
1360	C–C(=O)–O stretch of α,β-unsaturated ester

2.6.3 Nuclear magnetic resonance spectrometry

2.6.3.1 ^{1}H-NMR spectrum

The proton NMR spectrum of nimodipine was obtained using a Bruker Avance Instrument operating at 300, 400, or 500 MHz. Standard Bruker Software was used to obtain COSY and HETCOR spectra. The sample was dissolved in D_2O, and tetramethylsilane (TMS) was used as the internal standard. The proton NMR spectra are shown in Figures 5 and 6, the COSY ^{1}H-NMR spectra are shown in Figures 7 and 8, and the gradient HMQC-NMR spectrum is shown in Figure 9. Assignments for the ^{1}H-NMR resonance bands of nimodipine are given in Table 3.

2.6.3.2 ^{13}C-NMR spectrum

The carbon-13 NMR spectrum of nimodipine was obtained using a Bruker Avance Instrument operating at 75, 100, or 125 MHz. Standard

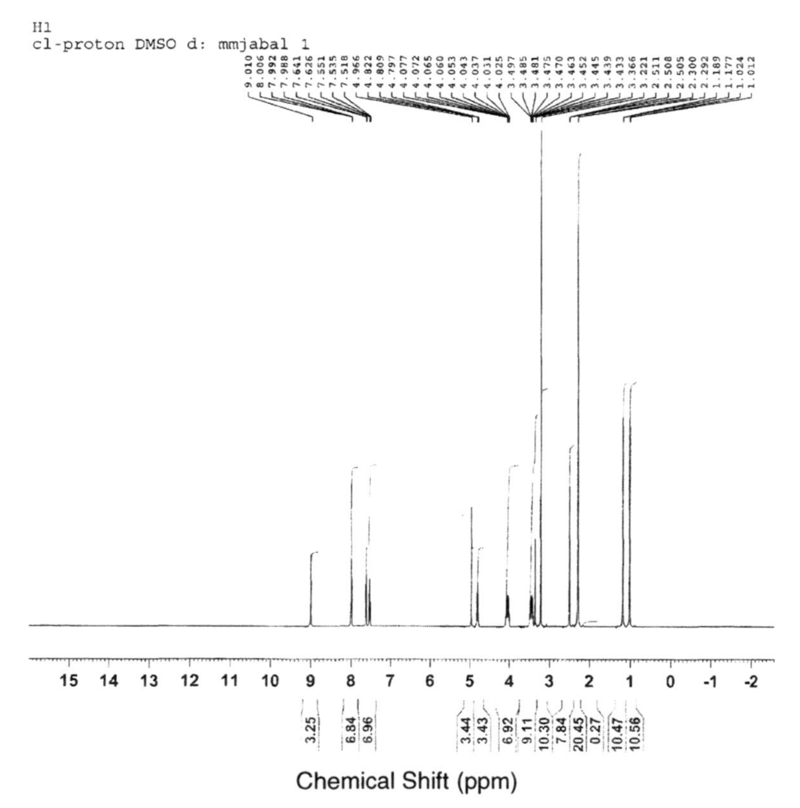

Figure 5. The ^{1}H-NMR spectrum of nimodipine in DMSO-d_6.

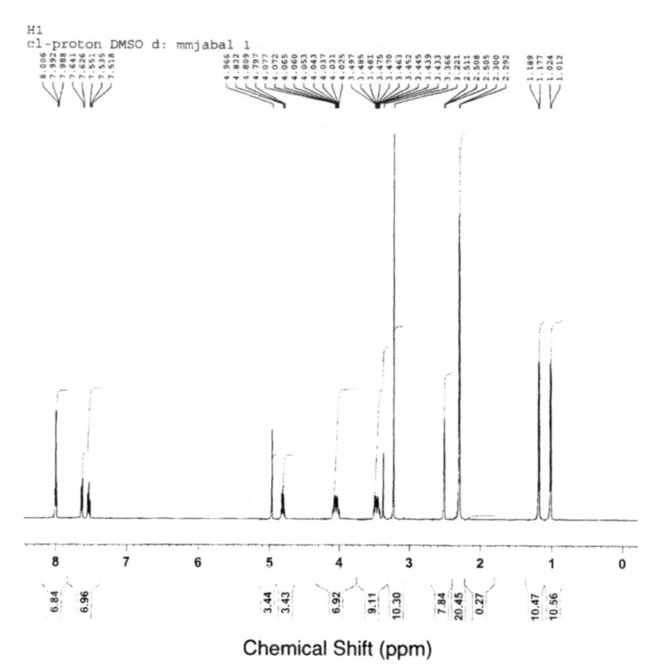

Figure 6. Expanded ^{1}H-NMR spectrum of nimodipine in DMSO-d_6.

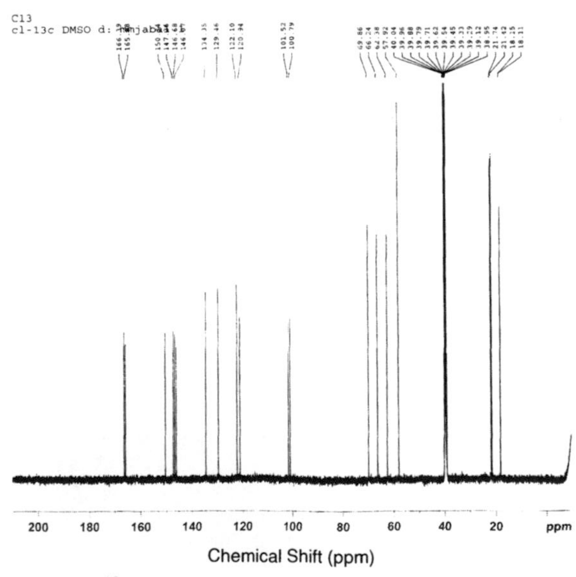

Figure 7. ^{13}C-NMR spectrum of nimodipine in DMSO-d_6.

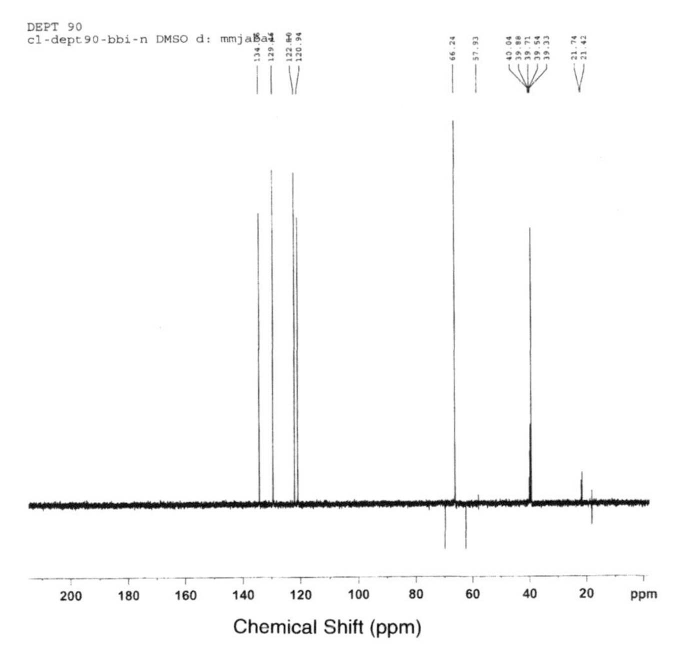

Figure 8. DEPT 90 ^{1}H-NMR spectrum of nimodipine in DMSO-d_6.

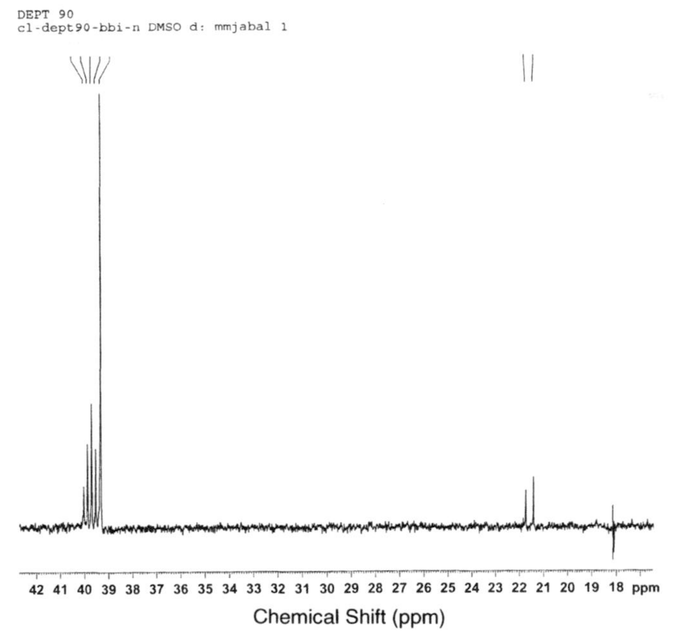

Figure 9. DEPT 90 ^{1}H-NMR spectrum of nimodipine in DMSO-d_6.

Table 3

Assignments of the ^{1}H-NMR Chemical Shifts of Nimodipine

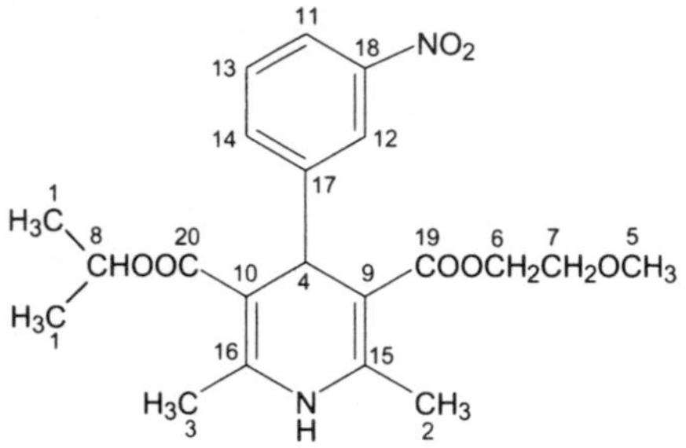

Proton atoms	Chemical shift (ppm relative to TMS)	Multiplicity*	Number of proton atoms
H-1	2.30	d	6
H-2, H-3	1.01, 1.02	Two d	6
H-5	3.36	s	3
H-6, H-7	3.43–3.49, 4.02–4.07	m	4
H-8	4.81	q	1
H-4	4.96	s	1
H-13	7.53	Two d	1
H-14	7.63	d	1
H-11	7.98	d	1
H-12	8.01	s	1
N–H	9.01	s	1

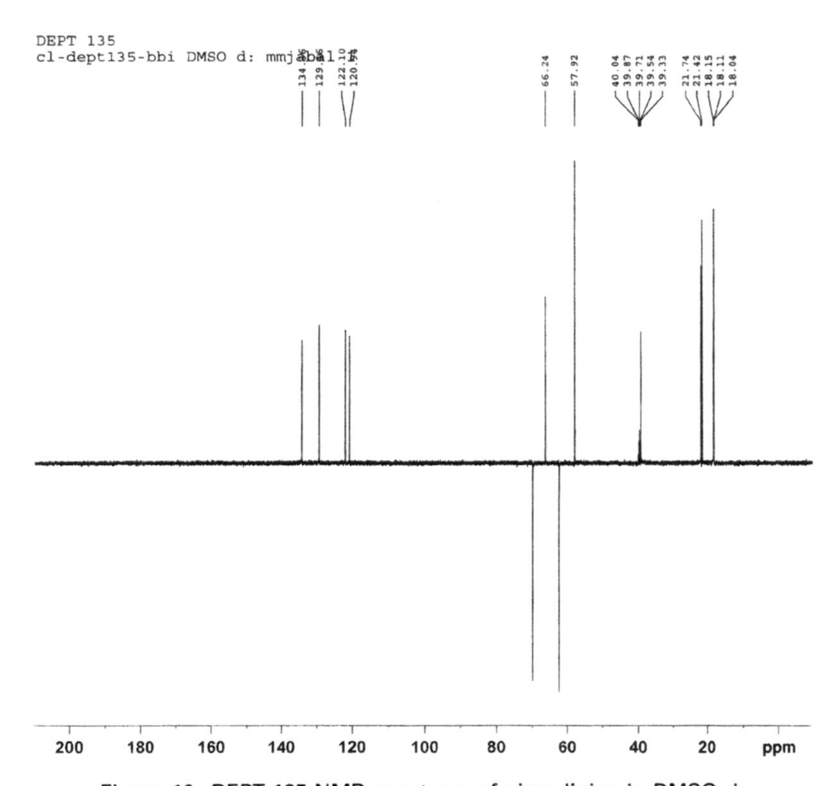

Figure 10. DEPT 135 NMR spectrum of nimodipine in DMSO-d_6.

Bruker Software was used to obtain DEPT spectra. The sample was dissolved in D_2O and tetramethylsilane (TMS) was used as the internal standard. The carbon-13 NMR spectrum of nimodipine is shown in Figure 10, and the DEPT-NMR spectra are shown in Figures 11 and 12. Assignments for the various carbon resonance bands of nimodipine are shown in Table 4.

2.7 Mass spectrometry

The electron impact (EI) spectrum of nimodipine shown in Figure 13 was recorded using a Shimadzu PQ-5000 GC–MS spectrometer. The spectrum shows a mass peak ($M^{\cdot}$) at $m/z = 418$, and a base peak at $m/z = 296$ resulting from the loss of the nitrophenyl group. These and other proposed fragmentation patterns of the drug are presented in Table 5.

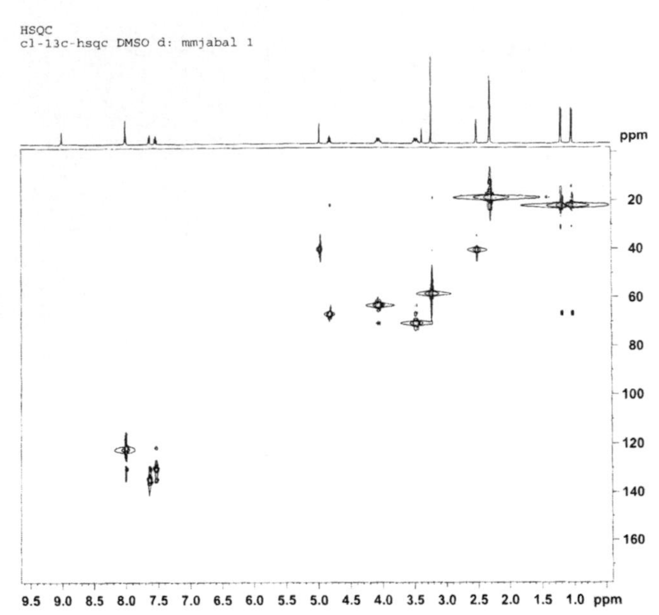

Figure 11. HSQC NMR spectrum of nimodipine in DMSO-d_6.

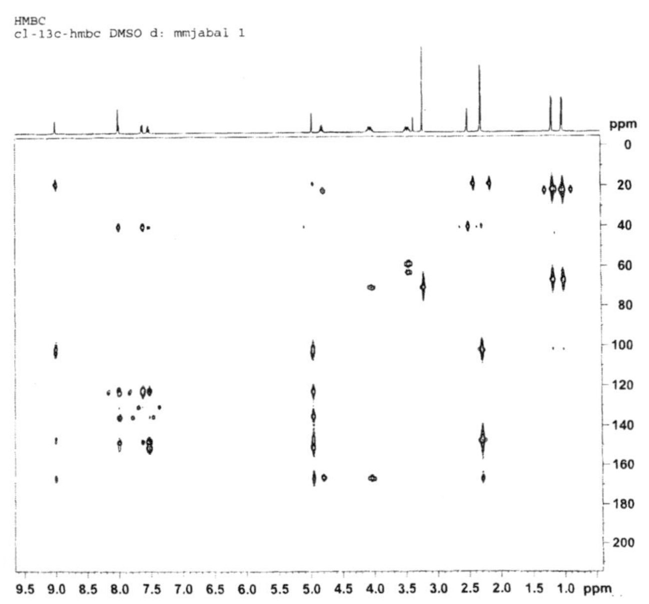

Figure 12. HMBC NMR spectrum of nimodipine in DMSO-d_6.

Table 4

ssignments of the ^{13}C-NMR Chemical Shifts of Nimodipine

Carbon atoms	Chemical shift (ppm relative to TMS)	Assignments at carbon number	Carbon atoms	Chemical shift (ppm relative to TMS)	Assignments at carbon number
C-1	18.13	2	C-12	122.10	1
C-2, C-3	21.42, 21.74	2	C-13	129.46	1
C-4	40.04	1	C-14	134.35	1
C-5	57.92	1	C-15, C-16	146.07, 146.68	2

(continued)

Table 4 (continued)

Carbon atoms	Chemical shift (ppm relative to TMS)	Assignments at carbon number	Carbon atoms	Chemical shift (ppm relative to TMS)	Assignments at carbon number
C-6, C-7	62.38, 69.86	2	C-17	147.26	1
C-8	66.24	1	C-18	150.34	1
C-9, C-10	100.79, 101.52	2	C-19, C-20	165.88, 166.39	2
C-11	120.94	1			

Table 5

Assignments for the Fragmentation Pattern Observed in the Mass Spectrum of Nimodipine

Mass number (*m*/*z*)	**Relative intensity (%)**	**Structural assignment**
418	6.86	$C_{21}H_{26}N_2O_7]^{++}$. mass peak (M·)
359	9.8	M–C_3H_7O
331	5.7	M–$C_4H_7O_2$
296	100	[M–$C_6H_4NO_2$] base peak
273	10.8	M–($C_3H_7O_2$ + CO + C_3H_6)
254	68.6	M–(C_3H_7O + C_2H_4O + NO_2 + CH_3)
227	4.9	M–($C_3H_7O_2$ + CO + C_3H_6 + NO_2)
196	53.9	M–($C_3H_7O_2$ + CO + C_3H_6 + NO_2 + 2CH_3 + H)
122	4.8	M–$C_{15}H_{22}NO_5$
67	5.8	$[CH_3–C≡C–C≡O]^+$
59	10.8	$[CH_3OCH_2CH_2]^+$

3. STABILITY AND STORAGE

Solutions of nimodipine are incompatible with some plastics, including polyvinyl chloride. The solutions are also light-sensitive, and should be protected from light [3]. The degradation half-lives of nimodipine are 16 and 56 h following exposure of aqueous solution of the drug to UV light

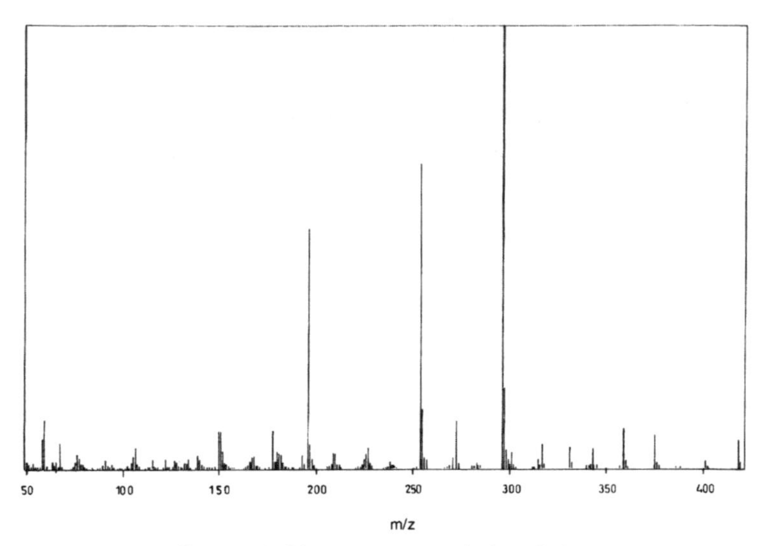

Figure 13. Mass spectrum of nimodipine.

at 360 nm and daylight, respectively [5]. Such solutions should be protected from freezing, and stored at 15–30°C.

4. REFERENCES

1. ***Remington's: Pharmaceutical Sciences***, 18th edn., A.R. Gennaro, ed., Mack Publishing Co., Pennsylvania, p. 855 (1990).
2. ***The Merck Index***, 12th edn., S. Budavari, ed., Merck and Co., NJ p. 1125 (1996).
3. ***Martindale, The Complete Drug Reference***, 33rd edn., S.C. Sweetman, ed., The Pharmaceutical Press, Chicago, p. 946 (2002).
4. ***Index Nominum 2000: International Drug Directory***, 17th edn., Swiss Pharmaceutical Society, Medpharm Scientific Publishers, p. 738 (2000).
5. ***Drug Information***, 95 edn., G.K. McEvoy, ed., American Society of Helth-System Pharmacists, p. 1259 (1995).

Nimodipine: Analytical Profile

Mohammed A. Al-Omar

Department of Pharmaceutical Chemistry
College of Pharmacy, King Saud University
P.O. Box 2457, Riyadh-11451
Kingdom of Saudi Arabia

PROFILES OF DRUG SUBSTANCES,
EXCIPIENTS, AND RELATED
METHODOLOGY – VOLUME 31
DOI: 10.1016/S0000-0000(00)00000-0

CONTENTS

1. **Compendial Methods of Analysis** 356
   1.1 Identification methods for the drug substance 356
   1.2 Tests 356
   1.3 Impurities and related substances 356
2. **Methods of Analysis Reported in the Literature** 359
   2.1 Spectrophotometric methods 359
   2.2 Electrochemical methods 360
   2.3 Thin-layer chromatography 361
   2.4 Gas chromatography 361
   2.5 Liquid chromatography 363
   2.6 Capillary electrophoresis (CE) 366
3. **References** 367

1. COMPENDIAL METHODS OF ANALYSIS

1.1 Identification methods for the drug substance

The European Pharmacopoeia [1] recommends the use of infrared absorption spectrophotometry for the identification of the pure drug. The procedure is as follows:

Examine by infrared absorption spectrophotometry comparing with the spectrum obtained with nimodipine CRS. If the spectra obtained in the solid state show differences, record further spectra using 20 g/L solutions in methylene chloride R and a 0.2-mm cell.

1.2 Tests [1]

1. Examine by infrared absorption spectrophotometry or comparing with the spectrum obtained with nimodipine CRS.
2. Dissolve 1 g in acetone and dilute to 20 mL with acetone. The solution should be clear.

1.3 Impurities and related substances [1]

The European Pharmacopoeia [1] described a liquid chromatographic method for the determination of three related substances, denoted as

A, B, and C:

Substance A: 2-methoxyethyl 1-methylethyl-2,6-dimethyl-4-(3-nitrophenyl) pyridine-3,5-dicarboxylate.

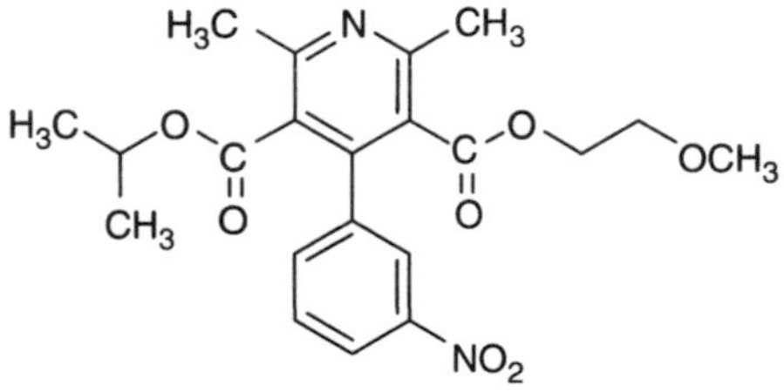

Substance B: In the structure below, $R = CH(CH_3)_2$. The systematic compound name is bis (1-methylethyl)-2,6-dimethyl-4-(3-nitrophenyl),1,4-dihydro-pyridine-3,5-dicarboxylate.

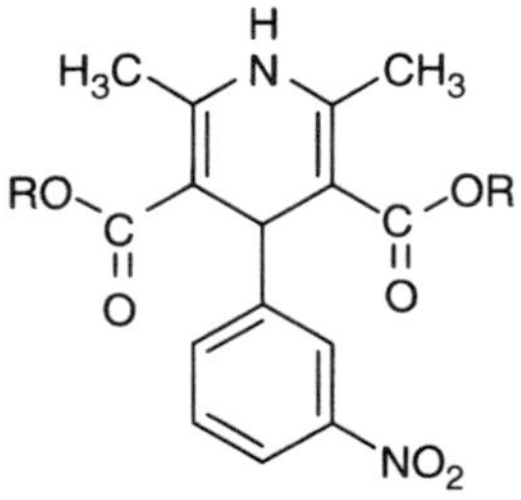

Substance C: In the structure above, $R = CH_2–CH_2–OCH_3$. The systematic compound name is bis (2-methoxyethyl)-2,6-dimethyl-4-(3-nitrophenyl)-1,4-dihydro pyridine-3,5-dicarboxylate.

The method requires preparation of the following three solutions.

Test Solution: Dissolve 40 mg of the substance to be examined in 2.5 mL of tetrahydrofuran and dilute to 25 mL with the mobile phase.

Reference Solution (a): Dilute 1 mL of the test solution to 100 mL with the mobile phase. Dilute 2 mL of the solution to 10 mL with the mobile phase.

Reference Solution (b): Dissolve 20 mg of nimodipine impurity A in 2.5 mL of tetrahydrofuran and dilute to 25 mL with the mobile phase. Dilute 1 mL of the solution to 20 mL with the mobile phase.

Reference Solution (c): Dilute 0.5 mL of the test solution to 20 mL with the mobile phase.

Reference Solution (d): Mix 1 mL of reference solution (b) and 1 mL of reference solution C, and dilute to 25 mL with the mobile phase.

The chromatographic procedure is carried out using a stainless steel column 12.5 cm long and 4.6 mm in internal diameter packed with octadecylsilyl silica gel (5 μm), which is maintained at 40°C. A mobile phase consisting of a mixture of 20 volumes of methanol, 20 volumes of tetrahydrofuran, and 60 volumes of water, is eluted at a flow rate of 2 mL/min. Detection is effected at 235 nm. When 20 μL of reference solution (d) is injected, the retention time of impurity A is about 7 min, and nimodipine elutes at about 8 min. Separately inject 20 μL of the test solution and 20 μL of reference solution (a). Record the chromatogram for the test solution for four times longer than the retention time of nimodipine. In the chromatogram obtained with the test solution, the area of the peak due to impurity A is not greater than the corresponding peak in the chromatogram obtained with reference solution (d) (i.e., 0.1%). None of the peaks (apart from the principal peak and the peak due to impurity A) has an area greater than the area of the principal peak in the chromatogram obtained with reference solution (a) (i.e., 0.2%). The sum of the areas of all peaks (apart from the principal peak) is not greater than 2.5 times the area of the principal peak in the chromatogram obtained with reference solution (a) (i.e., 0.5%). Disregard any peak due to the solvent and any peak with an area less than 0.5 times the area of the principal peak in the chromatogram obtained with reference solution (d).

Loss on drying

The permissible content is not more than 0.5% when determined on a 1.0 g sample.

Sulfated ash

The permissible sulfated ash content is not more than 0.1% when determined on 1.0 g sample.

Assay method for the bulk drug substance

The European Pharmacopoeia [1] recommends the following assay method. Dissolve 0.180 g with gentle heating in a mixture of 25 mL of

2-methyl-2-propanol and 25 mL of perchloric acid. Add 0.1 mL of ferroin, and titrate with 0.1 M ceric sulfate (titrate slowly toward the end of the titration). Carry out a blank experiment, and correct if necessary. Each milliliter of 0.1 M ceric sulfate is equivalent to 20.92 mg $C_{21}H_{26}N_2O_7$.

Assay method for tablets [1]

The assay method for nimodipine in tablets is performed according to the following procedure. Weigh and pulverize 20 tablets, and prepare the following solutions for use in the liquid chromatography method:

> *Solution (1)*: Add 50 mL of methanol to a quantity of the powdered tablets containing 60 mg of nimodipine, and mix with the aid of ultrasound for 5 min. Then add sufficient methanol to produce 100 mL, centrifuge, and use the supernatant liquid.
>
> *Solution (2)*: Prepare to contain 0.06% w/v of nimodipine BPCRS in methanol.
>
> *Solution (3)*: Prepare to contain 0.002% w/v of nimodipine BPCRS and 0.002% w/v of nimodipine impurity ABPCRS in methanol.

The chromatographic system and conditions described under *Related Substances* are used. The test is not valid unless in the chromatogram obtained with solution (3), the resolution factor between the peaks due to nimodipine and nimodipine impurity A is not less than 1.5, and the symmetry factor due to nimodipine is not less than 2.

Calculate the content of nimodipine in tablets using the declared content of $C_{21}H_{36}N_2O_7$ in nimodipine BPCRS. The content of nimodipine in the tablet should be 95–105% of the label claim.

2. METHODS OF ANALYSIS REPORTED IN THE LITERATURE

2.1 Spectrophotometric methods

Reddy *et al.* described two spectrophotometric methods for nimodipine in its dosage forms [2]. The first method involves the use of para dimethylaminocinnamaldehyde as a colorimetric reagent, and the measurement of the generated color at 510 nm. The second method is based on the use of the Folin–Ciocalteu reagent, and a colorimetric measurement at 640 nm. The Beer's law is obeyed over the ranges of 0.5–4 and 5–20 μg/mL, respectively.

Formation of a Schiff base product of the reduction product of nimodipine (using the Zn–HCl system) and 4-dimethylaminobenzaldehyde formed the basis of a spectrophotometric method for the determination of nimodipine in tablets [3]. A pink color exhibiting an absorption maximum at 580 nm was produced and remained stable for 30 min. The percent recovery was 99–101%, with coefficients of variation of 0.3–0.5.

Reduction of nimodipine to the corresponding aminoderivative (using Zn–HCl), followed by diazotization and coupling with the Bratton–Marshall reagent was reported [4]. The resulting color was measured at 550 nm, and the Beer's law was obeyed upto 40 μg/mL. Squella *et al.* determined nimodipine by measuring the absorbance of its solution in an acetate-phosphate buffer (pH 5) at 360 nm. The coefficient of variation for this method was 2.5% [5]. A similar method based on measuring the absorbance of nimodipine in ethanol at 237 nm was reported [6]. A linear relationship was observed between the concentration and the absorbance over the range of 0.14–6 μg/mL.

Belal *et al.* developed a fluorimetric method for the determination of nimodipine [7]. The method involves reduction of nimodipine with Zn–HCl, and measuring the fluorescence at 425 nm after excitation at 360 nm. The addition of triton X-100 was found to enhance the fluorescence. The useful concentration range for the fluorescence measurement is from 0.1 to 5.0 μg/mL, with a detection limit of 0.06 μg/mL (1.62×10^{-7} M). The method was applied to the analysis of commercial tablets, and was further applied to spiked human urine. The percent recovery in the latter study was 107.1 ± 2.54 ($n = 4$).

2.2 Electrochemical methods

The electrochemical behavior of nimodipine was studied in ammonia buffer containing 10% (v/v) ethanol [8]. A single-sweep oscillopolarographic method was then developed for nimodipine in tablets. The calibration graph (peak current at −0.73 V vs. concentration) was linear from 0.2 to 70 μM, and the detection limit was 10 μM. The same authors applied linear sweep voltammetry for the determination of nimodipine in tablets [9]. A reduction peak at −0.62V vs. the Ag/ACl reference electrode was obtained over the range 10 nM–0.8 μM with a detection limit of 1 nM. The method involved a preconcentration period of 60 s.

The anodic electrochemical behavior of various 1,4-dihydropyridines (including nimodipine) was studied in acetate-phosphate buffer (pH 8) containing 70% (v/v) ethanol [10]. An anodic peak due to a two-electron oxidation of the dihydropyridine ring to a pyridine derivative was obtained. Recoveries were 91.7–104%, with a RSD being less than 3%. A cathodic polarographic method for nimodipine in tablets was developed. The dependence of peak current on concentration in acetate-phosphate buffer of pH 5 containing 20% ethanol was linear over the range of 1 μM–1 mM, with a coefficient of variation of 0.2% [5].

2.3 Thin-layer chromatography

Enantiometric mixtures of nimodipine were separated on β-cyclodextrin-bonded silica gel plates. The plates were developed with light petroleum/ethyl acetate/methanol, methanol/1% triethylammonium acetate (pH 4.1), or methanol/acetonitrile/1% triethylammonium acetate as mobile phases. Spots were identified by illumination at 365 nm, or by exposure to iodine vapor [11].

Trace amounts of nimodipine could be determined using TLC on silica gel GF_{254} plates and a mobile phase consisting of chloroform/methanol/methylene chloride/*n*-hexane (26:12:10:5). A. TLC-scanner was used to measure the absorbance at 365 nm ($R_f = 0.48$). The calibration graph was linear from 5 ng to 1 μg, with a detection limit of 5 ng [12].

Nimodipine was determined using a HPTLC method in tablets after their extraction with methanol (paracetamol was used as an internal standard). The plates were developed with toluene/ethylacetate (1:1), and densitometric detection was effected at 240 nm. Calibration graphs were linear from 25 ng (the detection limit) to 200 ng. Recovery was 96.03%, and the relative standard deviation was 0.56% [13].

2.4 Gas chromatography

Metabolites of the dihydropyridines (including nimodipine) could be determined in urine through the use of gas chromatography–mass spectrometry after extractive methylation [14]. Derivatization was affected using methyl iodide. The derivative was dissolved in ethyl acetate, and was injected into a HP capillary (12 m × 0.2 mm;

cross-linked methylsilicone film thickness 330 nm) operated with temperature programming from 100°C and held for 3 min. Helium was used as a carrier gas (flow rate of 1 mL/min), and 70 eV EIMS were used. The overall recoveries ranged from 67 to 77%, with relative standard deviations less than 10%.

Mixtures of dihydropyridine derivatives (including nimodipine) were qualitatively and quantitatively determined by GC–MS [15]. Methanolic samples were injected into a HP column (12 m × 0.2 mm i.d.), operated with temperature programming to 280°C at 2°C/min, and then to 300°C at 10°C/min. Helium was used as the carrier gas (flow rate of 4.5 mL/min), and 70 eV EMIS was used for detection. The analysis time was 45–60 min. The recoveries were better than 99%, and the relative standard deviations were 2.11–2.46%.

Nimodipine was determined in tablets by GC after extraction with chloroform (using nifedipine as an internal standard) [16]. A column (3 ft × 2 mm) of 10% of OV-1 on Chromosorb G (AW), 80–100 mesh treated with DMCS and operated at 225°C, was used with nitrogen as the carrier gas. Flame ionization was used as the means of detection. The calibration graph was linear over the range 0.5–2 ng/mL, and the coefficient of variation was 2.28.

Nimodipine in plasma was determined by a GC method [17]. Plasma was treated with 2 M NaOH after addition of the internal standard, and then extracted with toluene. The column (10 m × 0.31 mm) consisted of cross-linked 5% phenylmethyl silicone, and the method used temperature programming from 90°C (held for 1 min) to 255°C at a heating rate of 25°C/min. Helium was used as a carrier gas, and the N-P detection mode was employed. The calibration graph was linear from 2 to 50 ng/mL, and the detection limit was 0.5 ng/mL.

The external factors influencing nimodipine concentrations during intravenous administration were studied using GC–electron-capture detection [18]. Nimodipine was extracted from plasma and analyzed using a column (1.8 m × 2 mm) packed with 3% of OV-17 on gas-Chrom Q (50–100 mesh). The column was operated at 255°C with nitrogen as the carrier gas (flow rate of 25 mL/min), and 63 Ni ECD. The calibration graph was linear for upto 1000 ng/mL, and the limit of detection was 0.5 ng/mL.

Nimodipine could be determined by GC in plasma and cerebrospinal fluid at concentrations as low as 1 ng/mL [19]. To avoid decomposition during the GC analysis, nimodipine was first oxidized to its pyridine derivative. A column (1.8 m × 3 mm) of 2% of OV-17 on Anakrom Q was operated at 250–260°C, with nitrogen as the carrier gas (flow rate of 25–30 mL/min) and ECD detection.

2.5 Liquid chromatography

The concentration–time profiles of nimodipine following intravenous and oral administration were followed using high-performance liquid chromatography (HPLC) after a solid phase extraction (SPE). A column of 3 μm Spherisorb ODS2 was eluted with a mobile phase (of 0.8 mL/min) consisting of 1 mM phosphate buffer (pH 5)/methanol (7:13), and UV detection was used. Diazepam was used as the internal standard, and the calibration graph was linear over the range of 5–200 ng/mL. Enantiometric ratios were determined on a 5 μm Ultron ES-OVM chiral column using a mobile phase (flow rate of 1 mL/min) consisting of ethanol/20 mM phosphate buffer (47:153). The data evaluation entailed using the peak area ratio of the two enantiomers [20].

The stability of nimodipine injectable formulations was studied using reversed phase HPLC, which made use of methyltestosterone as an internal standard. A YWG C_{18} column was eluted with methanol/water (13:7, v/v), and detection was effected at 238 nm. The calibration graph was linear over the range of 5.98–299 μM/L [21].

1,4-Dihydropyridines (including nimodipine) were determined in human plasma using HPLC coupled with electrochemical detection. The separation was performed using a Supelcosil LC-ABZ-Plus C_{18} column. A mobile phase consisting of methanol/H_2O (7:3), that contained 2 mM acetate buffer (pH 5), was eluted at a flow rate of 1 mL/min. The detector was equipped with a glassy carbon electrode that was operated at 1000 mV vs. an Ag/AgCl reference electrode in the direct current mode. The total elution time was less than 18 min [22].

Nimodipine, tarcine, and their metabolites were determined by HPLC in the plasma of patients with Alzheimer's disease [23]. After extraction and evaporation under nitrogen to dryness, the residue was reconstituted into acetonitrile/H_2O (1:3), and analyzed on a 5-μm Hypersil phenyl column (25 cm × 4.6 mm i.d.) equipped with a Nova Pak phenyl guard column and operated with acetonitrile/4.4 mM KH_2PO_4 buffer (39:61) as the mobile phase (flow rate of 1.5 mL/min). Detection was effected either fluorimetrically at 330 nm excitation and 360 nm emission, or by its UV absorbance at 239 nm. Calibration graphs were linear over the range of 2–500 μg/mL for nimodipine and two of its metabolites, and 4–500 μg/mL for the third metabolite.

Six drugs (including nimodipine) used together in a program of hypertension therapy were determined simultaneously by HPLC [24]. The tablets were extracted with methanol, and felodipne was added as an internal standard. A column (25 cm × 4 mm i.d.) of Jasco Metaphase ODS silica (5 μm) was used with a mobile phase (flow rate of 1.5 mL/min) of 10 mM phosphate buffer of pH 4.5/acetonitrile (1:1). Detection was effected at 250 nm, and the linear operational range was found to be 25–3200 ng/mL.

Five dihydropyridine enantiomers (including nimodipine) were separated by HPLC using cyclodextrin as the chiral selector [25]. Columns containing β-cyclodextrin covalently bonded to 5 μm silica (25 cm × 4 mm i.d.), or of (*R*)- or (*S*)-naphthyl ethyl carbomyl β-cyclodextrin covalently bonded to 5 μm silica (25 cm × 4.6 mm i.d.), were used. The mobile phases (flow rate of 0.8 mL/min) were mixtures of ethanol or acetonitrile, with 0.1% triethylamine, adjusted to pH 5 with acetic acid. Detection was at 239 nm.

Nimodipine was determined in human plasma using HPLC and methyltestosterone as an internal standard [26]. A Spherisorb ODS, 10 μm column (25 cm × 4.6 mm i.d.) with methanol/H_2O/*n*-butylamine (65000-35000-1) was used as the mobile phase (flow rate of 1.2 mL/min), with detection at 238 nm. The calibration graph was linear over the range of 5–100 ng/mL, the detection limit was 2 ng/mL, and the average recovery was 92%.

The enantiospecific determination of nimodipine in human plasma could be established by liquid chromatography–tandem mass-spectroscopy [27]. The separation was effected using a 8 μm Chiral OJ MOD column (25 cm × 2 mm i.d.), operated at 35°C with ethanol–n-heptane (1:4) containing 2 mM ammonium acetate as the mobile phase (flow rate of 300 μL/min) and MS detection. Calibration graphs were linear over the range of 0.5–75 ng/mL, with a detection limit of 0.25 ng/mL for each enantiomer.

Five chiral calcium antagonists (including nimodipine) were separated using β-cyclodextrin as the stationary phase in a Merck Chiradex column (25 cm × 4.6 mm i.d.) with 0.1% triethylamine containing 5–100% of methanol or acetonitrile (as modifiers) [28]. The solvent was adjusted to pH 5 with acetic acid, and eluted at a flow rate of 0.8 mL/min. UV absorbance was used for the detection.

The metabolism of nimodipine in rat liver microsomes was studied by HPLC [29]. The centrifugate of the reaction mixture was analyzed using a column (10 cm × 4 mm i.d.) of Spherisorb S3 ODS II at 40°C, with a mobile phase (flow rate of 0.5 mL/min) of 5 mM ammonium acetate buffer (pH 6.6)/methanol (2:3) and detection at 218 and 238 nm. Calibration graphs were linear over the range of 0.2–50 μM for nimodipine, and 0.2–20 μM for the three main metabolites with detection limits of 30–80 nM.

Zhang *et al.* described a HPLC method for the simultaneous determination of six different dihydropyridine calcium antagonists (including nimodipine) in plasma [30]. After extraction with ethyl ether/*n*-heptane (1:1) and centrifugation, the organic phase was heated at 50°C and the residue dissolved in aqueous 60% acetonitrile (the mobile phase). An Ultrasphere 5 μm ODS column (25 cm × 4.6 mm i.d.) was used as the means of separation. The mobile phase flow rate was 1 mL/min, and detection was effected at 238 nm. The calibration graphs were linear over the range of 50 ng/mL–6.4 μg/mL, with a detection limit of 5 ng/mL.

Qian and Gallo used HPLC to determine nimodipine in monkey plasma [31]. After extraction with ethyl acetate and evaporating the organic phase to dryness, the residue was dissolved in the mobile phase (aqueous 65% methanol). A Hypersil ODS column (15 cm × 4.6 mm i.d.) was used. The mobile phase flow rate was 1 mL/min, and detection was effected at 238 nm. The calibration graph was linear over the range of 0.01–2 μg/mL, and the recovery was 95–106%.

Jiang *et al.* studied optimization of the mobile phase composition for the HPLC separation of dihydropyridine drugs, including nimodipine [32]. These workers used a stainless steel column (15 cm × 4.6 mm) 10 μm operated at 30°C. The mobile phase was methanol-acetonitrile-water (24.8:27:48.2).

Wang *et al.* studied the stability of nimodipine by HPLC, using beclomethazone dipropionate as an internal standard [33]. The method used a stainless steel column (25 cm × 4 mm) containing YWG $C_{18}H_{37}$, which was eluted at a flow rate of 1 mL/min) with methanol–water–ethyl ether (35:15:4). Detection was effected at 238 nm. The calibration graph was linear over the range of 0.03–0.72 μg/mL, the recovery was 100.1%, and the coefficient of variation was 0.83%.

Jain and Jain reported a HPLC method for the microquantification of nimodipine in intravenous infusion fluids [34]. A column (5 cm × 4.6 mm i.d.) of Shim-pack FLC-ODS was eluted with aqueous 75% methanol as the mobile phase (flow rate of 1 mL/min). Detection was effected at 238 nm. The calibration graph was linear over the range of 5–25 μg/mL, and recoveries were from 98.31 to 100.6%.

HPLC was used to evaluate the enantiomeric resolution of dihydropepidine enantiomers (including nimodipine), using phenylcarbamates of polysaccharides as a chiral stationary phase [35]. A column (25 cm × 4.6 mm) packed with the arylcarbamate derivatives of amylase, cellulose, and xylem was used. Detection was effected using polarimetry at 435 nm. Using xylem bis-(3,5-dichlorophenylcarbamate) and a mobile phase (flow rate of 0.5 mL/min) of 0.1%, diethylamine in hexane-propan-2-ol (17:3) yielded separation of nimodipine.

Patil *et al.* utilized supercritical fluid chromatography for the simultaneous assay of three dihydropyridines, including nimodipine [36]. These workers used a 10 μm JASCO-RP C_{18} column (25 cm × 4.6 mm), operated at 45°C and 13.7 mPa, with 13.04% methanol in CO_2 as the mobile phase (flow rate of 2 mL/min), and detection at 230 nm. Calibration graphs were linear upto 10 μg/mL, with detection limit of 0.3 μg/mL. The same technique was used by Bhoir *et al.* for the separation of seven vasodilating agents [37]. A JASCO-RP-C_{18} column (25 cm × 4 mm i.d.) operated at 45°C was used with a mobile phase consisting of 14.2% methanol in CO_2 and a pressure of 10.13 mPa, and detection at 224 nm. The calibration graph was linear over the range of 0.2–10.8 μg/mL.

2.6 Capillary electrophoresis (CE)

Micellar electrokinetic chromatography was used as a fast screening method for the determination of 1,4-dihydropyridines, and was used for the separation of six drugs of this group including nimodipine [38]. A fused silica tube (58.5 cm × 75 μm i.d., 50 cm to detector) was operated at 25 kV with detection at 200–236 nm. The optimized buffer (pH 8.2) contained 20 mM SDS in 50 mM borate buffer modified with acetonitrile. Linear calibration graphs extended upto 70 μg/mL from lower limits of 6.7 μg/mL.

The same technique was used by Wang *et al.* for the separation of five dihydropyridine calcium channel blockers [39]. The method used a fused silica capillary (75 cm × 75 μm i.d.; 68 cm to detector) operated at 60°C with an applied voltage of 24 kV, and a running buffer of 40 mM borax/4 mM SDS at pH 8. Detection at 200 nm was used for quantitation.

Enantioseparation of dihydropyridine derivatives could be accomplished using capillary electrophoresis by means of neutral and negatively charged β-cyclodextrin derivatives [40]. The method used a fused silica capillary (47 cm × 75 μm; 40 cm to detector), operated at 25°C. The running phase was 50 mM phosphate buffer (pH 3) containing cyclodextrins and organic modifiers, and an applied voltage of 6–20 kV was applied. Detection was effected at 190–300 nm.

Gilar *et al.* have also applied CE to the separation of dihydropyridine calcium antagonists [25]. Their method used an uncoated capillary (76 cm × 75 μm i.d.) at 20 kV, a running buffer of 20–25 M phosphate-borate buffer (pH 9–9.5) containing 1% urea, and detection at 240 nm. The buffer also contained 10 mM β-cyclodextrin, and varying amounts of SDS.

3. REFERENCES

1. ***European Pharmacopoeia***, 3rd edn., Supplement 13, Council of Europe, Strasbourg, (2001).
2. M.N. Reddy, T.K. Murthy, K.V. Rao-Kanna, A.V. Gopal-Hara, and D.G. Sankar, *Indian Drugs*, ***38***, 140 (2001).
3. S.N. Bharathi, M.S. Prakash, M. Nagarajan, and K.A. Kumar, *Indian Drugs*, ***36***, 661 (1999).
4. K.P.R. Chowdary and G.D. Rao, *Indian Drugs*, ***32***, 548 (1995).
5. J.A. Squella, J.C. Strum, R. Leaue, and L.J. Nunez-Vergara, *Anal. Lett.*, ***25***, 281 (1992).
6. M. Liu, *Yaowu-Fenxi-Zazhi.*, ***10***, 171 (1990).
7. F. Belal, A. Al-Majed and S. Gulthof, *Pharmazie* (in press).
8. Y.H. Zeng and Y.L. Zhou, *Fenxi-Shiyanshi*, ***19***, 5 (2000).
9. Y.H. Zeng and Y.L. Zhou, *Fenxi-Huaxue.*, ***27***, 832 (1999).
10. A. Alvarez-Lueje, L.J. Nunez-Vergara, and J.A. Squella, *Electroanal.*, ***6***, 259 (1994).
11. Q.H. Zhu, P.X. Yu, Q.Y. Deng, and L.M. Zeng, *J. Planar Chromatogr. Mod. TLC*, ***14***, 137 (2001).
12. M.S. Zhou, P.L. Wang, and L.A. Chen, *Fenxi Ceshi Xuebao*, ***18***, 66 (1999).

13. V.M. Shinde, B.S. Desai, and N.M. Tendolkar, *Indian Drugs*, ***31***, 119 (1994).
14. H.M. Mauer and J.W. Arit, *J. Anal. Toxicol.*, ***23***, 73 (1999).
15. B. Marciniec and E. Kujawa, *Chem. Anal.*, ***40***, 511 (1995).
16. R.T. Sane, M.G. Gangrade, V.V. Bapat, S.R. Surve, and N.L. Chankar, *Indian Drugs*, ***30***, 147 (1993).
17. M.T. Rossel, M.G. Bogaret, and L. Hugghens, *J. Chromatogr. Biomed. Appl.*, ***98***, 224 (1990).
18. P. Jackbsen, E.O. Mikkelsen, J. Laursen, and F. Jensen, *J. Chromatogr. Biomed. Appl.*, ***47***, 383 (1986).
19. G.J. Krol, A.J. Noe, S.C. Yeh, and K.D. Ramesch, *J. Chromatogr. Biomed. Appl.*, ***30***, 105 (1984).
20. H. Wanner-Olsen, F.B. Gaarskaer, F.O. Mikkelsen, P. Jackobsen, and B. Voldby, *Chirality*, ***12***, 660 (2000).
21. Y.S. Hu, Q.H. Tang, and Q.Y. Du, *Sepu.*, ***18***, 376 (2000).
22. J.A. Lopez, V. Martinez, R.M. Alonso, and R.M. Jimenez, *J. Chromatogr.*, ***870***, 105 (2000).
23. G. Aymard, M. Cayre-Castel, C. Feruandez, L. Lacomblez, and B. Diquet, *Ther. Drug Monitor.*, ***20***, 422 (1998).
24. Y.P. Patel, S. Patil, I.C. Bhoir, and M. Sundaresan, *J. Chromatogr.*, ***828***, 283 (1998).
25. M. Gilar, M. Uhrova, and E. Tesarova, *J. Chromatogr. Biomed. Appl.*, ***681***, 133 (1996).
26. W. Lu, L.Z. Wang, G.L. Zhao, and Y.C. Sun, *Yaowu-Fenxi-Zazhi*, ***15***, 3 (1995).
27. W.M. Mueck, *J. Chromatogr.*, ***712***, 45 (1995).
28. E. Tesarova, M. Gilar, P. Hobza, M. Kabelac, Z. Degl, and E. Smolkova-Keulemansova, *J. High Resolut. Chromatogr.*, ***18***, 597 (1995).
29. P.M. Leuenberger, H.R. Ha, W. Pletscher, P.J. Meier, and O. Sticher, *J. Liq. Chromatogr.*, ***18***, 2243 (1995).
30. S.H. Zhang, Y.R. Zhen, L.C. Zhang, and S.B. Li, *Sepu.*, ***13***, 132 (1995).
31. M.X. Qian and J.M. Gallo, *J. Chromatogr. Biomed. Appl.*, ***1165***, 316 (1992).
32. J. Jiang, Y.U.Z. Wang, W.Y. Zhong, C.L. Zhang, Y.R. Zang, and D.Q. Zhang, *e.q. Zhang. Fenxi. Huaxue*, ***20***, 822 (1992).
33. S. Wang, W. Ma, and X. Wang, *Yaowu-Fenxi-Zashi*, ***11***, 81 (1991).
34. R. Jain and C.L. Jain, *Indian Drugs*, ***28***, 154 (1990).
35. Y. Okamoto, R. Aburatani, K. Hatada, M. Honda, J. Inotsume, and M. Nakano, *J. Chromatogr.*, ***513***, 375 (1990).

36. S.T. Patil, I.C. Bhoir, and M. Sundaresan, *Indian Drugs*, ***36***, 698 (1999).

37. I.C. Bhoir, B. Raman, M. Sandaresan, and A.M. Bhagwat, *J. Pharm. Biomed. Anal.*, ***17***, 539 (1998).

38. V. Martinez, J.A. Lopez, R.M. Alonso, and R.M. Jimenez, *J. Chromatogr.*, ***836***, 189 (1999).

39. J. Wang, J.F. Chen, and L.F. Zhao, *Fenxi. Huaxue*, ***27***, 82 (1999).

40. T. Christians and U. Holzgrabe, *Electrophoresis*, ***21***, 3609 (2000).

Nimodipine: Drug Metabolism and Pharmacokinetic Profile

Mohammed A. Al-Omar

Department of Pharmaceutical Chemistry
College of Pharmacy, King Saud University
P.O. Box 2457, Riyadh-11451
Kingdom of Saudi Arabia

PROFILES OF DRUG SUBSTANCES, EXCIPIENTS, AND RELATED METHODOLOGY – VOLUME 31
DOI: 10.1016/S0000-0000(00)00000-0

CONTENTS

1. Uses, Application, and Associated History 372
2. Absorption and Bioavailability 372
3. Metabolism 373
4. Excretion 374
5. References 375

1. USES, APPLICATION, AND ASSOCIATED HISTORY

Nimodipine is a dihydropyridine calcium channel blocker derivative which is used in the treatment of cerebrovascular disorders. Particularly, it is used as a prophylactic and in the treatment of ischemic neurological deficits that follow aneurysmal or traumatic subarachnoid hemorrhage [1]. It may also have an effect on patients with dementia of vascular or degenerative origin which is most commonly due to Alzheimer's disease and/or cerebrovascular disease [2, 3]. The drug acts mainly by dilating the cerebral blood vessels and improving the cerebral blood flow, and it may also prevent or reverse ischemic damage to the brain by limiting transcellular calcium influx [4]. It is well-known that calcium has a role in regulating various brain functions, which links membrane excitation to subsequent intracellular enzymatic responses. Nimodipine is an isopropyl calcium channel blocker which can easily cross the blood–brain barrier. It reduces the number of open calcium channels, and thus decreases the influx of calcium ions into the neurons [5].

The oral dose for the prophylaxis of neurological deficit is 60 mg every 4 h. Treatment should begin within 4 days of the onset of hemorrhage, and should continue for 12 days. If cerebral ischemia has already occurred, the neurological deficit may be treated by intravenous infusion of 1 mg per hour for 2 h as an initial dose. The starting dose should be increased to 2 mg per hour provided that no severe decrease in blood pressure is observed [1].

2. ABSORPTION AND BIOAVAILABILITY

Nimodipine is rapidly absorbed from the gastrointestinal tract following oral administration, but undergoes extensive first-pass metabolism in the liver [1]. The oral bioavailability is reported to be about 13%. Although the drug administration using capsules could cause one to reach peak levels faster, the bioavailability was not significantly different from that of the tablets [6]. The drug is highly bound to plasma protein (greater than 95%) [1, 7].

Mean cerebrospinal fluid concentrations were estimated to be 0.3 ± 0.2 μg/L in patients whose mean plasma concentrations were 76.9 ± 34 μg/L [8]. This indicates that nimodipine does not apparently bind to the cerebrospinal fluid proteins. The bioavailability of dihydropyridines following oral administration was found to increase after food intake, such as grapefruit juice [9]. After intravenous administration, nimodipine is distributed rapidly and uniformly. On the other hand, after oral administration, nimodipine is strongly differentiated, with high concentrations being found in the liver, kidneys, and fat, low concentrations in the brain and testes, and the rest in the plasma [10, 11]. Only small amounts of nimodipine pass through the placenta [11].

3. METABOLISM

Nimodipine is extensively metabolized in the liver, and undergoes extensive first-pass metabolism [1]. More than 18 metabolites have been detected and identified from the biotransformation of nimodipine. The drug undergoes different reactions before its excretion. These reactions include, dehydrogenation of the 1,4-dihydropyridine moiety, oxidation of the two ester groups, and oxidative demethylation which is followed by carboxylic acid formation via oxidation of the resulting alcohol. In addition to the oxidation reactions, hydroxylation of the methyl groups at the pyridine ring, hydroxylation of one methyl group at the isopropyl ester moiety, reduction of the aromatic nitro group, and glucuronidation can take place [12]. Nimodipine is mainly metabolized by cytochrome P450-3A (CYP3A), which is involved in the dehydrogenation and oxidation reactions [13]. The oxidation of 1,4-dihydropyridines is proposed to produce a reactive oxygen species that could be harmful to vulnerable tissues such as the brain [14]. Nimodipine is completely metabolized, and no unchanged drug was detected in either urine or feces [12].

Scherling *et al.* [12] have thoroughly investigated the biotransformation of nimodipine in rats, dogs, and monkeys, and have proposed the metabolic pathway scheme shown in Figure 1.

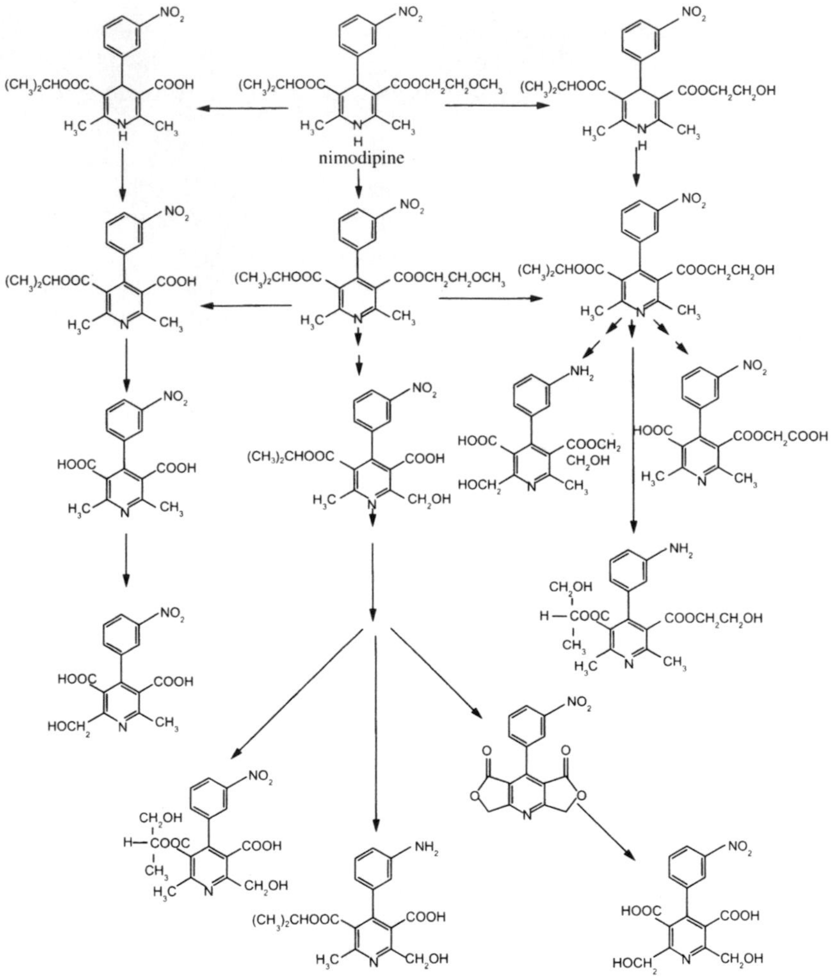

Figure 1. Metabolic pathway of nimodipine in rats, dogs, and in monkeys [12].

4. EXCRETION

Nimodipine is excreted in feces via the bile duct, and in urine *via* the glomular filtration, as metabolites [1, 10]. Fecal excretion is the major excretory route (greater than 67%) [11]. The parent compound and its metabolites were detected in breast milk [9]. The average elimination half-life for the drug is reported to be 9 h, but the initial decline in plasma is much more rapid, equivalent to a half-life of 1–2 h [1]. An average of 43% of bile-excreted nimodipine is subject to pronounced enterohepatic circulation, and 57% is excreted with the feces [15, 16]. Due to this cycle of excretion/reabsorption, the plasma concentration decreases and the increases. More than 80 and 19% of the reabsorbed quantity is excreted *via* bile and urine, respectively [15, 16]. The excretion routes detected in the fetus are essentially the same as that of the mother [11].

5. REFERENCES

1. ***Martindale, The Complete Drug Reference***, 33rd edn., S.C. Sweetman, ed., The Pharmaceutical Press, Chicago, p. 946 (2002).
2. J.M. López-Arrieta and J. Birks, *Cochrane Database Syst. Rev.*, Issue 3, CD000147 (2002).
3. J.M. López-Arrieta and J. Birks, *Cochrane Database Syst. Rev.*, Issue 1, CD000147 (2001).
4. T.N. Raju, E. John, R. Shankararao, L. Fornell, R. Vasa, and M. Abu-Harb, *Pharm. Res. Off. J. Ital. Pharm. Soc.*, ***33***, 5 (1996).
5. J.M. Gilsbach, *Acta Neurochir. Supp.*, ***45***, 41 (1988).
6. L.Q. Guo, H.S. Tan, and G. Chen, *Acta Pharmacol. Sinica*, ***14***, 295 (1993).
7. D. Maruhn, H.M. Siefert, H. Weber, K. Ramsch, and D. Suwelack, *Arzneimittelforschung*, ***35***, 1781 (1985).
8. K.D. Rämsch, G. Ahr, D. Tettenborn, and L.M. Auer, *Neurochiurgia*, ***28***, 74 (1985).
9. U. Fuhr, A. Maier-Brüggemann, H. Blume, W. Muck, S. Unger, J. Kuhlmann, C. Hoschka, M. Zaigler, S. Rietbrock, and A.H. Slaib, *Int. J. Clin. Pharmacol. Ther.*, ***36***, 126 (1998).
10. D. Maruhn, H.M. Siefert, H. Weber, K. Ramsch, and D. Suwelack, *Arzneimittelforschung*, ***35***, 1781 (1985).
11. D. Suwelack, H. Weber, and D. Maruhn, *Arzneimittelforschung*, ***35***, 1787 (1985).
12. D. Scherling, K. Bühner, H.P. Krause, W. Karl, and C. Wunsche, *Arzneimittelforschung*, ***41***, 392 (1991).
13. X.Q. Liu, Y.L. Ren, Z.Y. Qian, and G.J. Wang, *Acta Pharmacol. Sin.*, ***21***, 690 (2000).
14. C.B. äärnhielm and G. Hansson, *Biochem. Pharmacol.*, ***35***, 1419 (1986).
15. D. Suwelack and H. Weber, *Arzneimittelforschung*, ***10***, 231 (1985).
16. L. Parnetti, *Clin. Pharmacol.*, ***29***, 110 (1995).

Related Methodology Review Articles

Evaluation of the Particle Size Distribution of Pharmaceutical Solids

Harry G. Brittain

Center for Pharmaceutical Physics
10 Charles Road
Milford, NJ 08848, USA

PROFILES OF DRUG SUBSTANCES,
EXCIPIENTS, AND RELATED
METHODOLOGY – VOLUME 31
DOI: 10.1016/S0000-0000(00)00000-0

CONTENTS

1. **Introduction** 380
2. **Representations of Particle Shape, Size, and Distribution** . . . 381
   - 2.1 Particle shape 381
   - 2.2 Particle size 384
   - 2.3 Distribution of particle sizes. 386
3. **Sieving Analysis** 392
   - 3.1 Construction of sieves 394
   - 3.2 Performance of sieving analysis 397
     - 3.2.1 Summary of USP ⟨786⟩, Method I 397
     - 3.2.2 Summary of USP ⟨786⟩, Method II 398
   - 3.3 Validation of sieving methods 399
   - 3.4 Interpretation and presentation of sieving data . . . 399
   - 3.5 Sieving analysis of samples by polymodal distributions. 400
4. **Determination of Particle Size Distribution by Laser Light Scattering**. 404
   - 4.1 Background 406
   - 4.2 Photon correlation spectroscopy. 409
   - 4.3 Fraunhofer diffraction. 411
5. **Summary** 417
6. **References** 418

1. INTRODUCTION

The evaluation of particle size is an essential aspect of the physical characterization of a powdered solid, and such determinations are undertaken to obtain information about the size characteristics of an ensemble of particles. Since the particles of any given powdered solid will never be of exactly the same size, one needs to acquire data related to the average particle size and distribution of sizes about this average. Since very often the idea of a particle diameter proceeds from preconceived shape factors, the concept of particle size is irrevocably derived from aspects of particle shape and morphology.

It is well known to anyone who has studied powders that the various methods used to evaluate particle size distribution do not ordinarily yield comparable results. A key item of debate, and an almost unanswerable question, centers around what constitutes a "correct" method for the determination of particle size distributions. Without doubt, the best "correct" method is one whose sample can be obtained using an appropriate sampling procedure, one where the sample can be properly prepared and introduced into the instrument, and one where all instrumental parameters can be correctly used for the analysis. Therefore, the "correct" (but differing) particle size results obtained using different methodologies are all equally "correct", but each may simply be expressing its "correct" results in different terms.

It is most plausible to view the decision as to which particle size methodology is most appropriate for a given situation as a matter of accuracy versus precision. If absolute accuracy is most important, then rigorous research must be conducted to verify that the method finally adopted does indeed yield particle size results that are absolutely indicative of the characteristics of the bulk material. Very often, this will entail the use of a referee method to calibrate the method intended for more routine use. If, however, one is more interested in developing profiles of lot-to-lot variability, then the use of any of the available methods that yields "correct" results is appropriate. In the latter situation, the key is to adhere to a rigorously defined experimental protocol that enables the investigator to run exactly the same method each and every time.

2. REPRESENTATIONS OF PARTICLE SHAPE, SIZE, AND DISTRIBUTION

2.1 Particle shape

It is not possible to rationally discuss the size of a particle, or any distribution associated with the sizes of an ensemble of particles, without first considering the three-dimensional characteristics of the particle itself. This is because the size of a particle is expressed in terms of linear dimension characteristics derived from its shape, or in terms of its projected surface or volume. As will be shown, some methods of expressing particle size discard any concept of particle shape, and instead express the size in terms of a spherical equivalent to a chosen geometric shape.

An appropriate starting place for a discussion of particle shape can be found in general test ⟨776⟩ of the United States Pharmacopoeia [1]. In the performance of the shape aspect of this particular test procedure, the USP requires that, "For irregularly shaped particles, characterization of

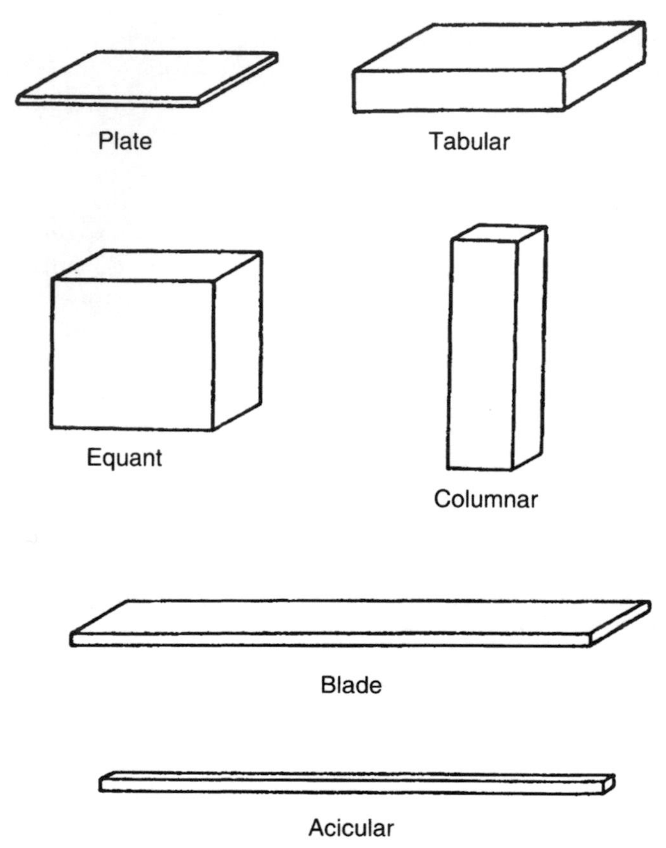

Figure 1. Descriptions of particle shape as defined by the United States Pharmacopoeia [1].

particle size must also include information on particle shape". The general method defines several descriptors of particle shape, which are illustrated in Figure 1. The USP definitions of these shape parameters are: (a) *Acicular* (slender, needle-like particle of similar width and thickness), (b) *Columnar* (long, thin particle with a width and thickness that are greater than those of an acicular particle), (c) *Flake* (thin, flat particle of similar length and width), (d) *Plate* (flat particles of similar length and width but with greater thickness than flakes), (e) *Lath* (long, thin, and blade-like particle), and (f) *Equant* (particles of similar length, width, and thickness; both cubical and spherical particles are included).

One rarely observes discrete particles but instead, is typically confronted with particles that have aggregated or agglomerated into more complex structures. The USP [1] provides a number of terms that serve to describe any degree of association: (a) *Lamellar* (stacked plates), (b) *Aggregate* (mass of adhered particles), (c) *Agglomerate* (fused or cemented particles), (d) *Conglomerate* (mixture of two or more types of particles), (e) *Spherulite* (radial cluster), and (f) *Drusy* (particle covered with tiny particles).

In addition, the particle condition can be described by another series of terms: (a) *Edges* (angular, rounded, smooth, sharp, fractured), (b) *Optical* (color, transparent, translucent, opaque), and (c) *Defects* (occlusions, inclusions). Furthermore, surface characteristics may be described as: (a) *Cracked* (partial split, break, or fissure), (b) *Smooth* (free of irregularities, roughness, or projections), (c) *Porous* (having openings or passageways), (d) *Rough* (bumpy, uneven, not smooth, and (e) *Pitted* (small indentations).

Of course, it is recognized that the pharmaceutical descriptors of particle shape are derived from the general concept of the crystallographic habit. The exact shape acquired by a crystal will depend on a variety of factors, such as the temperature, pressure, and composition of the crystallizing solution. Nevertheless, precipitation of a given compound generally results in a characteristic shape or outline. Since the faces of a crystal must reflect the internal structure of the solid, the angles between any two faces of a crystal will remain the same even if the crystal growth is accelerated or retarded in one direction or another. This behavior has been illustrated in Figure 2. Optical crystallographers will usually catalogue the various crystal faces and document the angles between these as they identify the crystal system to which the given particle belongs. When the particle is particularly well formed, a compilation of symmetry elements is also collected.

For many workers, however, the concept of qualitative shape descriptors has proven inadequate and this has necessitated the definition of more quantitatively defined shape coefficients [2]. For instance, Heywood [3] has defined the *Elongation Ratio* (n) as:

$$n = \frac{L}{B} \tag{1}$$

and the Flakiness Ratio (m) as:

$$m = \frac{B}{T} \tag{2}$$

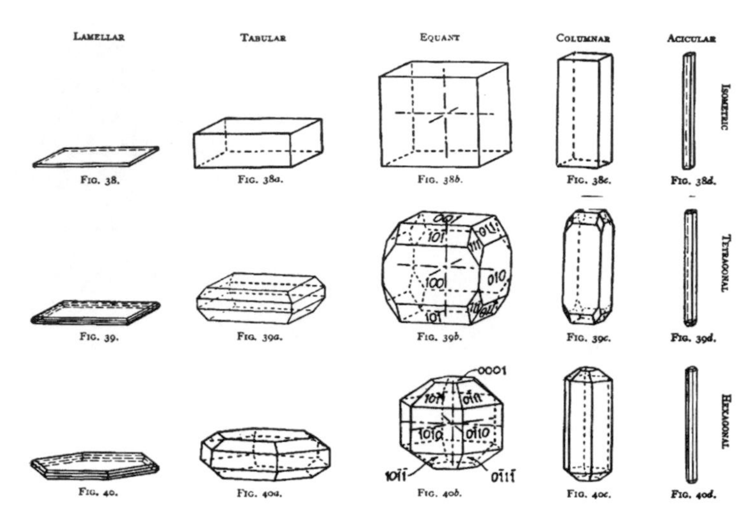

Figure 2. Illustration of how growth along certain crystal directions can profoundly alter the characteristic habit of various crystals.

where T is the particle thickness (the minimum distance between two parallel planes that are tangential to opposite surfaces of the particle), B is the breadth of the particle (the minimum distance between two parallel planes that are perpendicular to the planes defining the thickness), and L is the particle length (the distance between two parallel planes that are perpendicular to the planes defining thickness and breadth).

2.2 Particle size

It is not really possible to continue a discussion of particle shape or size without first developing definitions of particle diameter. This is, of course, rather trivial for a spherical particle since its size is uniquely determined by its diameter. For irregular particles, however, the concept of size requires definition by one or more parameters. It is often most convenient to discuss particle size in terms of derived diameters, such as a spherical diameter that is in some way equivalent to some size property of the particle. These latter properties are calculated by measuring a size-dependent property of the particle and relating this to a linear dimension.

Certainly the most commonly used measurements of particle sizes are the *length* (the longest dimension from edge to edge of a particle oriented parallel to the ocular scale) and the *width* (the longest dimension of the particle measured at right angles to the length). Intuitive as these properties may be, their definition is still best realized in the illustration of

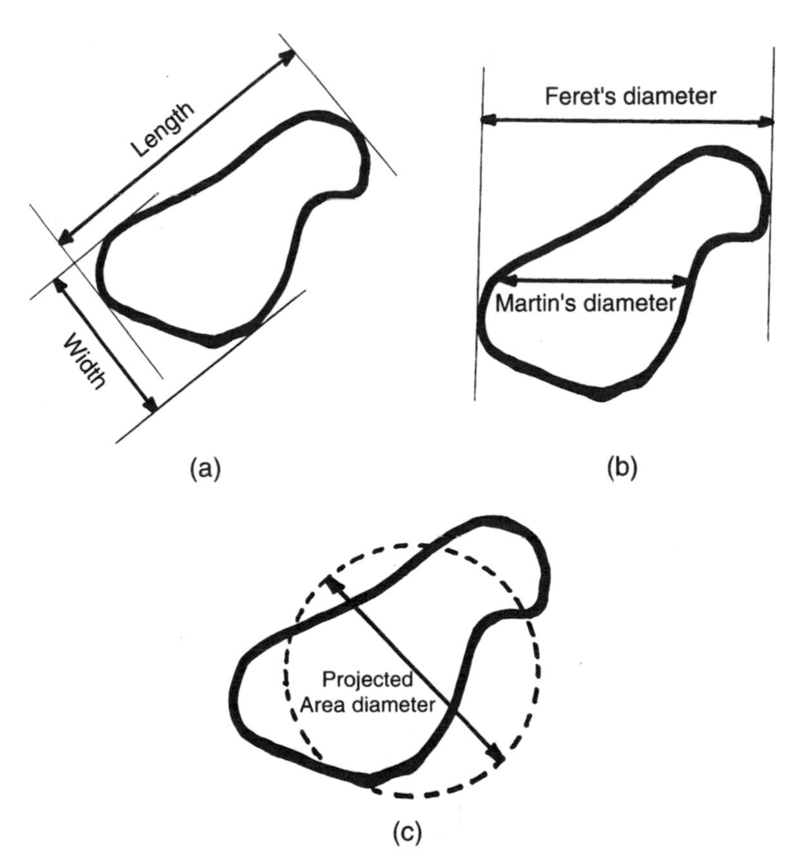

Figure 3. Some commonly used descriptors of particle size.

Figure 3a. Closely related to these properties are two other descriptors of particle size, *Feret's diameter* (the distance between imaginary parallel lines tangential to a randomly oriented particle and perpendicular to the ocular scale), and *Martin's diameter* (the diameter of the particle at the point that divides a randomly oriented particle into two equal projected areas). The definition of these quantities is illustrated in Figure 3b.

The coordinate system associated with the measurement is implicit in the definitions of length, width, Feret's diameter, and Martin's diameter, since the magnitude of these quantities requires some reference point. As such, these descriptors are most useful in discussing particle size as measured using microscopy, since one is working with immobilized particles. Defining spatial descriptors for freely tumbling particles is considerably more difficult, and hence has necessitated the definition of a series of derived particle descriptors. However, given the growing popularity of laser light scattering as a sizing methodology, derived statements of particle diameter are extremely useful.

The derived descriptors for particle size all begin with a homogenization of the length and width descriptors into either a circular or spherical equivalent, and make use of the ordinary geometrical equations associated with the derived equivalent. For instance, the *perimeter diameter* is defined as the diameter of a circle having the same perimeter as the projected outline of the particle. The *surface diameter* is taken as the diameter of a sphere having the same surface area as the particle, and the *volume diameter* is defined as the diameter of a sphere having the same volume as the particle. One of the most widely used derived descriptors is the *projected area* diameter, which is the diameter of a circle having the same area as the projected area of the particle resting in a stable position. The concept of projected area volume diameter is illustrated in Figure 3c.

Several other derived descriptors of particle diameter have been used for different applications. For instance, the *sieve diameter* is the width of the minimum square aperture through which the particle will pass. Other descriptors that have been used are the *drag diameter* (the diameter of a sphere having the same resistance to motion as the particle in a fluid of the same viscosity and at the same velocity), the *free-falling diameter* (the diameter of a sphere having the same density and the same free-falling speed as the particle in a fluid of the same density and viscosity), and the *Stokes' diameter* (the free-falling diameter of a particle in the laminar flow region).

2.3 Distribution of particle sizes

All analysts know that the particles making up real samples of powdered substances do not consist of any single type, but instead will generally exhibit a range of shapes and sizes. Particle size determinations are therefore undertaken to obtain information about the size characteristics of an ensemble of particles. Furthermore, since the particles under study are not of the exact same size, information is required regarding the average particle size and the distribution of sizes about this average.

One could imagine the situation where a bell-shaped curve is found to describe the distribution of particle sizes in a hypothetical sample, and this type of system is known as the *normal distribution*. Samples that conform to the characteristics of a normal distribution are fully described

Table 1

Particle Composition of a Hypothetical Sample Exhibiting a Normal Distribution

Size (μm)	Number in band	Number frequency	Percent less than	Percent greater than
5	50	1.67	1.67	98.33
10	90	3.00	4.67	95.33
15	110	3.67	8.33	91.67
20	280	9.33	17.67	82.33
25	580	19.33	37.00	63.00
30	600	20.00	57.00	43.00
35	540	18.00	75.00	25.00
40	360	12.00	87.00	13.00
45	170	5.67	92.67	7.33
50	120	4.00	96.67	3.33
55	60	2.00	98.67	1.33
60	40	1.33	100.00	0.00
Total	**3000**	**100**		

by a mean particle size and the standard deviation. An example of a sample exhibiting a normal distribution is provided in Table 1, where 3000 particles have been sorted according to an undefined determiner of their size. In the usual data representation, the number of particles in each size fraction is separated out, and from this one calculates the percentage of particles in each size fraction. This calculation yields the particle size histogram plotted in Figure 4a. The number frequency is ordinarily used to construct a cumulative distribution (shown in

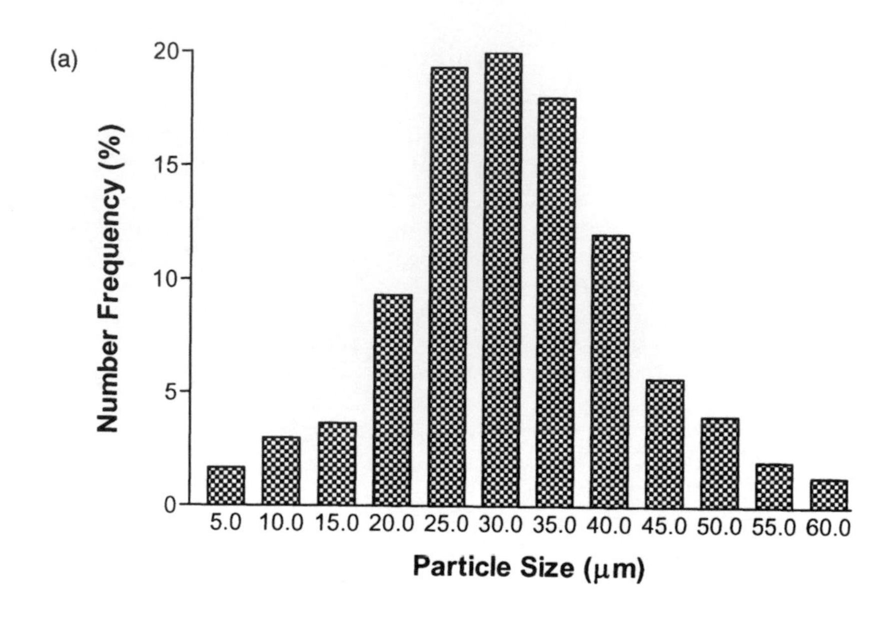

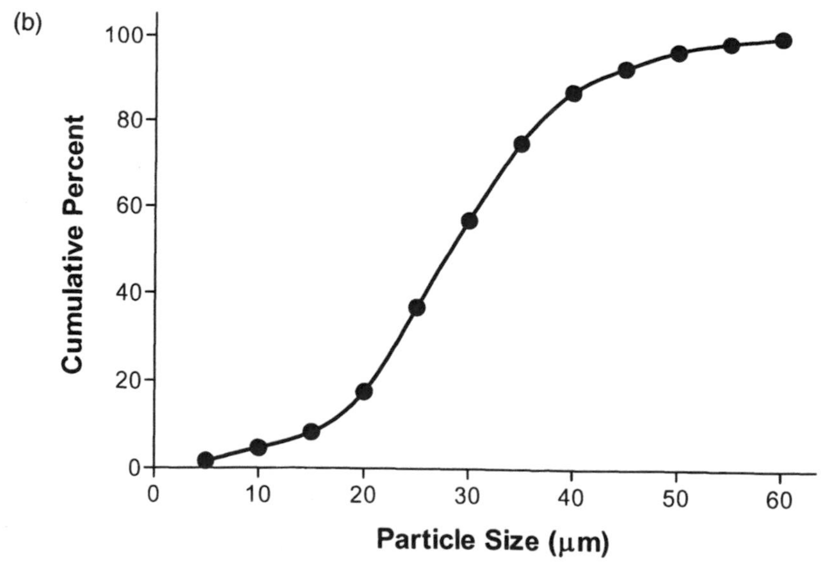

Figure 4. Particle size representations for a hypothetical normal distribution. Shown are (a) the frequency distribution and (b) the cumulative distribution.

Figure 4b), which can be ascending or descending depending on the nature of the study and what information is required.

The arithmetic *mean* of the ensemble of particle diameters is calculated using the relation:

$$d_{\mathrm{av}} = \frac{\sum n d_i}{\sum n} \tag{3}$$

where n is the number of particles having a diameter equal to d_i. The standard deviation in the distribution is then calculated using:

$$\sigma = \left[\frac{(d_{\mathrm{av}} - d_i)^2}{n}\right]^{1/2} \tag{4}$$

In the example of Table 1, one calculates that $d_{\mathrm{av}} = 30.2$ μm, and that $\sigma = 1.1$.

The most commonly occurring value in the distribution is the *mode*, and is the value at which the frequency representation is a maximum. The *median* divides the frequency curve into two equal parts, and equals the particle size at which the cumulative representation equals 50%. In a rigorous normal distribution, the mean, mode, and median have the same value. For a slightly skewed distribution, however, the following approximate relationship holds:

$$\text{mean} - \text{mode} = 3[\text{mean} - \text{median}] \tag{5}$$

It would be highly advantageous if powder distributions could be described by the normal function, since all of the statistical procedures developed for the Gaussian distributions could be used to describe the properties of the sample. However, unless the range of the particle sizes is extremely narrow, most powder samples cannot be adequately described by the normal distribution function. The size distribution of the majority of real powder samples is usually skewed toward the larger end of the particle size scale. Such powders are better described using the log-normal distribution type. When the particle distribution is plotted using the logarithm of the particle size, the skewed curve is transformed into one closely resembling a normal distribution, hence this terminology. This situation is illustrated in Figure 5.

The distribution in a log-normal representation can be completely specified by two parameters, the *geometric median* particle size (d_g) and

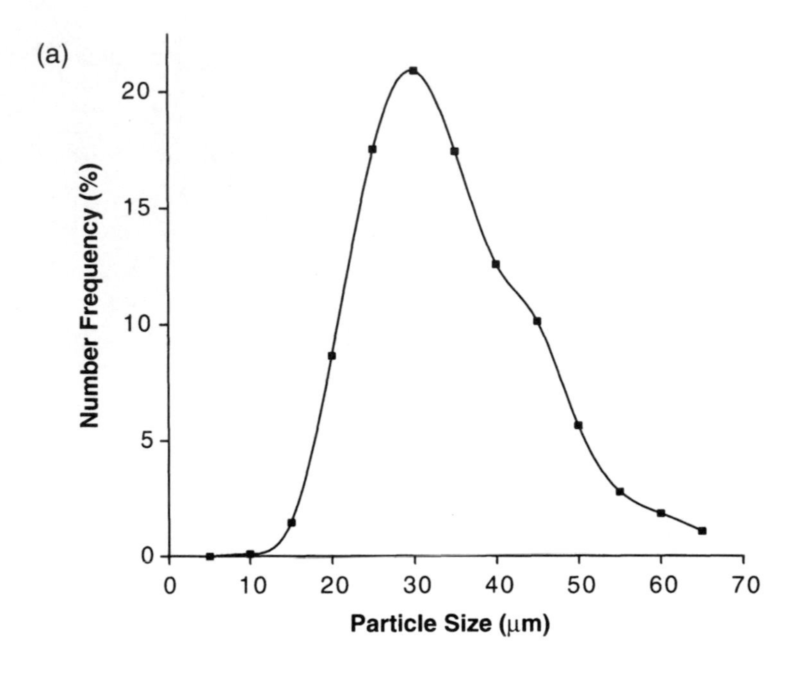

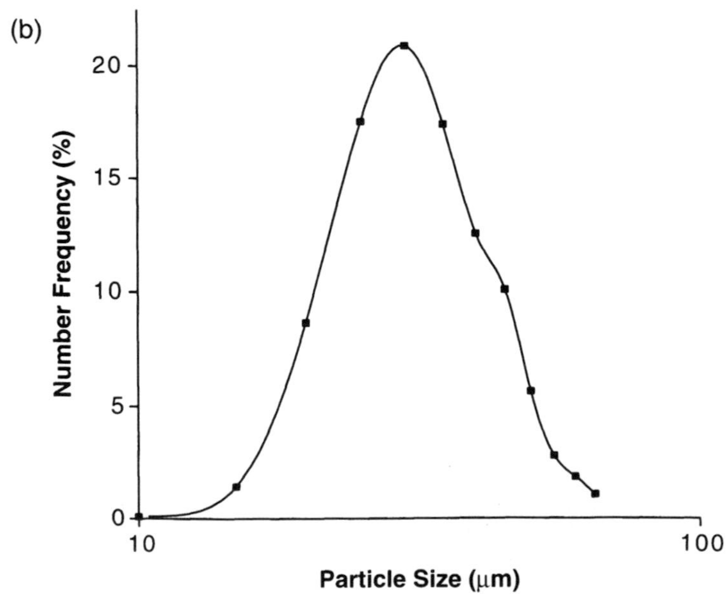

Figure 5. Particle size representations for a hypothetical log-normal distribution, plotted on a (a) linear scale and (b) logarithmic scale.

the standard deviation in the geometric mean (σ_g). The geometric median is the particle size pertaining to the 50% value in the cumulative distribution, and is calculated using:

$$d_g = \text{anti} - \log\left[\frac{\sum n \log(d_i)}{\sum n}\right] \tag{6}$$

where n is the mass of particles having the particle size equal to d_i. Two samples having identical d_g and σ_g values can be said to have been drawn from the same total population and exhibit properties or characteristics of the total population.

In many applications, the particle size results are processed by plotting the cumulative frequency data on a logarithmic scale. If a straight line is obtained, the particle size distribution is said to obey the log-normal function. One obtains the value of d_g as being equal to the 50% value of the distribution. The value of σ_g is obtained by dividing the 84.1% value of the distribution by the 50% value.

Although the distribution in the log-normal representation is completely specified by the geometric median particle size and the geometric mean standard deviation, a number of other average values have been derived to define useful properties. These are especially useful when the physical significance of the geometric median particle size is not clear. The *arithmetic mean* (d_{av}) particle size is defined as the sum of all particle diameters divided by the total number of particles, and is calculated using:

$$d_{\text{av}} = \frac{\sum nd_i}{\sum n} \tag{7}$$

The *surface mean* (d_s) particle size is defined as the diameter of a hypothetical particle having an average surface area, and is calculated using:

$$d_s = \left[\frac{\sum nd_i^2}{\sum n}\right]^{1/2} \tag{8}$$

The *volume mean* (d_v) particle size is the diameter of a hypothetical particle having an average volume, and is obtained from:

$$d_v = \left[\frac{\sum nd_i^3}{\sum n}\right]^{1/3} \tag{9}$$

The *volume–surface mean* (d_{vs}) particle size is the average size based on the specific surface per unit volume, and is calculated using:

$$d_{vs} = \frac{\sum nd_i^3}{\sum nd_i^2} \tag{10}$$

For the distribution plotted in Figure 5, one finds that $d_g = 32.91$ µm, $d_{av} = 34.42$ µm, $d_s = 35.93$ µm, $d_v = 37.43$ µm, and $d_{vs} = 40.62$ µm.

Various types of physical significance have been attached to the different expressions of particle size. For chemical reactions, the surface mean is important, while for pigments the volume mean value is the appropriate parameter. Deposition of particles in the respiratory tract is related to the weight mean diameter, and the dissolution of particulate matter is related to the volume–surface mean.

Particle size distributions can be sorted according to the mass (or volume) of the particles contained within a given size band, or to the number of particles contained in the same size band. With substances having real density values, the distribution of the same ensemble of particles can look quite different depending on how the data are plotted. This behavior has been illustrated in Figure 6, where the frequency and cumulative distribution plots are shown for the same sample, but where the data have been separately processed in terms of the mass and particle numbers.

Unfortunately not every powdered sample is characterized by the existence of a single distribution, and the character of real samples can be quite complicated. Recognizing the existence of multimodal distributions is not always straightforward, but their existence can often be detected through plotting the data on a log-probability paper. The existence of more than one particle population is indicated by a change in the slope of the line. This behavior has been illustrated in Figure 7, showing a single log-normal distribution and a multimodal sample consisting of two populations whose mean differs by about 50%. The break in the log plot is clearly evident, but if one were to simply plot the latter sample in either a frequency or cumulative view, one would not have been able to detect the existence of two particle size populations in the sample.

3. SIEVING ANALYSIS

One of the most conceptually simple methods used to deduce the particle size distribution of a powdered solid is that of analytical sieving.

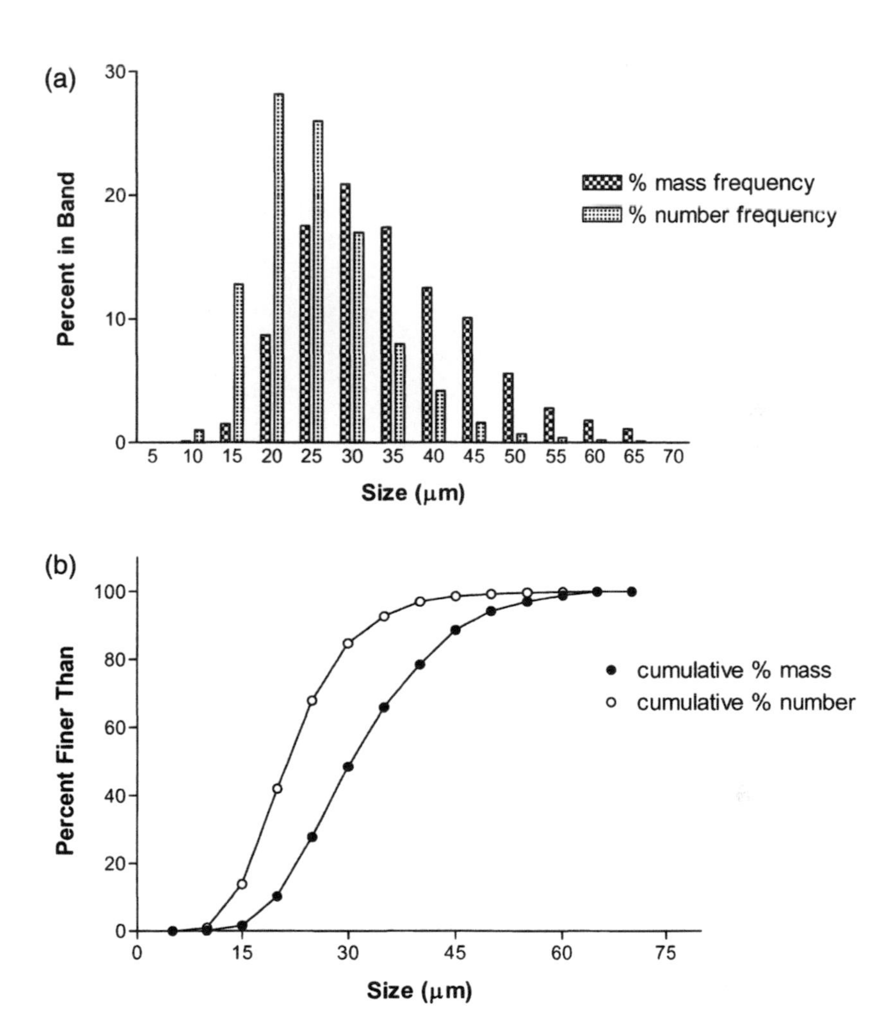

Figure 6. Particle size representations for a hypothetical log-normal distribution. Shown are (a) the frequency distribution and (b) the cumulative distribution, and each of these contains the difference obtained when processing the data in terms of either the particle number or the particle mass.

In sieving analysis, one simply passes the sample through wire meshes having openings of various sizes, and measures how much of the sample is retained on each sieve. Sieving is one of the fundamental methods for the classification of powders, and is the method of choice for determining the size distribution of coarse powders. Although sieving is most suitable for powders whose average particle size exceeds 25–50 μm, it can be used

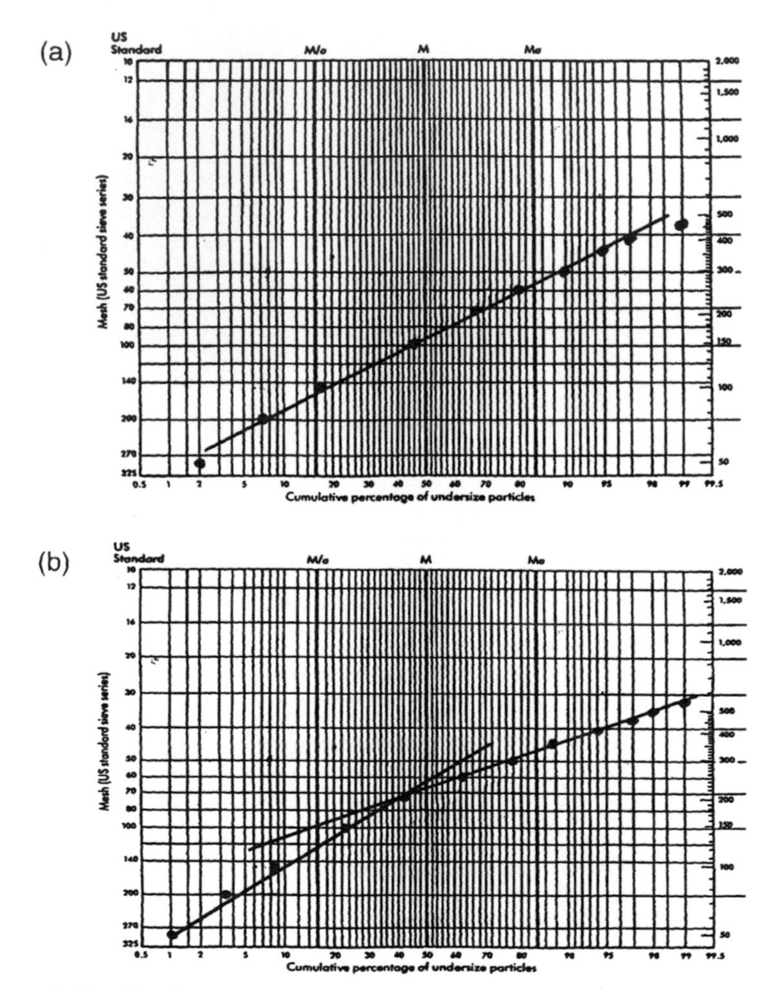

Figure 7. Particle size representations plotted on log-probability paper for (a) a single hypothetical log-normal distribution and for (b) a hypothetical sample containing two log-normal distributions whose average particle size differs by %.

for finer grades of powders if the method is properly validated and executed. Sieving analysis is difficult to perform for oily or cohesive powders, since they will tend to clog the sieve openings.

3.1 Construction of sieves

Sieves are constructed from wire mesh that is sealed into the base of an open cylindrical container, with the ideal sieve openings being nearly

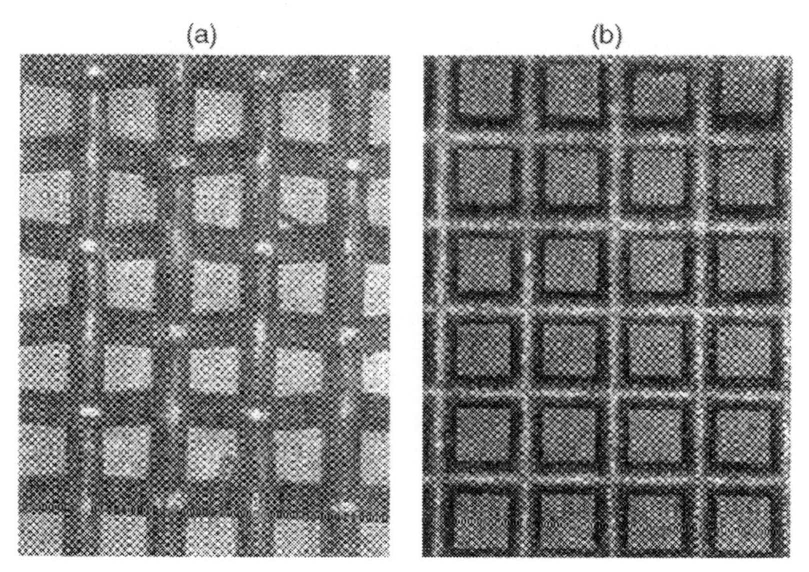

Figure 8. Photomicrographs of (a) woven-wire screen and (b) micromesh screen.

square apertures. Examples of the two most commonly encountered types of mesh screen are shown in Figure 8. The consequence of using wire mesh sieves is that the method of analytical sieving effectively provides a two-dimensional estimate of size, since the smallest lateral dimension of each particle dictates its ability to pass through a given sieve opening.

One widely used classification scheme for sieve sizes is denoted as the Tyler Standard Scale. In this system, the standard is based on a wire cloth having 200 openings per linear inch, which then defines a 200-mesh sieve. The diameter of the wire used for a 200-mesh screen is 53 μm, and the opening is 74 μm. The ratio between the adjacent sizes of the screen scale equals the square root of two. Thus, the areas of the openings of each sieve are twice those of the next finer sieve. In addition, the ratio between the widths of openings of alternate sieves in the series equals two. Closer sizing may be performed using screens that have a fixed opening width ratio equal to the fourth root of two.

Another system was defined by the National Bureau of Standards using the same ratio as the Tyler Standard Scale, but which was based on an opening of 1 mm. This system is known as the US Sieve Series. However, the differences between Tyler and US sieves are small, and they are often used interchangeably. However, USP general method ⟨786⟩ specifies the use of sieves within the US Series definition [4].

A summary of Tyler and US Standard sieve classifications, and the corresponding sieve openings, has been given [5], and the systems are compared in Table 2. It may be seen that for fine sieves, there is little effective difference between the two systems. However, the difference

Table 2

Sieve Classifications within the United States Standard and Tyler Systems

Mesh number	**Aperture opening**	
	United States Standard series (μm)	**Tyler series (μm)**
18	1000	991
20	840	833
25	710	701
30	590	589
40	420	417
45	350	351
50	297	295
60	250	246
70	210	208
80	177	175
100	149	147
120	125	124
140	105	104
170	88	88
200	74	74
230	62	61
270	53	53
325	44	43
400	37	38

Adapted from Ref. [5].

between the Tyler and US Standard sieves is more pronounced for the coarser sieves.

3.2 Performance of sieving analysis

The conduct of analytical sieving for the classification of pharmaceutical materials has been fully described in USP general method ⟨786⟩ [4]. The general method describes the use of both dry sieving (Method I) and wet sieving (Method II) procedures. In either method, the sieves are stacked on top of each other in ascending degrees of coarseness, and the powder to be tested is placed on the top sieve. The nest of sieves is completed by a well-fitting pan at the base and a lid at the top. Additional sources of information on the performance of sieving analysis may be found in the literature [6–9].

The nest of sieves is subject to a standardized period of agitation, which serves to cause the powder sample to distribute between the sieves. Agitation can be effected using vibration, rotation/tapping, or ultrasound. The horizontal sieve motion serves to loosen the powder packing, and permits subsieve particles to pass through. Vertical motion serves to mix the particles, and brings more of the subsieve particles to the screen surface. Although allowed by general method ⟨786⟩, the use of agitation by hand is not encouraged.

In a properly designed sieve test, the sample will be partitioned in approximately equal weights on each of five or six sieves, and on the bottom pan. The sieving analysis is considered to be complete when the weight on any of the test sieves does not change by more than 5% of the previous weight on that sieve. Performance of a sieving test yields the weight percentage of powder retained in each sieve size range.

3.2.1 Summary of USP ⟨786⟩, Method I

Method I in USP general method ⟨786⟩ gives the procedure to be followed when conducting the sieving analysis of dry powdered solids.

1. Tare each test sieve to the nearest 0.1 g.
2. Place an accurately weighed quantity of test specimen on the top (coarsest) sieve, and replace the lid.
3. Agitate the nest of sieves for 5 min.
4. Carefully remove each sieve from the nest without the loss of material.

5. Reweigh each sieve, and determine the weight of material on each sieve.
6. Determine the weight of the material in the collecting pan in a similar manner.
7. Reassemble the nest of sieves, and agitate for 5 min.
8. Remove and weigh each sieve as previously described.
9. Repeat these steps until the endpoint criteria (the weight on any of the test sieves does not change by more than 5% of the previous weight on that sieve) are met.

Upon completion of the analysis, the analyst is to reconcile the weights of material. The total losses must not exceed 5% of the weight of the original test specimen. If there is evidence that the particles retained on any sieve are aggregates (rather than single particles), the use of dry sieving is unlikely to give good reproducibility. At that point, the analyst could consider the use of Method II as an alternate technique.

3.2.2 Summary of USP ⟨786⟩, Method II

Method I in USP general method ⟨786⟩ gives the procedure to be followed when conducting the sieving analysis of wet or suspended solids.

1. Modify the lid and collecting pan of the sieve nest to permit addition of a liquid onto the surface of the top sieve and collection of the liquid from the pan.
2. Select a liquid in which the test specimen is insoluble, and modify the sieving method as follows.
3. Thoroughly disperse the dried test material in the liquid by gentle agitation, and pour this dispersion onto the top sieve.
4. Rinse the dispersion equipment with fresh liquid, and add the rinsings to the top sieve.
5. Feed the sieving liquid through a suitable pumping mechanism to the nozzle(s) in the lid, and collect the sieving liquid from the pan in a suitable container.
6. Continue the wet sieving process until the emerging liquid appears free of particles.
7. Remove each sieve from the sieve nest, and dry each sieve to a constant weight at the same temperature as that used above.
8. Determine the weight of dried material on each sieve.

The results are analyzed as those obtained using Method I.

3.3 Validation of sieving methods

One of the most important experimental parameters to determine in a sieving analysis is the time required to completely equilibrate the sample between all of the sieves. This time is typically determined by repeating the analysis with a fresh sample, and continuing the mechanical agitation for successively longer periods of time. The proper sieving time is the smallest amount of time which leads to conformance with the requirements for an endpoint determination. When this endpoint has been validated for a specific material, then a single fixed time of sieving may be used for future analyses, provided the particle size distribution does not change significantly. The determination of the proper sieving time is part of the robustness evaluation of the method.

The two most important analytical performance parameters to be determined during the validation of a sieving procedure are the precision and accuracy associated with the analysis. To evaluate precision, one repeats the particle size determination of a properly subdivided sample three to five times, and compares the percentages associated with each size fraction. The accuracy of a sieving analysis is usually evaluated using standard powders consisting of micron-sized glass spheres of known particle size distribution. The size openings in a sieve can be verified through the passage of the reference material, or through a microscopic study of the screens themselves.

It is not usually possible to determine the analytical performance parameters of specificity, limits of detection and quantitation, linearity, or range.

3.4 Interpretation and presentation of sieving data

In addition to the weights on the individual sieves and in the pan, the raw data must include the weight of the test specimen, the total sieving time, and the precise sieving methodology. The raw data are converted into a cumulative weight distribution, and if it is desired to express the distribution in terms of a cumulative weight undersize, the range of sieves used should include a sieve through which all the material passes.

In many cases, the particle size distribution of a real sample turns out to be adequately represented by a log-normal distribution. In that case, the distribution can be specified by the Geometric Median Particle Size (d_g) and the Geometric Mean Standard Deviation (σ_g). Two samples having identical d_g and σ_g values can be said to have been drawn from the same total population. The value of d_g is equal to the 50% value of the cumulative distribution, and the value of σ_g is obtained by dividing the 84.1% value of the distribution by the 50% value.

Sieving results are most commonly plotted on three-cycle log paper to compare the particle size with the cumulative percentage of undersize particles. If the plot is linear over the entire range, then the material is characterized by a log-normal distribution. If the line is curved, or consists of two or more linear segments, the distribution is polymodal.

For illustration purposes, the sieving results obtained on a powdered sample exhibiting a classic log-normal distribution are shown in Table 3. The typical fashion for presenting such data is to list out the data as a function of both the sieve mesh number and the associated sieve size opening (typically in units of μm). For each sieve in the nested series, one records the mass of sample retained on each sieve, the percentage of sample retained on each sieve, and usually the cumulative percentage of sample retained on each sieve. The cumulative percentage of sample passing through each sieve may also be summarized, although this quantity is not ordinarily used in the analysis.

To illustrate the sieving results, one may plot either the percentage of sample retained on each sieve in a histogram plot (Figure 9a) or the cumulative distribution as a function of sieve size (Figure 9b). The most illustrative mode of data representation is that of the three-cycle plot (Figure 9c), since any departure from linearity that would indicate the existence of a polymodal distribution is readily apparent.

For the distribution in Figure 9, one finds that the value of the geometric median particle size (d_g) equals 210 μm, and the geometric mean standard deviation (σ_g) is 1.57. However, as discussed in earlier sections, a number of other mean particle values can be calculated. For this distribution, one finds the surface mean particle size equal to 215 μm, the volume mean particle size equal to 231 μm, and the volume–surface mean particle size equal to 267 μm. Depending on the intended use for the particle size information, one mean value may have more significance to the analyst than another.

3.5 Sieving analysis of samples by polymodal distributions

When the particle size distribution of an analyzed sample is adequately characterized by a single (unimodal) distribution, then the method of

Table 3

Presentation of Analytical Sieving Results

Sieve mesh number	**Sieve size opening (μm)**	**Mass of sample retained on each sieve**	**Percentage of sample retained on each sieve**	**Cumulative percentage of sample retained on each sieve**	**Cumulative percentage of sample passing through each sieve**
40	425	7.49	5.2	5.2	94.8
50	300	13.55	9.4	14.7	85.3
60	250	21.38	14.9	29.5	70.5
70	212	32.87	22.9	52.4	47.6
100	150	41.32	28.8	81.2	18.8
140	106	22.47	15.6	96.9	3.1
270	53	3.26	2.3	99.1	0.9
Pan	n/a	1.24	0.9	100.0	0.0
	Total	**143.58**	**100**		

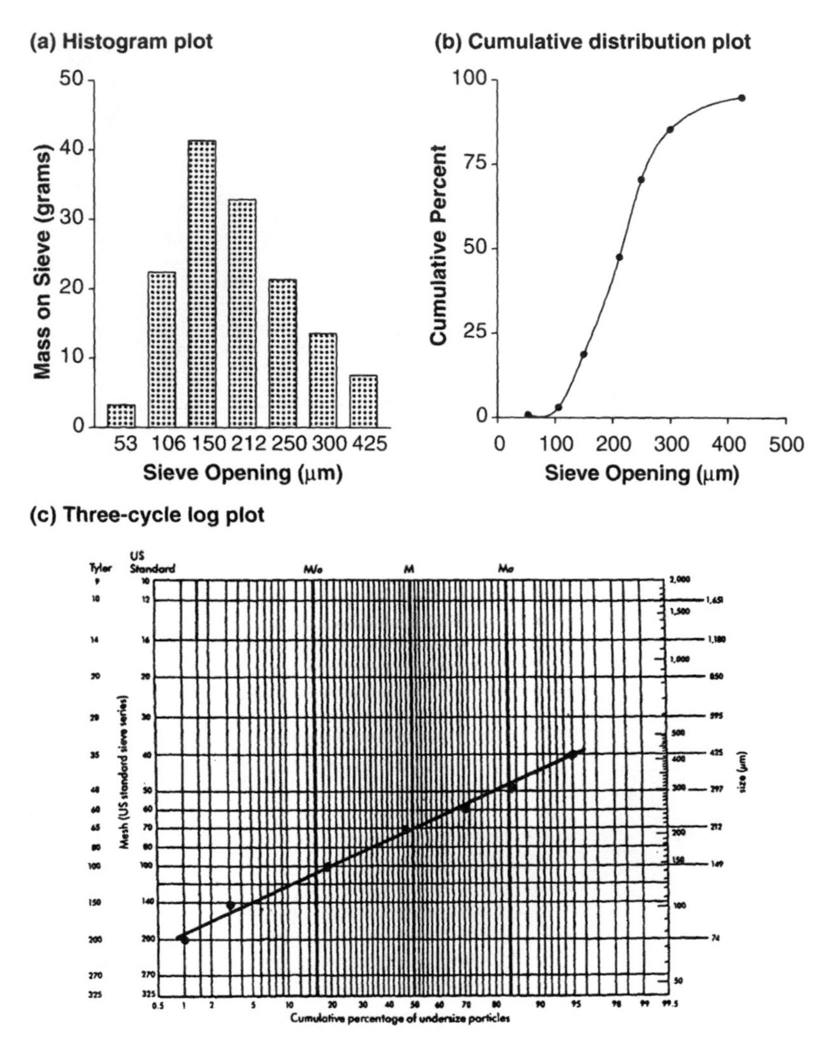

Figure 9. Representations of the particle size distribution of a sample consisting of a unimodal distribution.

data presentation of the preceding section works very well. It is equally important, however, to be able to recognize when the sample contains more than one population in its distribution.

To illustrate this situation, a series of model distributions consisting of two populations have been simulated, and the distributions analyzed according to the usual technique. Figure 10 shows the representation of a

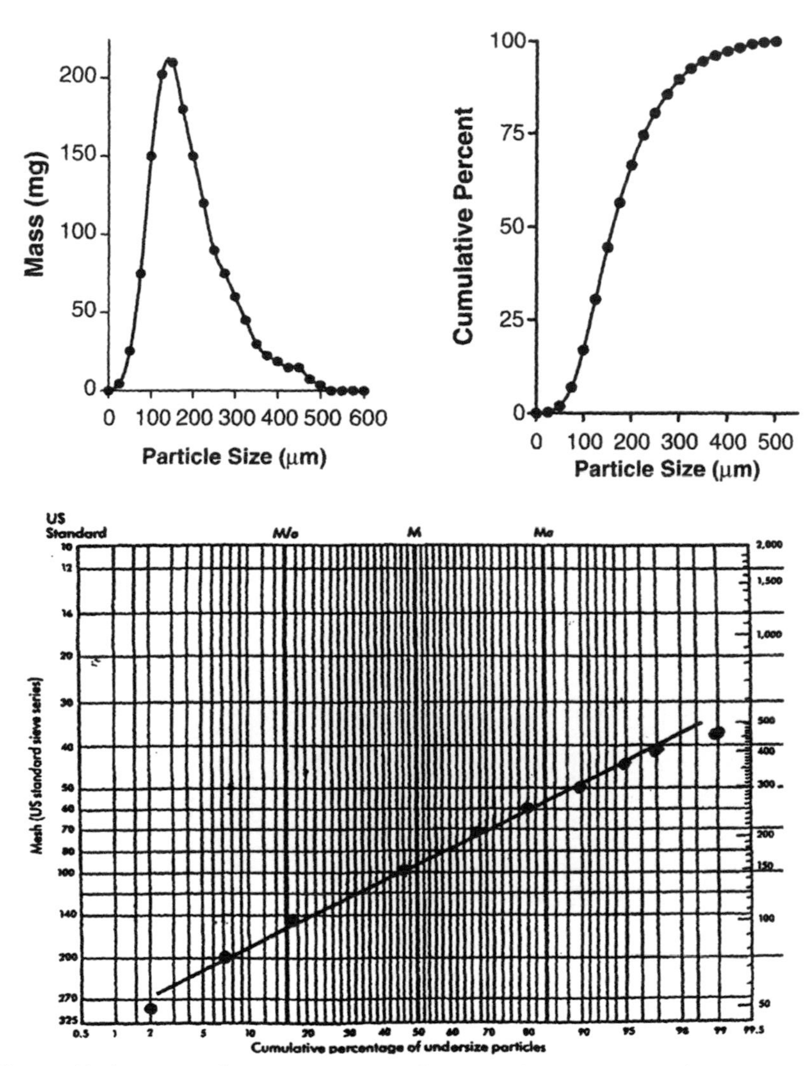

Figure 10. Log-normal representation of a sample consisting of a unimodal distribution.

unimodal distribution, while Figures 11–13 show the representations for samples consisting of bimodal distributions, with the modes differing by 100, 200, and 300 μm, respectively. The figures illustrate the power of the log plot, where the presence of the two size populations is far more discernable than in either the histogram or the cumulative distribution plots.

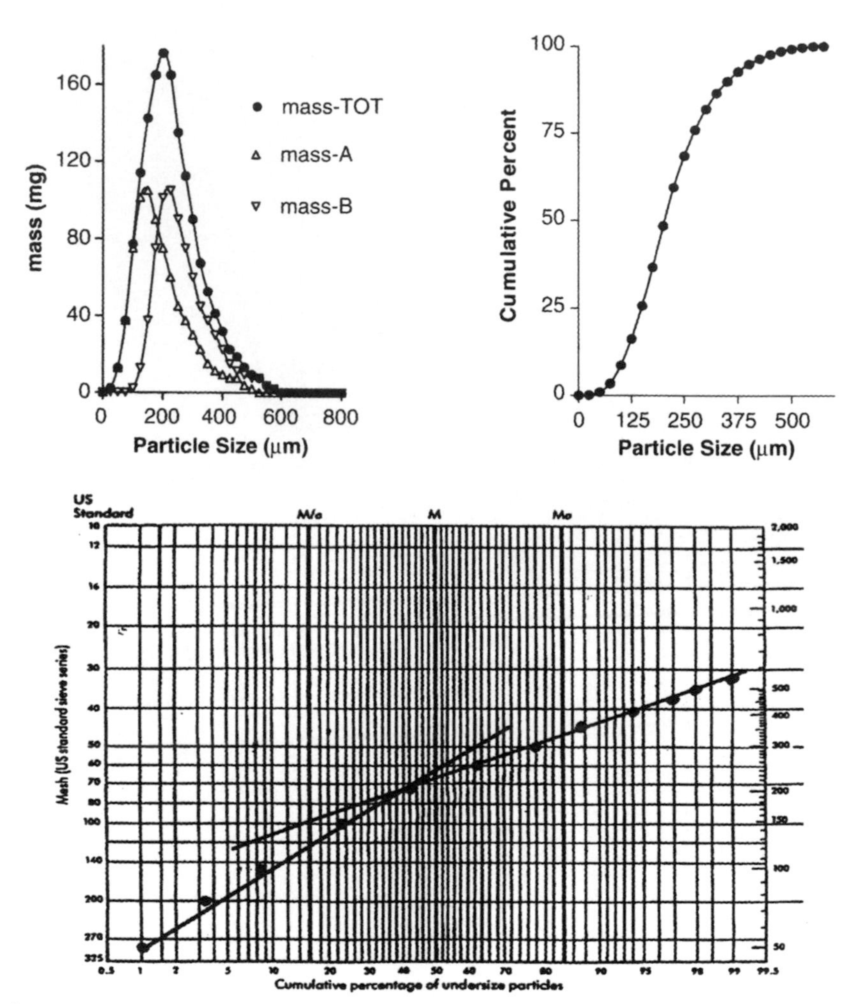

Figure 11. Representation of a sample consisting of a bimodal distribution, the modes of which differ by 100 μm.

4. DETERMINATION OF PARTICLE SIZE DISTRIBUTION BY LASER LIGHT SCATTERING

The use of laser light scattering for the determination of particle size distribution has become widespread. Owing to the nature of light scattering by particles, the technology is generally divided into two general methods. One of these uses high-angle light scattering, and is suited for determining the particle size of ultrafine particles. The other

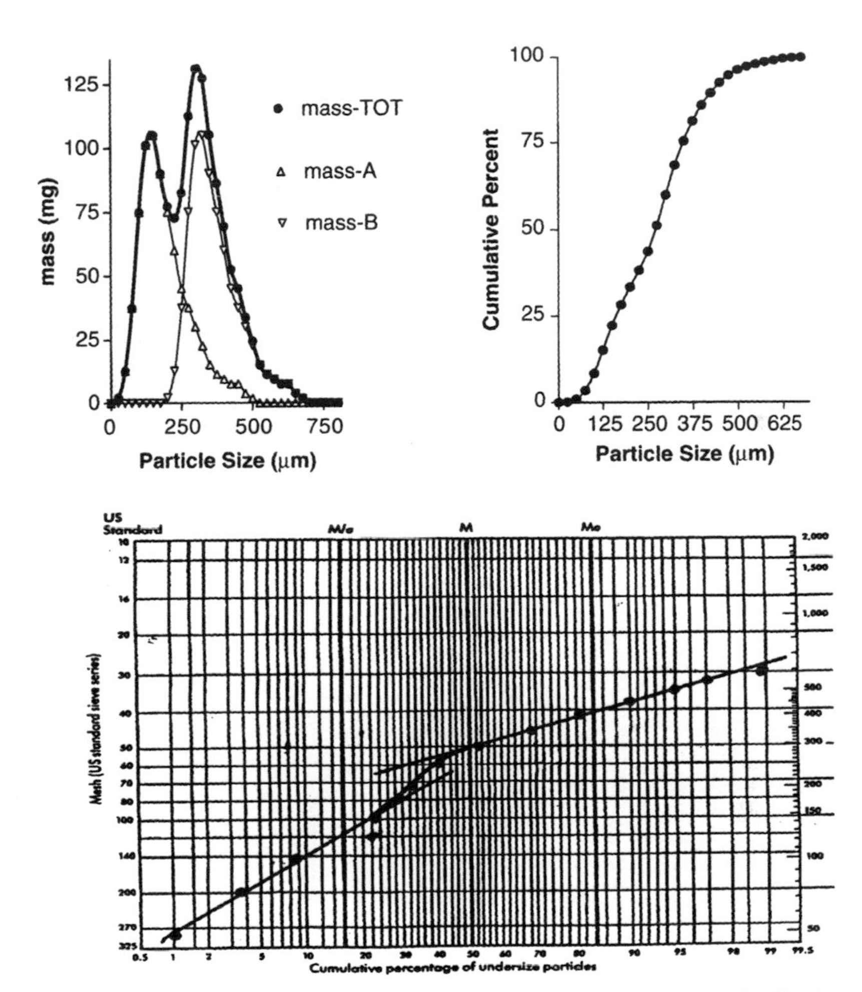

Figure 12. Representation of a sample consisting of a bimodal distribution, the modes of which differ by 200 μm.

technique makes use of low-angle scattering, and is appropriate for the characterization of larger particles. Since neither technique actually measures particle size directly, but instead interprets light scattering phenomena to deduce particle size information, the theory underlying each method will be outlined and then illustrated with applications.

As discussed earlier, it is generally recognized that analytical sieving is the most appropriate method to deduce the particle size of relatively coarse powders, and that particle size information on fine and ultrafine particles

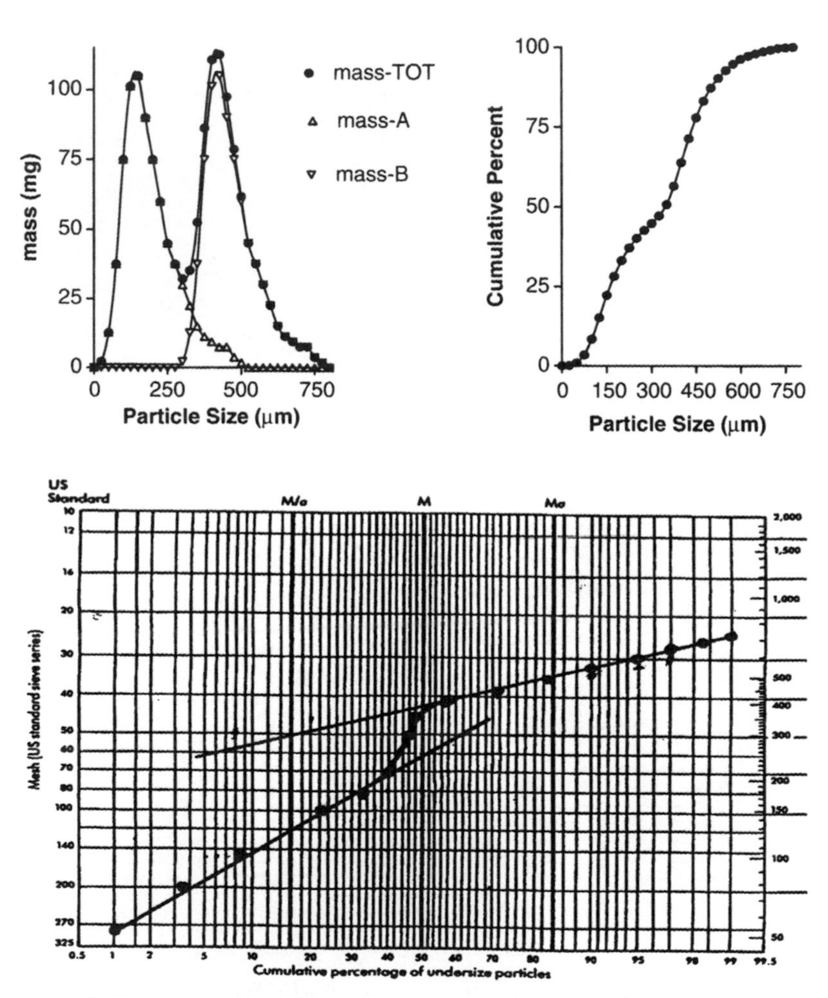

Figure 13. Representation of a sample consisting of a bimodal distribution, the modes of which differ by 300 μm.

should be obtained using laser light scattering. In this chapter, the basic precepts of laser light scattering for the evaluation of particle size information will be outlined, and more detailed expositions of the methodology are available in the literature [10–14].

4.1 Background

In the classical Rayleigh theory, the oscillating electric field in a beam of light passing through a dielectric medium induces oscillating electric dipoles in the suspended polarizable particles. These oscillating dipoles reradiate light that has the same wavelength as the incident (analogous to the antenna of a radio transmission), and are the source of the scattered light. The Rayleigh theory is therefore concerned with polarizability and the resultant induced dipole moment per particle. The theory is accurate for particles having a size that is much smaller than the wavelength of the light. Unfortunately, the theory breaks cannot be applied to the scattering in the condensed media owing to the destructive interference between scattered fields from different particles.

Smoluchowski and Einstein reformulated the Rayleigh theory into a form concerned with polarizability, and the dipole moment per unit volume of the dielectric medium. Since at any temperature there are thermal fluctuations in all properties of the medium, the polarizability per unit volume can be divided into a constant and a fluctuating part. The constant aspect of the polarizability per unit volume yields the effect of refraction, and the fluctuating part of the polarizability per unit volume yields scattering.

The intensity of the scattered light may be calculated using the fluctuation theory from the mean-square fluctuations in the density and entropy for a pure fluid, or from the concentration for a suspension or solution. For suspensions:

$$\langle \mathrm{Is}(q) \rangle = KNM^2 P(\theta) B(c) \tag{11}$$

In equation (11), $\langle \mathrm{Is}(q) \rangle$ is the time-averaged scattered light intensity, q is the wave vector amplitude of the scattering fluctuation, K is an optical constant, N is the number of particles contributing to the scattering, M is the mass of the particle, P is the particle form factor, θ is the scattering angle, B is the concentration factor, and c is the particle concentration.

The fluctuations may be decomposed into various frequencies, and at any angle the scattering is due to a particular fluctuation, represented by the wave vector (q) that has an amplitude given by:

$$q = \left[\frac{4\pi n}{\lambda}\right] \sin(\theta/2) \tag{12}$$

where n is the refractive index of the medium. This equation has the form of the Bragg equation for scattering, but here the scattering takes place from a thermally excited fluctuation having a wavelength equal to $2\pi/q$. Since the mass of a spherical particle is proportional to its volume (and therefore the cube of its diameter), the M^2 term gives rise to a d^6 factor in the scattering.

The concentration factor, $B(c)$, is a result of interparticle effects. In the limit where c approaches zero, $B(c)$ goes to unity. This feature requires the use of dilute suspensions in order to obtain accuracy in the results.

The particle form factor, $P(\theta)$, is a result of interference from interparticle interactions arising from finite-sized particles. Formulae for the $P(\theta)$ function exist for spherical shaped particles, as well as for the particle sizes lesser than certain values. For particles smaller than this (the usual occurrence for particles of pharmaceutical interest), when the relative indices of refraction of the particle and medium differ significantly, then the Lorenz–Mie theory must be used.

It should be noted that at small scattering angles (*i.e.*, where θ approaches zero), it is found that the $P(\theta)$ function goes to unity, and this situation favors the use of small angles of detection for studies on larger particles.

The theory for calculating scattering cross sections from first principles begins with Maxwell's electromagnetic equations, and culminates with the theory attributed to Lorenz and Mie. Although the Lorenz–Mie theory is exact, it does not yield simple analytical solutions relating particle size to optical measurements. The scattering cross section is ultimately related to a size parameter, the refractive index of the particle relative to the suspending medium (m), and the angle of observation.

A development of the Lorenz–Mie theory is beyond the scope of this chapter, but a number of situations having practical interest can be defined. If λ is the wavelength of light used for the study, and if d is the size of a scattering particle, then the Rayleigh theory is appropriate for those instances where d is much lesser than λ. Another way of expressing this is that if $d < \lambda/10$, then a study of the Rayleigh scattering will be appropriate. Alternatively if d is significantly greater than λ (or if $d > 4\lambda$), then the diffraction theory may be used to interpret the results. However, when d is approximately equal to λ, then only the full Lorenz–Mie theory can be relied upon.

The limiting cases are better defined when one accounts for both the size parameter and the relative index of refraction. For such definitions, Van de Hulst suggested the use of the parameter:

$$p = 2\pi d|m - 1|/\lambda \tag{13}$$

When the Van de Hulst parameter is calculated to be less than 0.3, then the Rayleigh scattering is the appropriate model. When the Van de Hulst parameter is much greater than 30, then one would use the diffraction theory. And finally, when the magnitude of the Van de Hulst parameter is approximately unity, one should use the full Lorenz–Mie theory.

4.2 Photon correlation spectroscopy

A collimated beam of light, or a laser beam, passing through a condensed medium undergoes a number of changes in its properties as a result of the interactions. Among other things, the temporal fluctuations that yield a broadening of the frequency of scattered light contain information regarding the motion of any particles suspended in the medium. In particular, the translational diffusion coefficient (D_T) can be calculated from the linewidth of the light beam, defined as the half-width at half-frequency. There are well-known relationships between D_T and the particle size associated for particles having simple shapes.

If one examines the intensity of light scattered from a medium containing suspended particles, one finds that the time dependence of the intensity contains a fluctuating component owing to the Brownian motion of the scattering particles. This behavior is illustrated in Figure 14. The scattered electric field is a function of particle position, and is therefore continually changing. Since the intensity of scattered light is proportional to the square of the electric field, it too must fluctuate over time. The diffusion coefficient (which is the parameter of interest) is contained in the fluctuations, and these are described by a time-dependent correlation function.

It can be shown that the autocorrelation function from a system of rigid, monodisperse, compact particles has the form of an exponential decay:

$$C(t) = \langle I \rangle^2 [1 + b \exp(-2\Gamma_t)] \tag{14}$$

In equation (14), I is the total light intensity, b is an experimental constant, and $\Gamma_t = D_T q^2$. In empirical practice, I is determined

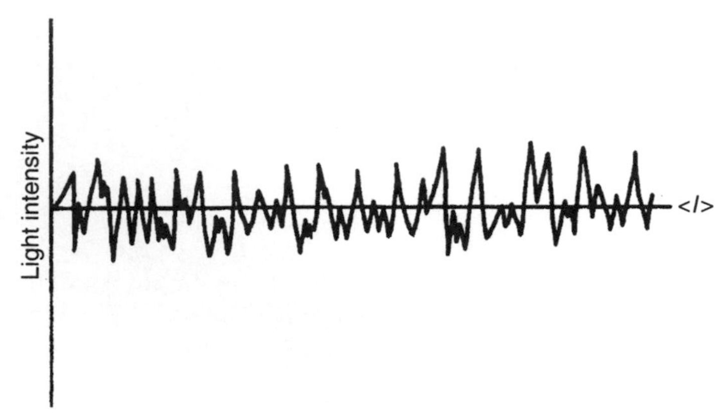

Figure 14. Representation of the scattered light intensity arising from the passage of a beam of light through a medium containing suspended particulate matter.

experimentally, b and Γ_t are determined through curve fitting, and D_T is then calculated.

If the shape of the scattering particle is known or assumed, one can calculate a particle size using information derived from the time dependence of Brownian motion-induced light scattering. For instance, the diffusion constant of a sphere according to the Stokes–Einstein equation is given by:

$$D_T = kT/3\pi\eta d \tag{15}$$

where k is the Boltzmann constant, T is the absolute temperature, and η is the viscosity of the medium through the light passes. Since D_T is determined experimentally, the particle size (d) can be calculated.

It should be noted that the particle size calculated using photon correlation spectroscopy is a hydrodynamic diameter, reflecting both the size of the particle and its encapsulating solvent shell. In addition, unless one has some type of knowledge regarding the actual shape of the particle (*i.e.*, ellipsoid or something else, use of the Stokes–Einstein equation yields a particle size that can only represent the effective spherical diameter of the scattering particle.

Nevertheless, for very small particles (100–750 nm), large-angle PCS is the method of choice for particle sizing, although the interpretation of data is considerably more difficult for polydisperse samples. Since the technique does not involve the counting of single particles, the size distribution must be obtained from the deconvolution of the sum over all of the single exponentials contributing to the measured autocorrelation function. Although the deconvolution is difficult, numerical techniques for the process exist. An excellent summary of the use of PCS spectroscopy in particle sizing is available [15].

Submicron emulsions have been used as colloidal drug carriers for the intravenous administration of lipophilic drugs, where it has been recognized that the control of particle size distribution is one of the most important characteristics [16]. The optimal size for emulsion droplets is 50 nm–1 μm, and particles larger in size than 5 μm are deemed clinically unacceptable as they cause the formation of pulmonary emboli [17]. When employed in the context of other characterization methodology, PCS spectroscopy can yield important information on the structure and properties of fat emulsions.

However, one must exercise some caution during the conduct of PCS spectroscopic studies. As discussed earlier, the experimental observables can only be interpreted correctly when working with a dilute suspension, which ordinarily requires that one substantially dilute down the original suspension to be analyzed. For example, it was found that the measured size of 19 nm latex spheres was strongly influenced by the concentration of particles in the suspension used for the determination [18]. As illustrated in Figure 15, errors as large as 25% were encountered at the highest particulate concentration. These errors arise from effects associated with multiple scattering, and it has been reported that the effect of concentration is most pronounced for studies conducted on particles smaller than 200 nm [19].

4.3 Fraunhofer diffraction

Contrary to the popular belief, the determination of particle size information using small angle light scattering can only be effected if one employs a suitable theoretical model to interpret the scattering data. One method that has achieved a great deal of popularity entails the use of diffraction theory to deduce the requisite particle size information.

The passage of a beam of light through an aperture results in the phenomenon of diffraction, which can be described using the Fresnel–Kirchhoff diffraction integral. When the size of the aperture is small

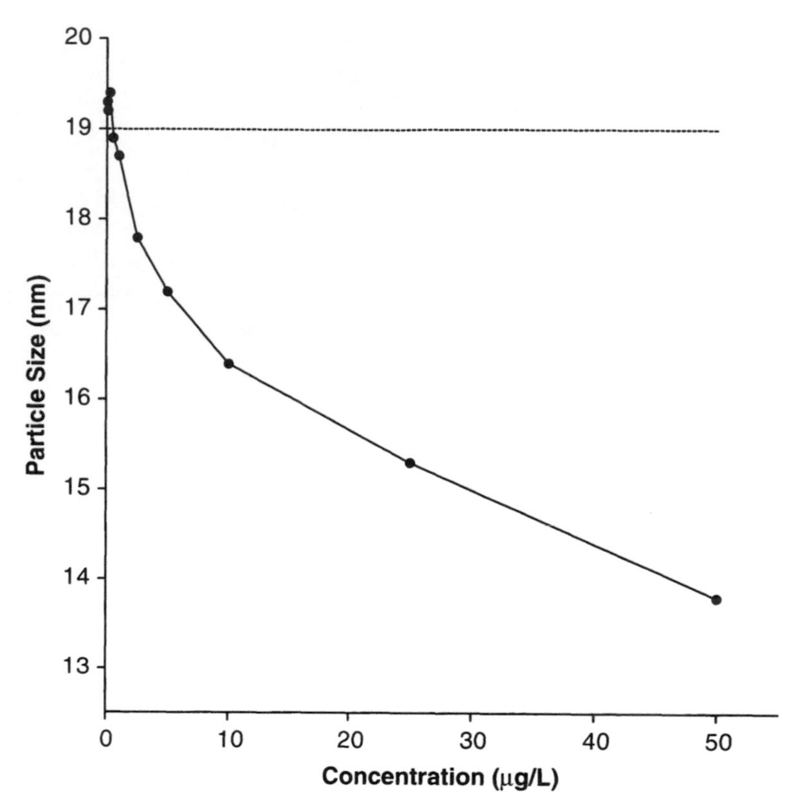

Figure 15. Dependence of the size determined for 19 nm latex spheres as a function of their total concentration. The figure was obtained by plotting data that were provided in [18].

relative to the dimensions of observation and the distance from the source, and the point of observation and the source are very close to the direction of propagation (*i.e.*, small angles of observation), the condition of Fraunhofer diffraction holds. This situation is illustrated in Figure 16.

Strictly speaking, the theory of Fraunhofer diffraction applies only to diffraction caused by the edges of an aperture. However, through the use of the Babinet's theorem, one may deduce that the diffraction pattern of an opaque object whose cross-sectional area and shape are equal to that of the aperture will be the same as that of the aperture itself. However, with this connection, one may use the diffraction theory of light passing through an aperture to obtain estimates of the size of particles whose size is comparable to that aperture.

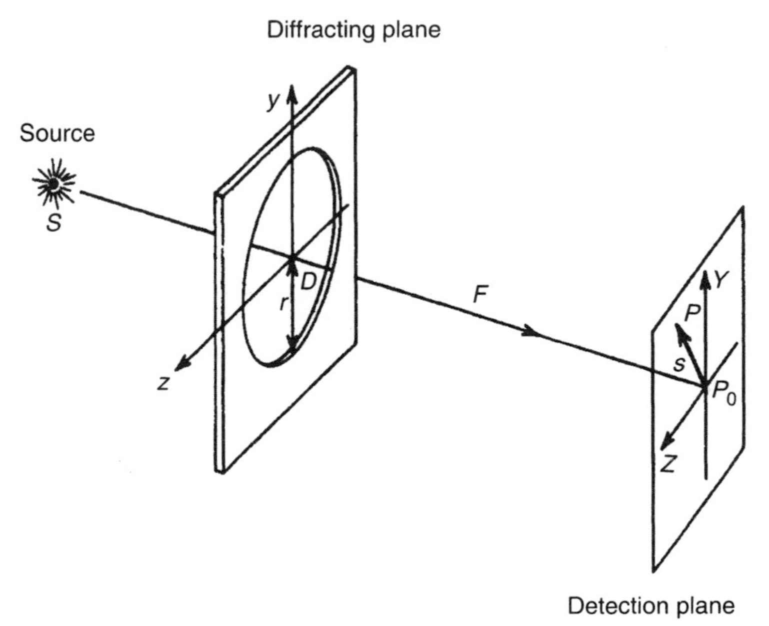

Figure 16. Experimental arrangement appropriate for the observation of Fraunhofer diffraction. The area of the aperture is small relative to the quantities illustrated as S–D and F, both of which are comparable to the wavelength of the light used.

The intensity of the diffraction pattern for an aperture, disk, or sphere of radius r is given by the Airy equation:

$$I = I_o\left[\frac{2J_1(x)}{x}\right]^2 \tag{16}$$

where I_o is the intensity at the center of the pattern, $J_1(x)$ is the first-order spherical Bessel function, and its function is defined by:

$$x = \frac{2\pi rs}{\lambda F} \tag{17}$$

In equation (17), s is the distance in the detection plane and F is the focal length of the lens.

The diffraction pattern will consist of a series of bright and dark concentric rings surrounding the bright central disk. Measuring the intensity pattern to deduce a size distribution is difficult, but measuring and analyzing the light energy distributed over a finite area of a detector is possible. For this purpose, the equation is integrated to obtain the fraction of the light contained within a circle of radius, s, on the detector:

$$L = 1 - J_0(x)^2 - J_1(x)^2 \quad (18)$$

The detectors used for the detection of diffracted light by suspended particles consist of a series of concentric rings, each of which is capable of reporting a signal proportional to the light intensity impinging on its surface. It is therefore more appropriate to determine the light energy falling between two radii, s_l and s_2, associated with diffraction by a single particle of radius r:

$$L(s_1{:}s_2) = C\pi r^2\{[J_0(x)^2 - J_1(x)^2]_{S1} - [J_0(x)^2 - J_1(x)^2]_{S2}\} \quad (19)$$

where πr^2 is the particle cross-sectional area, and C is an optical constant.

Equation (19) pertains to the light scattering of a single particle. For an ensemble of N particles, each of radius r, the relation between light intensity and number distribution is given by:

$$L(s_1{:}s_2) = C\pi \sum N_i r_i^2\{[J_0(x)^2 - J_1(x)^2]_{S1} - [J_0(x)^2 - J_1(x)^2]_{S2}\} \quad (20)$$

The summation in equation (20) is performed over all appropriate size classes. Using a particle density function, one can transform this expression into a relation between light energy and weight distribution.

The minimal experimental configuration that can yield a size distribution will consist of a ring detector having at least 15 concentric rings, and hence the light–size relationship consists of at least 15 simultaneous equations. To analytically solve for the size distribution derived from a 15-element detector requires the manipulation of a 225-element matrix. The complexity of the analysis is usually even greater than this since many detectors use over 30 rings. For instance, a 31-ring detector would require the solving of a 961-element matrix.

Owing to the degree of difficulty in processing the data, the most commonly used method of data processing uses least-squares criteria. A number of models have been proposed for this purpose, each of which begins with the basic assumptions and then works toward a minimalization of the differences between the ideal and the empirical. The size distribution is ultimately derived from the modifications required by the method the least-squares minimalization used, as will be illustrated in the following three examples.

In the **unimodal model-dependent method**, one begins by assuming some sort of model (such as the log-normal representation) for the size distribution. After that, one simulates the detector response expected for the assumed distribution, and then optimizes the model distribution by minimizing the sum of squared deviations from the measured detector response. When the desired degree of convergence is achieved, the reported distribution is taken as being the particle size distribution of the dispersed particles responsible for the scattering.

In the **unimodal model-independent method**, one begins by assuming that the actual particle size distribution actually consists of a finite number of fixed size classes. After that, one simulates the detector response that would be expected for this distribution, and then one optimizes the weight fractions in each size class by minimizing the sum of squared deviations from the measured detector response. When the minimization process is complete, the reported particle size distribution will reflect the optimized sequence of weight fractions.

In the **multimodal model-independent method**, one begins by simulating diffraction patterns for known size distributions. Then one superimposes random noise on the pattern, and calculates the expected element responses for the detector configuration being used. After that, one inverts the patterns by the same minimization algorithm used by the system, and then compares the inverted patterns with the known size distribution. After the preset degree of quantitative correctness is attained, the final responses are reported as the particle size distributions.

However, with most commercial systems, the choice of model and algorithm are transparent to the user. For this reason, the most useful results are obtained when particle size measurements are made using the same laser light scattering system, following the same method of sample preparation, and obtained using the same system parameters. Within the bounds of a rigorously specified range of conditions, it is possible to obtain good information regarding the particle size differences between different lots of material.

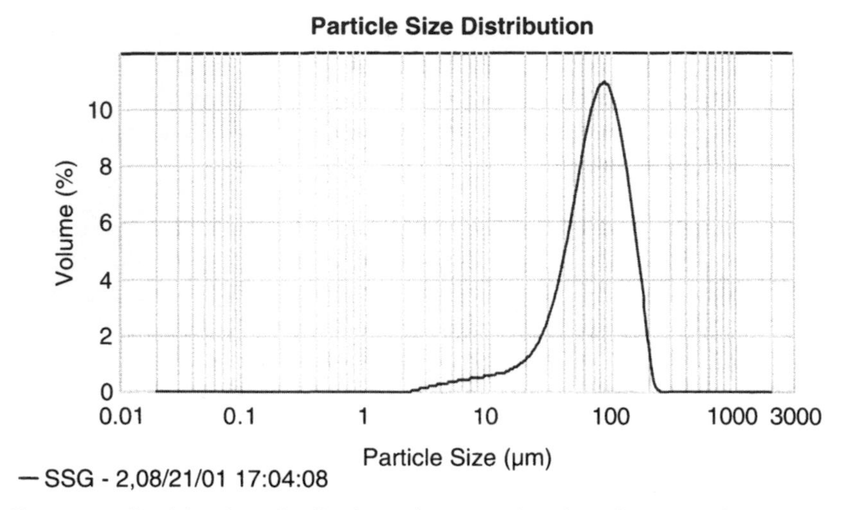

Figure 17. Particle size distribution of a sample of sodium starch glycolate suspended in water.

Assuming the instrument is working properly, the questions to address when determining the quality of a particle size entail an evaluation as to whether the sample was obtained using appropriate sampling procedures, whether the sample was properly prepared and introduced into the instrument, and whether all instrumental parameters were correctly used for the analysis [20]. When these conditions are met, then the derived particle size distributions are judged to be "correct", and useful for evaluation of the system under study.

As a simple example of a typical particle size analysis, a sample of sodium starch glycolate was suspended in water and its particle size obtained using a Malvern Mastersizer 2000 system. The distribution is graphed in Figure 17, where it may be observed that the substance was composed of essentially a single population of particles. The volume weighted mean particle size of this sample was found to be 82.4 μm, with 10% of the particles being smaller than 28.8, 50% of the particles being smaller than 77.0 μm, and 90% of the particles being smaller than 144.2 μm.

It is generally understood that disintegrating agents contained in tablet formulations exert their function as a result of the swelling of individual particles that takes place upon contact with water. To study this phenomenon, the particle size distribution of the sample of sodium starch glycolate dispersed in water was followed over time. As shown in

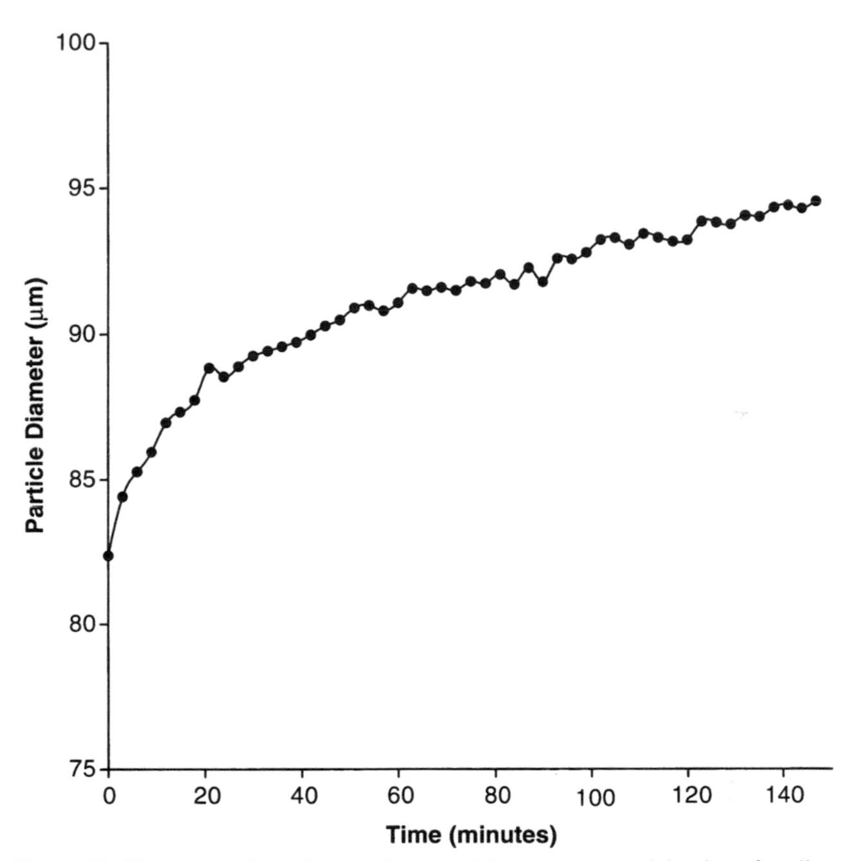

Figure 18. Time evolution of the volume-weighted mean particle size of sodium starch glycolate suspended in water.

Figure 18, the average particle size of the granules increased by approximately 15% over a 2-h time period.

5. SUMMARY

Sieving remains one of the fundamental methods for the classification of powders, and is the method of choice for coarse powders. The general consensus is that sieving is most suitable for granular solids, or powders whose average particle size exceeds 25–50 μm. It is quite important to remember that the particle diameter information obtained using analytical sieving represents the minimum square aperture through which the particle can pass. Details of particulate shape will influence the separation of particles by sieving, since particles will pass through openings based on their cross-sectional diameter. Nevertheless, analytical sieving, properly performed, is a valuable method for determining the size distribution of granular and coarse powders.

There is little doubt that owing to its relative simplicity of use, light scattering will assume an ever-increasingly important role in the size evaluation of pharmaceutical materials. This aspect has been recognized by the United States Pharmacopeia, which is in the process of developing a general test method for this purpose. Presumably, once the method reaches official status, the light scattering techniques will become as firmly ensconced in characterization studies as sieving analyses have been in the past.

6. REFERENCES

1. General Test ⟨776⟩, Optical Microscopy, ***United States Pharmacopoeia 27***, U.S. Pharmacopoeial Convention, Rockville, MD, pp. 2332–2334 (2004).
2. T. Allen, ***Particle Size Measurement***, 3rd edn., Chapman and Hall, London, pp. 107–120 (1981).
3. H. Heywood, *J. Pharm. Pharmacol.*, ***S15***, 56T (1963).
4. General Test ⟨786⟩, "Particle Size Distribution Estimation by Analytical Sieving," ***United States Pharmacopoeia 27***, U.S. Pharmacopoeial Convention, Rockville, MD, pp. 2335–2338 (2004).
5. R.E. Gordon, T.W. Rosanske, D.E. Fonner, N.R. Anderson, and G.S. Banker. "Granulation Technology and Tablet Characterization", in: ***Pharmaceutical Dosage Forms: Tablets,*** Chapter 5, H.A. Lieberman, L. Lachman, and J.B. Schwartz, eds., Marcel Dekker, New York, p. 265 (1990).
6. J.M. Dellavalle, ***Micromeritics***, 2nd edn., Pitman Publishing Corp., New York, pp. 96–122 (1948).
7. R.D. Cadle, ***Particle Size Determination***, Interscience Publishers, New York, pp. 175–196 (1955).
8. R.R. Irani and C.F. Callis, ***Particle Size: Measurement, Interpretation, and Application***, John Wiley & Sons, New York, pp. 107–122 (1963).
9. T. Allen, ***Particle Size Measurement***, 3rd edn., Chapman and Hall, London, pp. 165–186 (1981).
10. R.E. Gordon, T.W. Rosanske, D.E. Fonner, N.R. Anderson, and G.S. Banker. "Granulation Technology and Tablet

Characterization", in: ***Pharmaceutical Dosage Forms: Tablets,*** Chapter 5, H.A. Lieberman, L. Lachman, and J.B. Schwartz, eds., Marcel Dekker, New York, p. 265 (1990).

11. T. Allen, ***Particle Size Measurement***, 3rd edn., Chapman and Hall, London, (1981).

12. B.H. Kaye, ***Direct Characterization of Fine Particles***, John Wiley & Sons, New York, (1981).

13. H.G. Barth, ***Modern Methods of Particle Size Analysis***, John Wiley & Sons, New York, (1984).

14. T. Provder, ***Particle Size Distribution: Assessment and Characterization***, American Chemical Society, Washington, DC (1987).

15. W. Tscharnuter. "Photon Correlation Spectroscopy in Particle Sizing", in: ***Encyclopedia of Analytical Chemistry***, R.A. Meyers, ed., John Wiley & Sons, Chichester, pp. 5469–5485 (2000).

16. S. Benita and M.Y. Levy, *J. Pharm. Sci.*, ***82***, 1069–1079 (1993).

17. W.R. Burnham, P.K. Hansrani, C.E. Knott, J.A. Cook, and S.S. Davis, *Int. J. Pharm.*, ***13***, 9–22 (1983).

18. B.W. Muller and R.H. Muller, *J. Pharm. Sci.*, ***73***, 915–918 (1984).

19. Y. Tian and L.C. Li, *Drug Dev. Indust. Pharm.*, ***24***, 275–280 (1998).

20. H.G. Brittain and G. Amidon, *Am. Pharm. Rev.*, ***6***, 68–72 (2003).

CUMULATIVE INDEX

Bold numerals refer to volume numbers.

Acebutolol, **19**, 1
Acetaminophen, **3**, 1; **14**, 551
Acetazolamide, **22**, 1
Acetohexamide, **1**, 1; **2**, 573; **21**, 1
Acetylcholine chloride, **31**, 3, 21
Acyclovir, **30**, 1
Adenosine, **25**, 1
Allopurinol, **7**, 1
Amantadine, **12**, 1
Amikacin sulfate, **12**, 37
Amiloride hydrochloride, **15**, 1
Aminobenzoic acid, **22**, 33
Aminoglutethimide, **15**, 35
Aminophylline, **11**, 1
Aminosalicylic acid, **10**, 1
Amiodarone, **20**, 1
Amitriptyline hydrochloride, **3**, 127
Amobarbital, **19**, 27
Amodiaquine hydrochloride, **21**, 43
Amoxicillin, **7**, 19; **23**, 1
Amphotericin B, **6**, 1; **7**, 502
Ampicillin, **2**, 1; **4**, 518
Apomorphine hydrochloride, **20**, 121
Arginine, **27**, 1
Ascorbic acid, **11**, 45
Aspartame, **29**, 7
Aspirin, **8**, 1
Astemizole, **20**, 173
Atenolol, **13**, 1
Atropine, **14**, 325
Azathioprine, **10**, 29
Azintamide, **18**, 1
Aztreonam, **17**, 1

Bacitracin, **9**, 1
Baclofen, **14**, 527
Benazepril hydrochloride, **31**, 117
Bendroflumethiazide, **5**, 1; **6**, 597
Benperidol, **14**, 245
Benzocaine, **12**, 73
Benzoic acid, **26**, 1
Benzyl benzoate, **10**, 55
Betamethasone diproprionate, **6**, 43
Bretylium tosylate, **9**, 71
Brinzolamide, **26**, 47
Bromazepam, **16**, 1
Bromcriptine methanesulfonate, **8**, 47
Bumetanide, **22**, 107
Bupivacaine, **19**, 59
Busulphan, **16**, 53

Caffeine, **15**, 71
Calcitriol, **8**, 83
Camphor, **13**, 27
Captopril, **11**, 79
Carbamazepine, **9**, 87
Carbenoxolone sodium, **24**, 1
Cefaclor, **9**, 107
Cefamandole nafate, **9**, 125; **10**, 729
Cefazolin, **4**, 1
Cefixime, **25**, 39
Cefotaxime, **11**, 139

Cefoxitin sodium, **11**, 169
Ceftazidime, **19**, 95
Ceftriaxone sodium, **30**, 21
Cefuroxime sodium, **20**, 209
Celiprolol hydrochloride, **20**, 237
Cephalexin, **4**, 21
Cephalothin sodium, **1**, 319
Cephradine, **5**, 21
Chloral hydrate, **2**, 85
Chlorambucil, **16**, 85
Chloramphenicol, **4**, 47; **15**, 701
Chlordiazepoxide, **1**, 15
Chlordiazepoxide hydrochloride, **1**, 39; **4**, 518
Chloropheniramine maleate, **7**, 43
Chloroquine, **13**, 95
Chloroquine phosphate, **5**, 61
Chlorothiazide, **18**, 33
Chlorpromazine, **26**, 97
Chlorprothixene, **2**, 63
Chlortetracycline hydrochloride, **8**, 101
Chlorthalidone, **14**, 1
Chlorzoxazone, **16**, 119
Cholecalciferol, **13**, 655
Cimetidine, **13**, 127; **17**, 797
Ciprofloxacin, **31**, 163, 179, 209
Cisplatin, **14**, 77; **15**, 796
Citric Acid, **28**, 1
Clarithromycin, **24**, 45
Clidinium bromide, **2**, 145
Clindamycin hydrochloride, **10**, 75
Clioquinol, **18**, 57
Clofazimine, **18**, 91; **21**, 75
Clomiphene citrate, **25**, 85
Clonazepam, **6**, 61
Clonfibrate, **11**, 197
Clonidine hydrochloride, **21**, 109
Clorazepate dipotassium, **4**, 91
Clotrimazole, **11**, 225
Cloxacillin sodium, **4**, 113
Clozapine, **22**, 145
Cocaine hydrochloride, **15**, 151
Codeine phosphate, **10**, 93
Coichicine, **10**, 139
Cortisone acetate, **26**, 167
Crospovidone, **24**, 87
Cyanocobalamin, **10**, 183
Cyclandelate, **21**, 149
Cyclizine, **6**, 83; **7**, 502
Cyclobenzaprine hydrochloride, **17**, 41
Cycloserine, **1**, 53; **18**, 567
Cyclosporine, **16**, 145
Cyclothiazide, **1**, 65
Cypropheptadine, **9**, 155

Dapsone, **5**, 87
Dexamethasone, **2**, 163; **4**, 519
Diatrizoic acid, **4**, 137; **5**, 556
Diazepam, **1**, 79; **4**, 518
Dibenzepin hydrochloride, **9**, 181
Dibucaine, **12**, 105
Dibucaine hydrochloride, **12**, 105
Diclofenac sodium, **19**, 123
Didanosine, **22**, 185
Diethylstilbestrol, **19**, 145
Diflunisal, **14**, 491
Digitoxin, **3**, 149; **9**, 207
Dihydroergotoxine methanesulfonate, **7**, 81
Diloxanide furoate, **26**, 247
Diltiazem hydrochloride, **23**, 53
Dioctyl sodium sulfosuccinate, **2**, 199; **12**, 713
Diosgenin, **23**, 101
Diperodon, **6**, 99
Diphenhydramine hydrochloride, **3**, 173
Diphenoxylate hydrochloride, **7**, 149
Dipivefrin hydrochloride, **22**, 229
Dipyridamole, **31**, 215
Disopyramide phosphate, **13**, 183
Disulfiram, **4**, 168
Dobutamine hydrochloride, **8**, 139
Dopamine hydrochloride, **11**, 257
Dorzolamide hydrochloride, **26**, 283; **27**, 377
Doxorubicine, **9**, 245
Droperidol, **7**, 171

Echothiophate iodide, **3**, 233
Econazole nitrate, **23**, 127
Edetic Acid (EDTA), **29**, 57
Emetine hydrochloride, **10**, 289
Enalapril maleate, **16**, 207
Ephedrine hydrochloride, **15**, 233
Epinephrine, **7**, 193
Ergonovine maleate, **11**, 273
Ergotamine tartrate, **6**, 113
Erthromycin, **8**, 159
Erthromycin estolate, **1**, 101; **2**, 573
Estradiol, **15**, 283
Estradiol valerate, **4**, 192
Estrone, **12**, 135
Ethambutol hydrochloride, **7**, 231
Ethynodiol diacetate, **3**, 253
Etodolac, **29**, 105
Etomidate, **12**, 191
Etopside, **18**, 121
Eugenol, **29**, 149

Fenoprofen calcium, **6**, 161
Fenoterol hydrobromide, **27**, 33
Flavoxoate hydrochloride, **28**, 77
Flecainide, **21**, 169
Fluconazole, **27**, 67
Flucytosine, **5**, 115
Fludrocortisone acetate, **3**, 281
Flufenamic acid, **11**, 313
Fluorouracil, **2**, 221; **18**, 599
Fluoxetine, **19**, 193
Fluoxymesterone, **7**, 251
Fluphenazine decanoate, **9**, 275; **10**, 730
Fluphenazine enanthate, **2**, 245; **4**, 524
Fluphenazine hydrochloride, **2**, 263; **4**, 519
Flurazepam hydrochloride, **3**, 307
Flutamide, **27**, 115
Fluvoxamine maleate, **24**, 165
Folic acid, **19**, 221
Furosemide, **18**, 153

Gadoteridol, **24**, 209
Gentamicin sulfate, **9**, 295; **10**, 731
Glafenine, **21**, 197
Glibenclamide, **10**, 337
Gluthethimide, **5**, 139
Gramicidin, **8**, 179
Griseofulvin, **8**, 219; **9**, 583
Guaifenesin, **25**, 121
Guanabenz acetate, **15**, 319
Guar gurn, **24**, 243

Halcinonide, **8**, 251
Haloperidol, **9**, 341
Halothane, **1**, 119; **2**, 573; **14**, 597
Heparin sodium, **12**, 215
Heroin, **10**, 357
Hexestrol, **11**, 347
Hexetidine, **7**, 277
Histamine, **27**, 159
Homatropine hydrobromide, **16**, 245
Hydralazine hydrochloride, **8**, 283
Hydrochlorothiazide, **10**, 405
Hydrocortisone, **12**, 277
Hydroflumethaizide, **7**, 297
Hydroxyprogesterone caproate, **4**, 209
Hydroxyzine dihydrochloride, **7**, 319
Hyoscyamine, **23**, 155

Ibuprofen, **27**, 265
Imipramine hydrochloride, **14**, 37
Impenem, **17**, 73
Indapamide, **23**, 233
Indinivar sulfate, **26**, 319
Indomethacin, **13**, 211
Iodamide, **15**, 337
Iodipamide, **2**, 333
Iodoxamic acid, **20**, 303
Iopamidol, **17**, 115
Iopanoic acid, **14**, 181
Ipratropium bromide, **30**, 59
Iproniazid phosphate, **20**, 337
Isocarboxazid, **2**, 295
Isoniazide, **6**, 183
Isopropamide, **2**, 315; **12**, 721

Isoproterenol, **14**, 391
Isosorbide dinitrate, **4**, 225; **5**, 556
Isosuprine hydrochloride, **26**, 359
Ivermectin, **17**, 155

Kanamycin sulfate, **6**, 259
Ketamine, **6**, 297
Ketoprofen, **10**, 443
Ketotifen, **13**, 239
Khellin, **9**, 371

Lactic acid, **22**, 263
Lactose, anhydrous, **20**, 369
Lansoprazole, **28**, 117
Leucovorin calcium, **8**, 315
Levallorphan tartrate, **2**, 339
Levarterenol bitartrate, **1**, 149; **2**, 573; **11**, 555
Levodopa, **5**, 189
Levothyroxine sodium, **5**, 225
Lidocaine, **14**, 207; **15**, 761
Lidocaine hydrochloride, **14**, 207; **15**, 761
Lincomycin, **23**, 275
Lisinopril, **21**, 233
Lithium carbonate, **15**, 367
Lobeline hydrochloride, **19**, 261
Lomefloxacin, **23**, 327
Lomustine, **19**, 315
Loperamide hydrochloride, **19**, 341
Lorazepam, **9**, 397
Lovastatin, **21**, 277

Mafenide acetate, **24**, 277
Malic Acid, **28**, 153
Maltodextrin, **24**, 307
Mandelic Acid, **29**, 179
Maprotiline hydrochloride, **15**, 393
Mebendazole, **16**, 291
Mebeverine hydrochloride, **25**, 165
Mefenamic acid, **31**, 281
Mefloquine hydrochloride, **14**, 157
Melphalan, **13**, 265
Meperidine hydrochloride, **1**, 175
Meprobamate, **1**, 207; **4**, 520; **11**, 587
Mercaptopurine, **7**, 343
Mesalamine, **25**, 209; **27**, 379
Mestranol, **11**, 375
Metformin hydrochloride, **25**, 243
Methadone hydrochloride, **3**, 365; **4**, 520; **9**, 601
Methaqualone, **4**, 245
Methimazole, **8**, 351
Methixen hydrochloride, **22**, 317
Methocarbamol, **23**, 377
Methotrexate, **5**, 283
Methoxamine hydrochloride, **20**, 399
Methoxsalen, **9**, 427
Methylclothiazide, **5**, 307
Methylphenidate hydrochloride, **10**, 473
Methyprylon, **2**, 363
Metipranolol, **19**, 367
Metociopramide hydrochloride, **16**, 327
Metoprolol tartrate, **12**, 325
Metronidazole, **5**, 327
Mexiletine hydrochloride, **20**, 433
Minocycline, **6**, 323
Minoxidil, **17**, 185
Mitomycin C, **16**, 361
Mitoxanthrone hydrochloride, **17**, 221
Morphine, **17**, 259
Moxalactam disodium, **13**, 305

Nabilone, **10**, 499
Nadolol, **9**, 455; **10**, 732
Nalidixic acid, **8**, 371
Nalmefene hydrochloride, **24**, 351
Nalorphine hydrobromide, **18**, 195
Naloxone hydrochloride, **14**, 453
Naphazoline hydrochloride, **21**, 307
Naproxen, **21**, 345
Natamycin, **10**, 513; **23**, 405
Neomycin, **8**, 399
Neostigmine, **16**, 403
Nicotinamide, **20**, 475
Nifedipine, **18**, 221
Nimesulide, **28**, 197

Nimodipine, **31**, 337, 355, 371
Nitrazepam, **9**, 487
Nitrofurantoin, **5**, 345
Nitroglycerin, **9**, 519
Nizatidine, **19**, 397
Norethindrone, **4**, 268
Norfloxacin, **20**, 557
Norgestrel, **4**, 294
Nortriptyline hydrochloride, **1**, 233; **2**, 573
Noscapine, **11**, 407
Nystatin, **6**, 341

Ondansetron hydrochloride, **27**, 301
Ornidazole, **30**, 123
Oxamniquine, **20**, 601
Oxazepam, **3**, 441
Oxyphenbutazone, **13**, 333
Oxytocin, **10**, 563

Pantoprazole, **29**, 213
Papaverine hydrochloride, **17**, 367
Particle Size Distribution, **31**, 379
Penicillamine, **10**, 601
Penicillin-G, benzothine, **11**, 463
Penicillin-G, potassium, **15**, 427
Penicillin-V, **1**, 249; **17**, 677
Pentazocine, **13**, 361
Pentoxifylline, **25**, 295
Pergolide Mesylate, **21**, 375
Phenazopyridine hydrochloride, **3**, 465
Phenelzine sulfate, **2**, 383
Phenformin hydrochloride, **4**, 319; **5**, 429
Phenobarbital, **7**, 359
Phenolphthalein, **20**, 627
Phenoxymethyl penicillin potassium, **1**, 249
Phenylbutazone, **11**, 483
Phenylephrine hydrochloride, **3**, 483
Phenylpropanolamine hydrochloride, **12**, 357; **13**, 767
Phenytoin, **13**, 417
Physostigmine salicylate, **18**, 289
Phytonadione, **17**, 449
Pilocarpine, **12**, 385
Piperazine estrone sulfate, **5**, 375
Pirenzepine dihydrochloride, **16**, 445
Piroxicam, **15**, 509
Polythiazide, **20**, 665
Polyvinyl alcohol, **24**, 397
Polyvinylpyrollidone, **22**, 555
Povidone, **22**, 555
Povidone-Iodine, **25**, 341
Pralidoxine chloride, **17**, 533
Praziquantel, **25**, 463
Prazosin hydrochloride, **18**, 351
Prednisolone, **21**, 415
Primidone, **2**, 409; **17**, 749
Probenecid, **10**, 639
Pyrazinamide hydrochloride, **4**, 333; **28**, 251
Procaine hydrochloride, **26**, 395
Procarbazine hydrochloride, **5**, 403
Promethazine hydrochloride, **5**, 429
Proparacaine hydrochloride, **6**, 423
Propiomazine hydrochloride, **2**, 439
Propoxyphene hydrochloride, **1**, 301; **4**, 520; **6**, 598
Propyl paraben, **30**, 235
Propylthiouracil, **6**, 457
Pseudoephedrine hydrochloride, **8**, 489
Pyrazinamide, **12**, 433
Pyridoxine hydrochloride, **13**, 447
Pyrimethamine, **12**, 463

Quinidine sulfate, **12**, 483
Quinine hydrochloride, **12**, 547

Ranitidine, **15**, 533
Reserpine, **4**, 384; **5**, 557; **13**, 737
Riboflavin, **19**, 429
Rifampin, **5**, 467
Rutin, **12**, 623

Saccharin, **13**, 487
Salbutamol, **10**, 665
Salicylamide, **13**, 521

Salicylic acid, **23**, 427
Scopolamine hydrobromide,
19, 477
Secobarbital sodium, **1**, 343
Sertraline hydrochloride, **24**, 443
Sertraline lactate, **30**, 185
Sildenafil citrate, **27**, 339
Silver sulfadiazine, **13**, 553
Simvastatin, **22**, 359
Sodium nitroprusside, **6**, 487;
15, 781
Solasodine, **24**, 487
Sorbitol, **26**, 459
Sotalol, **21**, 501
Spironolactone, **4**, 431; **29**, 261
Starch, **24**, 523
Streptomycin, **16**, 507
Strychnine, **15**, 563
Succinycholine chloride, **10**, 691
Sulfacetarnide, **23**, 477
Sulfadiazine, **11**, 523
Sulfadoxine, **17**, 571
Sulfamethazine, **7**, 401
Sulfamethoxazole, **2**, 467; **4**, 521
Sulfasalazine, **5**, 515
Sulfathiazole, **22**, 389
Sulfisoxazole, **2**, 487
Sulfoxone sodium, **19**, 553
Sulindac, **13**, 573
Sulphamerazine, **6**, 515
Sulpiride, **17**, 607

Talc, **23**, 517
Teniposide, **19**, 575
Tenoxicam, **22**, 431
Terazosin, **20**, 693
Terbutaline sulfate, **19**, 601
Terfenadine, **19**, 627
Terpin hydrate, **14**, 273
Testolactone, **5**, 533
Testosterone enanthate, **4**, 452
Tetracaine hydrochloride, **18**, 379
Tetracycline hydrochloride,
13, 597
Theophylline, **4**, 466
Thiabendazole, **16**, 611
Thiamine hydrochloride, **18**, 413
Thiamphenicol, **22**, 461
Thiopental sodium, **21**, 535
Thioridazine, **18**, 459
Thioridazine hydrochloride,
18, 459
Thiostrepton, **7**, 423
Thiothixene, **18**, 527
Ticlopidine hydrochloride, **21**, 573
Timolol maleate, **16**, 641
Titanium dioxide, **21**, 659
Tobramycin, **24**, 579
α-Tocopheryl acetate, **3**, 111
Tolazamide, **22**, 489
Tolbutamide, **3**, 513; **5**, 557;
13, 719
Tolnaftate, **23**, 549
Tranylcypromine sulfate, **25**, 501
Trazodone hydrochloride, **16**, 693
Triamcinolone, **1**, 367; **2**, 571; **4**, 521;
11, 593
Triamcinolone acetonide, **1**, 397;
2, 571; **4**, 521; **7**, 501; **11**, 615
Triamcinolone diacetate, **1**, 423;
11, 651
Triamcinolone hexacetonide, **6**, 579
Triamterene, **23**, 579
Triclobisonium chloride, **2**, 507
Trifluoperazine hydrochloride,
9, 543
Triflupromazine hydrochloride,
2, 523; **4**, 521; **5**, 557
Trimethaphan camsylate, **3**, 545
Trimethobenzamide hydrochloride,
2, 551
Trimethoprim, **7**, 445
Trimipramine maleate, **12**, 683
Trioxsalen, **14**, 705
Tripelennamine hydrochloride,
14, 107
Triprolidine hydrochloride, **8**, 509
Tropicamide, **3**, 565
Tubocurarine chloride, **7**, 477
Tybamate, **4**, 494

Valproate sodium, **8**, 529
Valproic acid, **8**, 529
Verapamil, **17**, 643
Vidarabine, **15**, 647
Vinblastine sulfate, **1**, 443; **21**, 611
Vincristine sulfate, **1**, 463; **22**, 517
Vitamin D3, **13**, 655

Warfarin, **14**, 423

X-Ray Diffraction, **30**, 271
Xylometazoline hydrochloride, **14**, 135

Yohimbine, **16**, 731

Zidovudine, **20**, 729
Zileuton, **25**, 535
Zomepirac sodium, **15**, 673

Printed in the United Kingdom by
Lightning Source UK Ltd., Milton Keynes
140370UK00001B/84/A